Chemistry Research and Applications

Physics Research and Technology

nova science publishers
www.novapublishers.com

Chemistry Research and Applications

Properties and Uses of Antimony
David J. Jenkins (Editor)
2022. ISBN: 979-8-88697-081-4 (Softcover)
2022. ISBN: 979-8-88697-088-3 (eBook)

The Chemistry of Calcium Carbonate
George R. Hood (Editor)
2022. ISBN: 979-8-88697-442-3 (Hardcover)
2022. ISBN: 979-8-88697-448-5 (eBook)

The Science of Carbamates
Güllü Kaymak (Editor)
2022. ISBN: 978-1-68507-708-2 (Softcover)
2022. ISBN: 978-1-68507-872-0 (eBook)

Deep Eutectic Solvents: Properties, Applications and Toxicity
Carlos Eduardo de Araújo Padilha, PhD, Everaldo Silvino dos Santos, PhD, Francisco Canindé de Sousa Júnior, PhD, Nathália Saraiva Rios, PhD (Editors)
2022. ISBN: 978-1-68507-719-8 (Hardcover)
2022. ISBN: 978-1-68507-799-0 (eBook)

Polycyclic Aromatic Hydrocarbons: Sources, Exposure and Health Effects
Warren L. Gregoire (Editor)
2022. ISBN: 978-1-68507-626-9 (Softcover)
2022. ISBN: 978-1-68507-685-6 (eBook)

Cyanide: Occurrence, Applications and Toxicity
Bill M. Torres (Editor)
2022. ISBN: 978-1-68507-619-1 (Softcover)
2022. ISBN: 978-1-68507-670-2 (eBook)

More information about this series can be found at
https://novapublishers.com/product-category/series/chemistry-research-and-applications/

Luc Thomas Ikelle

Introduction to Multidisciplinary Science in an Artificial-Intelligence Age

The Matter in our Universe, Biological Cells, and Plate Tectonics

nova
science publishers

www.novapublishers.com

https://doi.org/10.52305/VAZP3077

Library of Congress Cataloging-in-Publication Data

ISBN: 978-1-68507-992-5

Published by Nova Science Publishers, Inc. † New York

This book is dedicated to Larissa, Kevin-Luc Matiba, and Darrell-Thomas Udoh, my gifts of 4D life.

Luc Thomas Ikelle

Contents

Preface

Higher science education is about preparing people to address the scientific and technological challenges of our times. The list of these challenges includes, but is not limited to, feeding the world's growing population, improving the length and quality of lives on Earth, improving short- and long- distance communication and transportation, predicting of, adapting to, and mitigating of natural hazards, and exploring space to improve our lives on Earth and for the survival of our civilization. This book series prepares students to understand and contribute, to address these challenges. Because these challenges, even in most of their narrow focuses, transcend the current college curriculums or programs. The Earth's climate change is an example of the narrow focus of the large topic of natural hazards. It spans almost all aspects of modern sciences, including biology, chemistry, and physics. Moreover, the span of these challenges may also be telling us that we have to rethink higher sciences education because the current splintering of science education into endless disciplines may not be the best or a unique way to prepare minds to address these challenges. We may be locking talented young minds to a certain viewpoint forever, an unintended indoctrination. At least some universities and colleges must start moving away from the monolithic way of delivering higher science education or to create a multidisciplinary science as a separate program and hopefully unleash a new generation of super engineers and scientists who are speaking scientific language. Moreover, this approach may optimize the higher education time for some people.

The splintering of science education may not even be the most effective way of preparing young minds for the jobs of tomorrow.

With the huge progress being made in the field of artificial intelligence (AI), the new jobs are about innovation. Broadly speaking, AI is a set of techniques that are aimed at approximating some aspects of human or animal cognition by inserting computer software into robots and other related systems. AI systems can nowadays perform many traditionally human tasks with high efficiency and without errors. Yet they are very far from being able to create. Very soon AI systems will become central to all science disciplines and the central platform for multidisciplinary studies. They are modern versions of computers, which extend beyond programming a set of scientific rules for a discipline-specific dataset. AI systems can automatically learn these rules, improve them, and even create new ones. We can easily input different kinds of datasets into them and even introduce rules for outcomes. So, while specialization and expertise are especially critical in graduate schools and adult life, a broad horizon is critical for expertise and innovation and to command and create AI systems. Moreover, how can one teach AI without having a background in the biology of life, chemistry, and physics? Likewise, how can one teach the biology of life, chemistry, or physics without incorporating AI?

Therefore, the objective of this book series is to provide an example of a text for undergraduate science education that prepares students to contribute to the challenges of improving the length and quality of lives on Earth, improving short- and long-distance communication and transportation, predicting of, adapting to, and mitigating natural hazards, and exploring space, to improve our lives on Earth and for the survival of our civilization. The main topics covered here are vacuum and matter, heat, chemical and nuclear reactions, porosity, permeability, viscosity, immersibility, wettability, gravity, electricity, and magnetism, acoustic and elastic waves, the biology of life, and tomography. A basic background in matter, heat, chemical and nuclear reactions, gravity, electricity, and magnetism, and the biology of lives is central for present and future jobs in healthcare. Along with electromagnetic wave propagation in a vacuum and matter and elastic wave propagation in matter, one can develop a solid background for understanding energy flow, which is an essential tool in healthcare, communication, and transportation, natural hazard prediction and mitigation, and space exploration. Tomography and spectroscopy experiments are the basis of most technologies in medicine and mineral exploration. We demonstrate in the last part of this book series that the basic equations of physics and chemistry are the same everywhere in our solar system, and probably in our entire universe. Biological equations, including equations for life, are probably also the same everywhere in the universe, although some of them are still unknown. Because of the multidisciplinary approach adopted in this text, the description of the atmosphere, hydrosphere, biosphere, and the lithosphere, along with solar energy, which is essential to the function of each of these spheres and their interactions, readily follows from these equations. The interconnectivity and interactions of these spheres are difficult to be taught in current programs, yet they are central to our livelihoods and come naturally in this text.

This text covers the materials of multiple current programs in a short space because these programs include huge duplications that we have avoided here. They also unnecessarily over-elaborate on some materials, which are actually just examples, exercises, or applications of a given topic, another issue avoided in this text. No chapter is divided into biology, chemistry, and physics. We define and motivate the importance of a topic or problem. Then we address it, both qualitatively and quantitatively, starting with basic principles followed by a series of applications. I also mentioned various terminologies of the same quantity to accommodate readers with a fixed background from many current sciences and engineering disciplines. Finally, the book series contains analytical problems, as well as computational problems, that include MATLAB software and Mathematica software developed especially for the classes associated with this book series. The computational exercises focus on numerical modeling, imaging, and statistics. In some chapters, essays on topics of considerable broad interest and debate are discussed at the end of these chapters based on the background developed in this book series. These essays may help students see how the background that they have acquired plays a central role in understanding and addressing the major scientific and technological questions of our times.

The book series is intended for multi-semester course work. The first two chapters of the first book are broad. They intend to provide a broad perspective on some basic notions that science students are expected to be familiar with. They also motivate the rest of the book series and provide a clear indication of the cross-fertilization of disciplines in the 21st century. In the other chapters, we motivate the topic, followed by a focus presentation of the topic, and then a broad set of examples of applications. For example, in Chapter 2 of the fourth book, we discuss heat radiation, conduction, and convection. We then describe their applications in climate change, in determining the solid Earth temperature gradient. In addition, we describe the Earth's mantle convection, radiation of energy from the core of the Sun to its surfaces, and bioheat transfer. We also introduce the second law of thermodynamics and describe its application to efficiently develop heat and pump engines, including wind turbines.

The chapters are also closed related with a clear progression. For example, in Chapter 1 of the first book, we introduced oxygen in the atmosphere as essential for some lives on Earth. In Chapter 2, we described oxygen isotopes. In Chapter 3, we introduce subsurface oxides and minerals. In Chapter 1 of the second book, we distinguish living things that require atmospheric oxygen for their survival. We also present photosynthesis, rock burial in the ground, and other examples of chemical reactions involving oxygen and oxygen compounds, and so on. It does not make sense for people to call themselves engineers or scientists if they are not familiar with the basic features of oxygen gas and electromagnetic waves around us.

The book series is divided into five volumes. This first book has three chapters. We start in Chapter 1 with an overview of the Earth, our solar system, and our universe. We also describe the early Earth's atmospheres, the occurrence of oxygen in our atmosphere and carbon-based life on Earth, and some of the past mass-extensions on Earth. In the next two chapters we introduce, with significant details, some of the core fundamental notions that underpin modern science and engineering education and scientific training. These notions are (1) organic and inorganic matter, (2) radiative matter, (3) chemical bonds, (4) molecular vibrations, (5) living things, (6) DNA and RNA, (7) rock type and rock cycle, (8) rock weathering, (9) geologic hotspots, (10) geologic faults, (11) earthquakes and volcanoes, and (12) mantle convection. We then illustrate, with applications across disciplines, the importance of these notions in our lives and in understanding the world around us. Some of these applications have to do with the origin of chemical elements and of a significant amount of molecules, the structures and compositions of biological cells, cell reproduction mechanisms, genetics and gene mutations, gene manipulation, the origin of the first biological cells, plate tectonics in our solar system, prediction of large earthquakes in Chile and Japan, and conventional and unconventional energy resources.

Chapter 1

A View of Earth, Our Solar System, and Our Universe

Where do we live? We are on planet Earth, a part of our solar system, which is a small part of the Milky Way galaxy—the Milky Way galaxy is just one of the 2 trillion, or so, galaxies in our universe. It turns out that all these astronomical objects are connected to our daily lives. For example, our lives are totally dependent on the presence of liquid water on Earth, on the correct amount of oxygen in the atmosphere, on solar energy, on the Earth's core magnetic field that protects us from the sun (a star) radiation, on space, weather, and human satellites and spacecrafts in space for communication of all kinds on Earth, and on the characteristics of living things as the key inspiration for medicines, vaccine and pharmacological testing, car, boat, and airplane transportation, electricity storage, and artificial intelligence. How did these connections happen, how did they start, and how can we use and/or modify them to enhance human lives on Earth? These age-old questions, among others, are still the central questions of modern science. We will elaborate on these questions in the next chapters, as well as how we draw inspiration from the natural phenomena of our universe to develop modern technology. Our objective in this chapter is to provide an overview of the Earth, our solar system, and, in a limited form, our universe, with its galaxies.

Key sentences: *Everything that we are looking for on other planets of our solar system is here at some scale or was here on Earth at a certain point in time and space at some scale. Everything that happened or is happening to our other planets is happening or happened here on Earth at a certain point in time and space at some scale. Answers are around you and/or in Earth's history.*

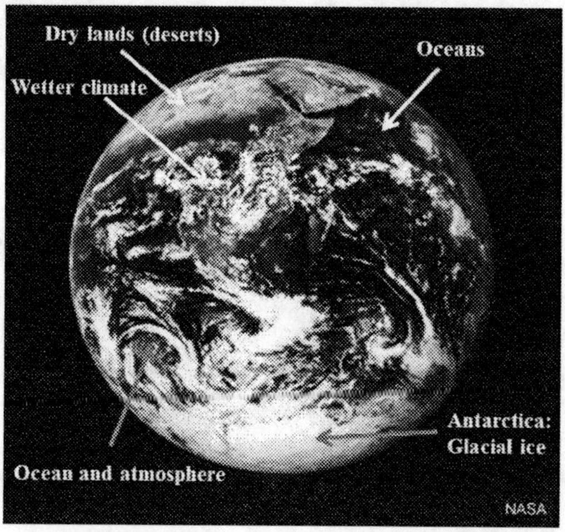

Figure 1.1. A view of the Earth from space through Apollo 17—taking a full photograph of the Earth from space takes some doing. In 1972, the crew of Apollo 17 took a camera to the Moon in order to bring the full sphere into view.

1.1. A View of Earth: The Four Spheres

Our objective in this section is to recall some of the basic components of the Earth, including its composition. Let us start with a view of the Earth from space, as seen through the Apollo program, more precisely Apollo 17, as depicted in Figure 1.1. Notice that in the Saharan dry land and the wet sub-Saharan, water is almost everywhere on Earth, including oceans and glacial ices, as we will discuss

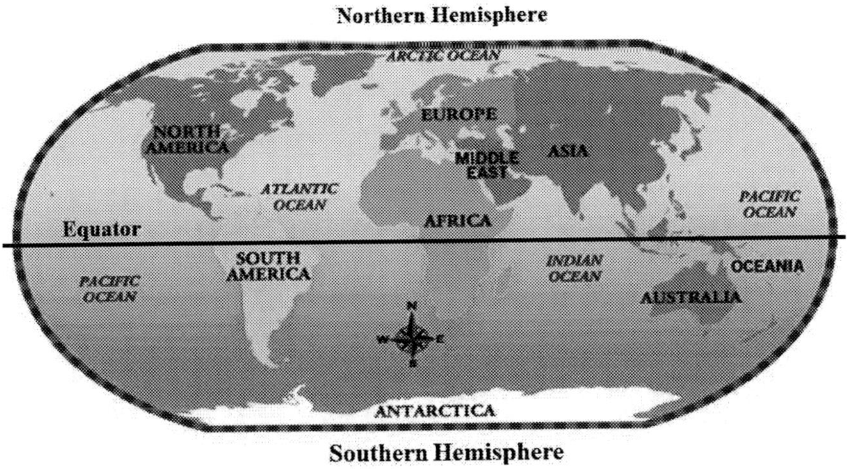

Figure 1.2. Another view of the Earth showing the seven continents (Africa, Antarctica, Asia, Australia, Europe, North America, and South America) and the five oceans (the Antarctic Ocean, the Arctic Ocean, the Atlantic Ocean, the Indian Ocean, and the Pacific Ocean).

later. A close look at the Earth's surface (Figure 1.2) shows that 61 percent of the Northern Hemisphere is covered by water and 81 percent of the Southern Hemisphere is covered by water. However, on a planetary scale, the ocean is small. In fact, there is more water within the Earth's interior than in its oceans and atmosphere—note that Earth's surface is not flat; it has topography. Ignoring oceans, the Earth's surface is dominated by two distinct elevations: Most land is 0-2 km above sea level, and most of the seafloor is 3-5 km below sea level (see Figure 1.3).

As we go through this chapter and this book's subsequent chapters, it is important that you re-familiarize yourselves with the major features of the Earth's surface, namely, the seven continents (Africa, Antarctica, Asia, Australia, Europe, North America, and South America), the five oceans (the Antarctic Ocean, the Arctic Ocean, the Atlantic Ocean, the Indian Ocean, and the Pacific Ocean), and the major mountain ranges (the Himalayan mountains in Asia, the Alps in Europe, the Andes in South America, the Rockies and Mount St. Helens in North America, Mount Pinatubo in the Philip-

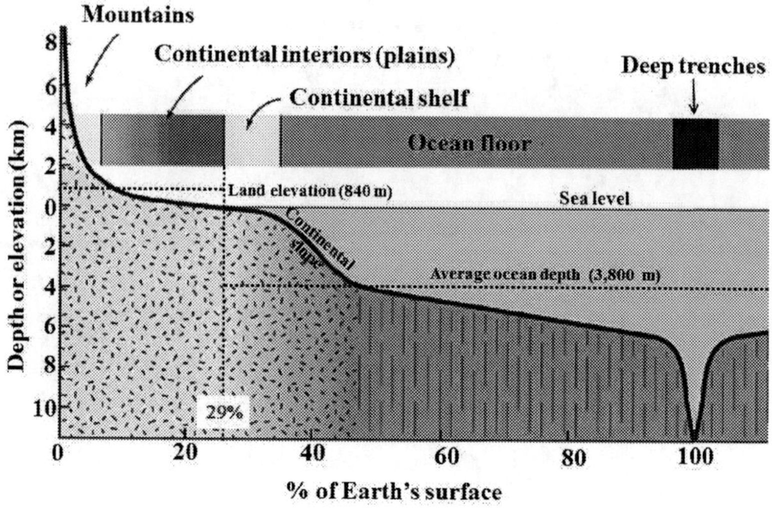

Figure 1.3. A schematic illustration of the relative heights of the major features of the Earth's surface. The horizontal axis is the percentage of the Earth's surface, whereas the vertical axis shows the elevation above or below sea level. Sea level is the average level of the surface of the Earth's oceans. This illustration also shows that the Earth has two types of crust, continental and oceanic. Deep sea trenches are long, narrow depressions on the sea bottom. They are associated with maximum ocean depths of about 10 km. In 2021, explorer Victor Vescovo reached the deepest point in the Mariana Trench (about 10.93 km)—the deepest known point in all Earth's oceans. (Adapted from Encyclopedia Britannica, 2010).

pines, and Mount Kea and Mauna Loa in Hawaii) on the Earth's surface—other useful statistics to keep in mind are that the average land elevation is 840 meters, the average ocean depth is 3,800 meters, the average ocean temperature is 3.9 degrees Celsius (39 degrees Fahrenheit, or 277 degrees Kelvin), and the average global temperature is 14.6 degrees Celsius (58.3 degrees Fahrenheit, or 287.8 degrees Kelvin)—about 50 percent of the Earth's population lives less than 240 km from an ocean. In Japan, by contrast, 96 percent of the human population lives less than 100 km from the ocean. Tibet (see Figure 1.4) and Antarctica (see Figure 1.5) have two of the most hostile environments on the Earth's surface. The Tibetan plateau is one of the most striking features of continental tectonics, developed following the Indian continental subduction and collision with the Asian continent (see Chapter 3). It is the largest mountain mass on Earth. Most of the plateau—an area larger than Texas—has an average elevation of 5 km. It has the world's 26 of the 100 highest mountains clustered together in a single place. The plateau is bordered to the south by the inner Himalayan range and to the north by the Kunlun Mountains. Antarctica, with very cool surface temperatures of between -10 degrees and -60 degrees Celsius, is a cold, dark, and remote place, located in the Southern Hemisphere.

It is a continent where almost all the bedrock is covered by the ice sheet; Antarctica is a land mass surrounded by an ocean. Sea ice can move freely in the open ocean. During the winter, up to 18 million square kilometers of ocean are covered by sea ice, but by the end of summer, only about 3 million square kilometers of sea ice remain. Because the Arctic and Antarctic seas are extremely cold and remote, we often think that these two regions are nearly identical with respect to climate-associated living conditions. Actually, they are significantly different. Polar bears live only in the Arctic, and penguins live only in the Antarctic. There are about 4 million indi-

(a)

(b)

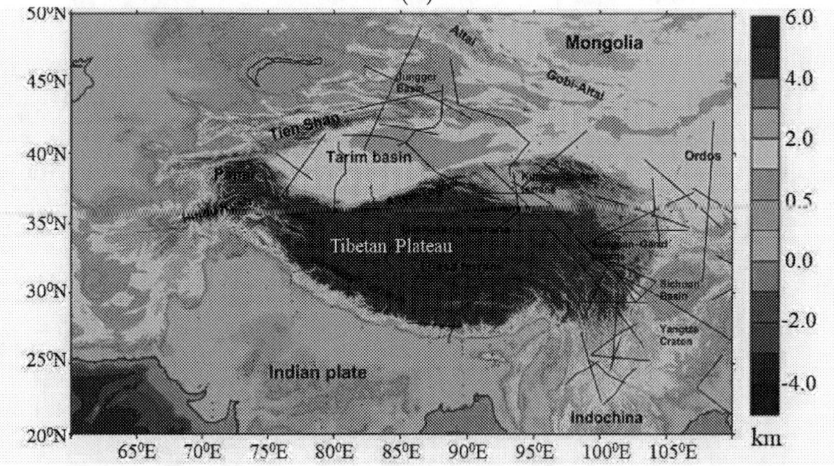

(c)

Figure 1.4. (a) An aerial view of the Tibetan plateau snapped by crew on the International Space Station (Image credit: NASA). Mountains higher than 8000 meters include Makalu (8462 m), Everest (8850 m), Lhotse (8516 m) and Cho Oyu (8201 m). View of a small portion of the Tibetan plateau for detailed analysis. (c) A topographic map of the Tibetan plateau.

genous peoples (e.g., the Inuit, Chukchi, Sami, Yupik, and Inupiat) who live around the Arctic, while Antarctica has never had any native people living there. Antarctica has only ever been inhabited by

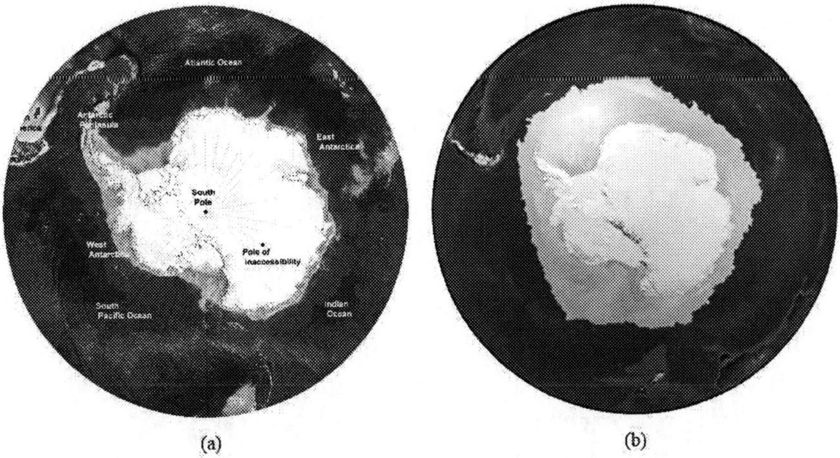

Figure 1.5. (a) A map of the Antarctic region. The Antarctic is a region located at the South Pole. The access points are Argentina, South Africa, New Zealand, and Australia. (b) A map of Antarctic sea ice extent in the winter. Antarctica is 98 percent covered in ice. [Adapted and credited: National Snow and Ice Data Center, University of Colorado, Boulder, CO.]

scientific stations and their personnel who only stay for a year or two. There are ships that bring tourists each summer for six to 14 days. There is still nobody who was born there and has always lived there. Actually, Antarctica is the part of the Earth's surface that most resembles some part of Mars's surface. It is almost thermally isolated from the rest of the planet by the oceans and the atmospheric polar vortex, whereas the Arctic is primarily affected by seasonal atmospheric transport and river flows from surrounding continents (see Figure 1.6). Both regions hugely influence the global oceans. The key difference between the Arctic and Antarctica, as far as climate change is concerned, is that the Antarctic sea ice extent has been increasing steadily at a rate of about 1.8 percent a decade over the last 40 years (Comiso et al., 2017), while the Arctic sea ice extent has been decreasing at a rate of over 4 percent a decade (Vaughan et al., 2013). In other words, the Arctic is not as cold as the Antarctic. What are these differences in sea ice between the two regions? This question is important because the Arctic and Antarctic influence climate over a significant part of the globe through changes in sea level, decreased oceanic heat transport, and increased emissions of greenhouse gases from thawing permafrost. We address this question in Ikelle (2022d).

There are more than seventy lakes about four kilometers beneath the Antarctic ice sheet. The discovery and/or confirmation of these lakes have been made possible by GPR (ground penetration radar) data-acquisition surveys and GPR data imaging, especially that water from none of the lakes has not been sampled directly yet because the lakes are subject to high pressure (about 355 bars), low temperatures (about -3° degrees Celsius), and permanent darkness.

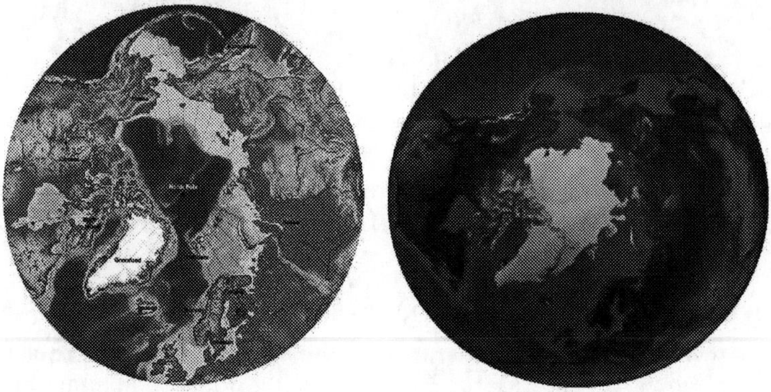

(a) (b)

Figure 1.6. (a) A map of the Arctic region. The Arctic is centered on the North Pole sea, which is surrounded by land. The red dots are access points. (b) A map of Arctic sea ice extent in the winter. [Adapted and credited: National Snow and Ice Data Center, University of Colorado, Boulder, CO].

We generally divide the Earth into four spheres: (i) the atmosphere, (ii) the hydrosphere, (iii) the biosphere, and (iv) the lithosphere. The atmosphere is a thin gaseous envelope surrounding the Earth. The hydrosphere is water, primarily the oceans. The biosphere includes all living things on the planet. The lithosphere is the rocky outer shell of the Earth. All these spheres—the lithosphere, the hydrosphere, the atmosphere, and the biosphere—are linked through flows of mass, energy, and life. Their interactions have affected the development of our planet. Solar energy is essential to the function of each of these spheres and for the entire system. Note that the word *cryosphere* is often used to describe the frozen water in the Earth's system and includes frozen parts of the oceans, such as the waters surrounding Antarctica and the Arctic, which are here considered part of the hydrosphere. Let us now describe in a little more detail the atmosphere, the hydrosphere, the biosphere, and the lithosphere

1.1.1. Atmosphere

An atmosphere is a layer of gas that surrounds a world. Atmospheres form from out-gassing from volcanoes, evaporation of surface liquid, sublimation of surface ice, impacts by the solar wind and comets, and the planet's early magnetic field trapping gases in the solar system. All these mechanisms will be described in the coming chapters.

Composition. The atmosphere is unique in the solar system because it provides the air we breathe and protects us from the sun's intense heat. It is thin (90 percent of it occupies 16 km) and tenuous. It is composed of 78 percent nitrogen[1] (N_2), 21 percent oxygen (O_2), 0.9 percent argon (Ar), and minor amounts of carbon dioxide, methane, and water vapor. In other words, three gases—nitrogen, oxygen, and argon—represent 99.9 percent of the gases in the atmosphere. Yet some of the gases that make up the remaining 0.1 percent (such as carbon dioxide, methane, and water vapor) can have huge effects on the Earth's climate and therefore on our living conditions. We further address this issue in Ikelle (2022a). Note that about 3 billion years ago, oxygen was not present, and therefore there was no life as we know it today. Photosynthesis of microbes and plants, which began about 3 billion years ago, put oxygen into the atmosphere. Photosynthesis is a process used by plants and other organisms to convert light energy, usually from

[1]Nitrogen is a colorless, tasteless, odorless gas that is slightly soluble in water. It is a non-toxic and inert gas that neither burns nor supports combustion and, at room temperature, does not react with any substance. On heating, however, it combines with oxygen to produce nitric oxide, with hydrogen to produce ammonia, with silicon to form silane, and with metals such as lithium, calcium, barium, magnesium and aluminum to form the nitrides.

the sun, into chemical energy which can then be released to fuel the organisms' activities (see Ikelle, 2022a). This process produces oxygen and removes carbon dioxide (CO_2) and methane (CH_4) from the atmosphere. It is remarkable that photosynthesis of plants puts oxygen in the atmosphere at the right percentage, 21 percent. Why not less—say, 15 percent? If so, there would not be human life on Earth as we know it today. Note also that the content of CO_2 in the atmosphere is very small, about 0.03 percent, and yet the total amount of CO_2 is roughly 2 trillion tons in the atmosphere surrounding the Earth.

A layer model. The atmosphere is generally divided into several layers, as depicted in Figure 1.8a. This division is essentially based on the temperature profile. The ground is shown in dark color at the bottom of this figure. The approximate altitude in kilometers is given on the right-hand scale. The approximation here is caused by the fact that the altitudes of the layer boundaries vary with the topography of the Earth's surface. We cannot expect these boundaries to have the same altitudes above the highest peak of the Himalayas and above sea level. The mountain's highest point is about 8 kilometers above sea level.

Let us now briefly describe the layers of the atmosphere: (i) The troposphere is the lowest atmospheric layer, as we can see in Figure 1.8a. It is about 11 km thick. Most clouds are found in the troposphere. The troposphere is thinner at the poles (averaging about 8 km thick) and thicker at the equator (averaging about 16 km thick). The temperature in the troposphere decreases with altitude. (ii) The stratosphere ranges from 11 to 48 kilometers above the Earth's surface. The stratosphere contains the ozone layer, which absorbs most of the harmful ultraviolet radiation from the sun. The temperature in the stratosphere increases slightly with altitude. The highest temperature in this region is about zero degrees Celsius (i.e., 32 degrees Fahrenheit, or 273 degrees Kelvin). (iii) The mesosphere is obviously above the stratosphere, between about 45 and 95 km altitude. It is characterized by a rapid decrease in temperature with altitude. The temperature is about -90 degrees Celsius (-130 degrees Fahrenheit, or 183 degrees Kelvin) at the top of the mesosphere. (iv) The thermosphere starts about 95 km above the Earth's surface. This temperature is quite high. Note that its temperature is not measured with a thermometer but by looking at the motion and speed of the rarefied gases in this layer, which are very energetic, but would not affect a thermometer. Temperatures in this layer may be as high as thousands of degrees Celsius. (v) The exosphere is the layer beyond

Computational exercise: Scripps Institution of Oceanography has data describing the variations of CO_2 volume and the ratio of oxygen and nitrogen volume in the atmosphere (O_2/N_2) from ten stations around the world: Alert, NWT, Canada; Barrow, Alaska; Cold Bay, Alaska; Cape Kumukahi, Hawaii; La Jolla Pier, California; Mauna Loa Observatory, Hawaii; American Samoa; Cape Grim, Australia; Palmer Station, Antarctica; South Pole. The data were obtained using the methods described by Keeling et al. (1998). These data can be downloaded for free from scrippso2.ucsd.edu/data.html. We used the monthly average for the last 20 years (from 1991 to 2020). These data are also available at imodeedupack.com/ims/chp1. Figure 1.7 shows the scatterplot of these data. Use the following Matlab commands:

```
load co2ff; co2 = co2ff(:, 2); load oxygff; oxyg = oxygff(:, 2);
scatter(co2, oxyg)
```

to reproduce Figure 1.7. Verify that

$$O_2/N_2 \approx -10.07\,CO_2 + 3462\ , \tag{1.1}$$

when O_2/N_2 is measured in per meg and CO_2 in ppm, where ppm stands for parts per million (i.e., 1 ppm = 0.0001 percent) and 4.8 per meg is equivalent to 1 ppm (i.e., 1μmole per mole of dry air). Just as percent means "out of a hundred," ppm means "out of a million." The use of ppm and even ppb (parts per billion) is another way of expressing how small the amounts of some gases like O_2 are in the Earth's atmosphere today. Notice that when O_2/N_2 rises, CO_2 falls in an approximately linear way.

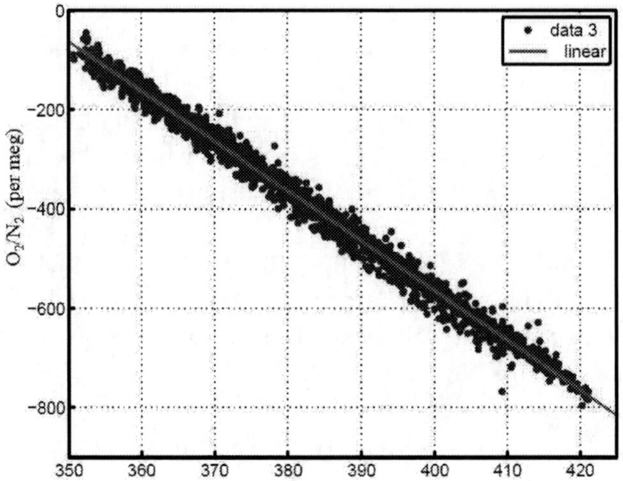

Figure 1.7. The scatterplot of the variations of atmospheric oxygen-nitrogen ratio versus atmospheric carbon dioxide from 1991 to 2020. We used data of Scripps Institution of Oceanography.

the thermosphere. (vi) The ionosphere overlaps the other atmospheric layers above the Earth. The air is ionized[2] by the sun's ultraviolet light. These ionized layers affect the transmittance and reflectance of radio waves. We will discuss this issue in the climate-change chapter. Note that the exosphere and the ionosphere are not shown in Figure 1.8a. Pressure and density in the atmosphere decrease with increasing altitude. Basically, pressure is caused by the weight of any overlying material. The upper atmosphere has less material above it; therefore, its pressure is lower. However, the temperature has a more complicated profile with altitude, as we can see in Figure 1.8.

Some other atmospheres in the solar system. In addition to Earth, many of the other planets in the solar system have atmospheres (Figure 1.9), including Mars and Venus. Venus's atmosphere is composed mostly of carbon dioxide. It also contains minor

[2]Ionization here is the process by which electrically neutral atoms or molecules are converted to electrically charged atoms or molecules, ions). In general, ionization occurs whenever sufficiently charged particles or radiant energy travel through gases, liquids, or solids.

amounts of nitrogen and other trace elements, including compounds based on hydrogen, nitrogen, sulfur, carbon, and oxygen. Though shallower than that of Earth, the atmosphere of Venus is much hotter and denser than Earth's. As greenhouse gases (carbon dioxide and methane) warm the lower atmosphere, they cool the upper atmosphere, leading to a compact thermosphere. Based on the temperature profile of Venus's atmosphere, as shown in Figure 1.8b, we can see that Venus has no stratosphere. Its troposphere extends to an altitude of 65 km (an altitude at which the mesosphere has already been reached on Earth). At the top of the troposphere, temperature and pressure reach Earth-like levels. The Martian atmosphere is very thin and composed mainly of carbon dioxide, along with some nitrogen and argon. The thin atmosphere also increases the variability of the planet's temperature (Figure 1.8c). Mars' surface temperatures vary from lows of approximately -140 degrees Celsius (-220 degrees

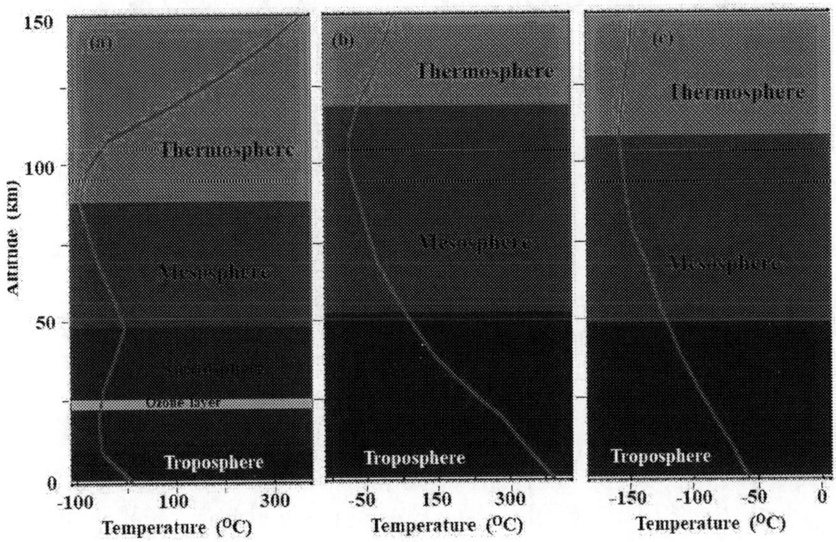

Figure 1.8. (a) This drawing depicts the layers of the Earth's atmosphere. (b) This drawing shows the layers of Venus's atmosphere. Venus has almost the same mass as Earth but a thicker atmosphere, which is composed of 96 percent carbon dioxide. (c) This drawing shows the layers of Mars's atmosphere. The atmosphere on Earth is a hundred times thicker.

Fahrenheit, or 133 degrees Kelvin) during the polar winters to highs of up to 20 degrees Celsius (68 degrees Fahrenheit, or 293 degrees Kelvin) in the summers. The study of extraterrestrial atmospheres is useful because it allows us to gain more insight into the Earth's atmosphere.

Figure 1.9. A comparison of three terrestrial planets: Earth, Mars, and Venus. Mars is much smaller than Earth and Venus. Mars is very cold because it is too far from the sun. Venus is too hot because it is close to the sun, and Earth is just right because it is at an intermediate distance from the Sun.

In addition to Earth, many of the other planets in the solar system have atmospheres (Figure 1.9), including Mars and Venus. Venus's atmosphere is composed mostly of carbon dioxide. It also contains minor amounts of nitrogen and other trace elements, including compounds based on hydrogen, nitrogen, sulfur, carbon, and oxygen. Though shallower than that of Earth, the atmosphere of Venus is much hotter and denser than Earth's. As greenhouse gases (carbon dioxide and methane) warm the lower atmosphere, they cool the upper atmosphere, leading to a compact thermosphere. Based on the temperature profile of Venus's atmosphere, as shown in Figure 1.8b, we can see that Venus has no stratosphere. Its troposphere extends to an altitude of 65 km (an altitude at which the mesosphere has already been reached on Earth). At the top of the troposphere, temperature and pressure reach Earth-like levels. The Martian atmosphere is very thin and composed mainly of carbon dioxide, along with some nitrogen and argon. The thin atmosphere also increases the variability of the planet's temperature (Figure 1.8c). Mars' surface temperatures vary from lows of approximately -140 degrees Celsius (-220 degrees Fahrenheit or 133 degrees Kelvin) during the polar winters to highs of up to 20 degrees Celsius (68 degrees Fahrenheit or 293 degrees Kelvin) in the summers. The study of extraterrestrial atmospheres is useful because it allows us to gain more insight into Earth's atmosphere.

Quiz: Where does an atmosphere end?

Box 1.1. Early Atmosphere and Oxygen Impact on Living Things

Atmospheric oxygen is a basic element of life and a necessary element for the survival of all living organisms, except for some microbes. Any change in its concentration impacts our health. In fact, the Earth's atmosphere had no free oxygen (O_2) until about 2.4 billion years ago and there was no life before then, except some microbes that consumed organic components in the oceans. This early atmosphere (4.5 to 2.5 billion years ago) is known as the Archean atmosphere. It had more vigorous volcanism than today and volcanic gases dominated the Archean atmosphere which was mainly composed of nitrogen (N_2) and carbon dioxide (CO_2) and some traces of water (H_2O), hydrogen gas (H_2), ammonia (NH_3), methane (CH_4), hydrogen sulfide (H_2S), as illustrated in Figure 1.10 (see Schlesinger, 1997; Abe, 1993; Kasting, 1993; Allègre and Schneider, 1994; Morel and Reinfelder, 1995; and Kump et al., 1991; for details). The surface temperatures of the Archean Earth were around 20 to 25 degrees Celsius (see Chapter 3), despite the fact that the heat output by the Archean sun was probably about 25 percent lower than today, but because of its carbon dioxide-rich atmosphere. The period between 2.5 billion and 570 million years ago, known as the Proterozoic Earth, marked the appearance of free oxygen in the Earth's atmosphere. Because of the coupling of the oxygen and carbon dioxide cycles in photosynthesis (Ikelle, 2022a), as oxygen rises, carbon dioxide falls. Like the Archean, the Proterozoic was for most of its duration dominated by microbes. However many of them who did not rely on oxygen to live died.

The passage from the Archean period to the Proterozoic, marked not only a major shift in the atmosphere composition, but also in the world climate with the first known major ice age about 2.3 billion years before the present. The cause of this ice age is the combination of methane and oxygen in reactions induced by sunlight (see Ikelle, 2022a). Also a number of large meteorites hit the Earth during the Proterozoic period, and some of them might have been large enough to wipe out as much as 60 percent of surface life.

A biological revolution occurred at the beginning of Phanerozoic Earth (i.e., the period from 570 million years to the present). The first millions of years of this period saw an explosion of multi-celled animals. The arrival of these new inhabitants along with the spread of large vascular plants accompanied the rise in atmospheric oxygen to about 35 percent of atmospheric volume between 375 and 275 million years ago. The sudden drop in O_2 below 15 percent of atmospheric volume about 275 million years ago may be due to wildfires. Such wildfires can be ignited by volcanic activity, meteorite impacts, and/or lightning. The disappearance of a large size of insects (more than ten-fold larger than those in similar groups alive today) (Shear and Kukalova-Peck, 1990) and amphibians about 275 million years ago can be attributed to these wildfires. Our modern oxygen level of 21 percent reduces the occurrence of wildfires. Below about 15 percent, fires, even if ignited, will not propagate (Scott, 2000; Glasspool et al., 2004 and 2010; and Belcher et al., 2013).

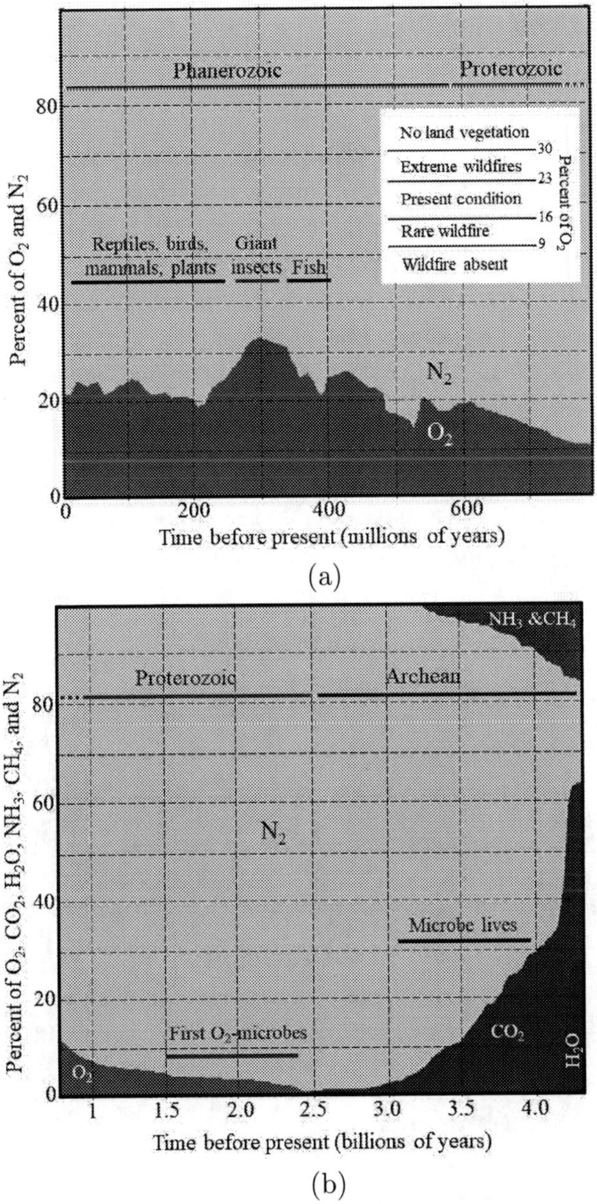

Figure 1.10. (a) The variations in atmospheric composition over time. The layered model indicates the amount of O_2 in the atmosphere that can cause wildfires. (b) Prior to 3.5 billion years ago, O_2 was absent.

1.1.2. Hydrosphere

The hydrosphere refers to the total mass of water on or near the Earth's surface. Water in its purest form is the compound H_2O; water contains two hydrogen (H) atoms bonded to one oxygen (O) atom. Generally it is rare to find pure water; there are usually other compounds, ions, and/or particles mixed with it. Though water may move and carry other substances with it, the small water molecules make life on Earth possible. In fact, liquid water makes the Earth a special place. Our planet has a temperature range that allows water to remain in a liquid state. In a cold place like Pluto [which has an average surface temperature of -229 degrees Celsius (-380 degrees Fahrenheit, or 44 degrees Kelvin)], all of the water would be permanently frozen solid. However, on a very hot planet, all of the water would be in a gaseous state. Water vapor and solid water are relatively useless to Earth's organisms.

Figure 1.11. Water is everywhere on the planet. The volume of water in the planet's ice caps, glaciers, rivers, lakes, aquifers and oceans is about 1.4 million trillion m^3 and has not changed in millennia, and it will not be changing in the foreseeable future.

Saltwater and freshwater. Water covers 71 percent of the Earth's surface. Current estimates are that there are about 1400 million km^3 of water on Earth. This figure includes water in liquid and frozen forms in groundwater, glaciers, oceans, lakes, and streams. Therefore, by the law of averages, it is more likely to find minerals below the ocean than on land. Of the water covering the Earth's surface, 97.5 percent is in the oceans (i.e., saltwater accounts for 97.5 percent of this amount), and 2.5 percent is in glaciers, groundwater, lakes, and streams (i.e., freshwater accounts for only 2.5 percent of the amount of water covering the Earth's surface). Let us examine the freshwater partition. Of this water, 68.7 percent is in the form of ice and permanent snow in the Arctic, the Antarctic, and mountainous regions. Next, 29.9 percent exists as fresh groundwater. Only 0.26 percent of the total amount of freshwater on the Earth is easily accessible. It is found in lakes, reservoirs, and river systems. In other words, not only is the amount of freshwater on Earth small, but also the amount of accessible fresh water is tiny.

The water cycle. In addition to the limited amount of freshwater available for use, water appears everywhere on Earth (see Figure 1.11). Water is in the air, on the land, between the rocks (rocks are aggregates of minerals, with igneous, sedimentary, and metamorphic

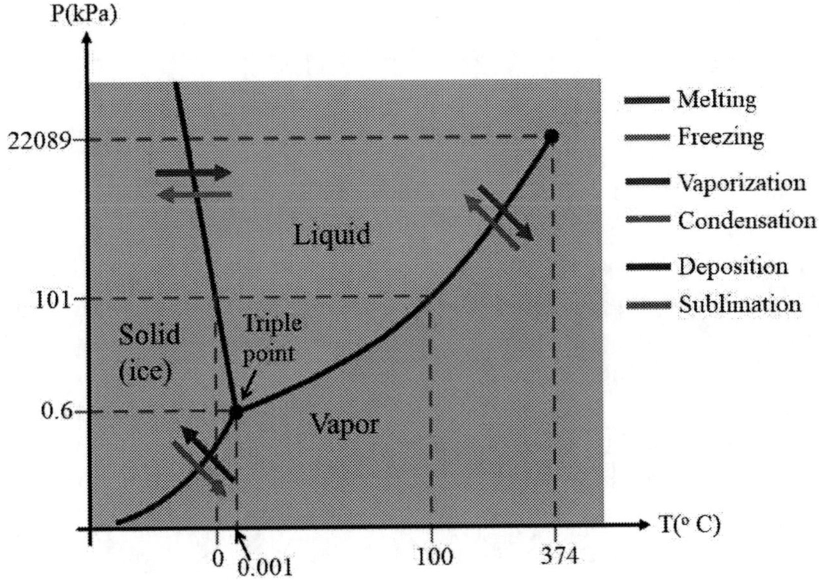

(a)

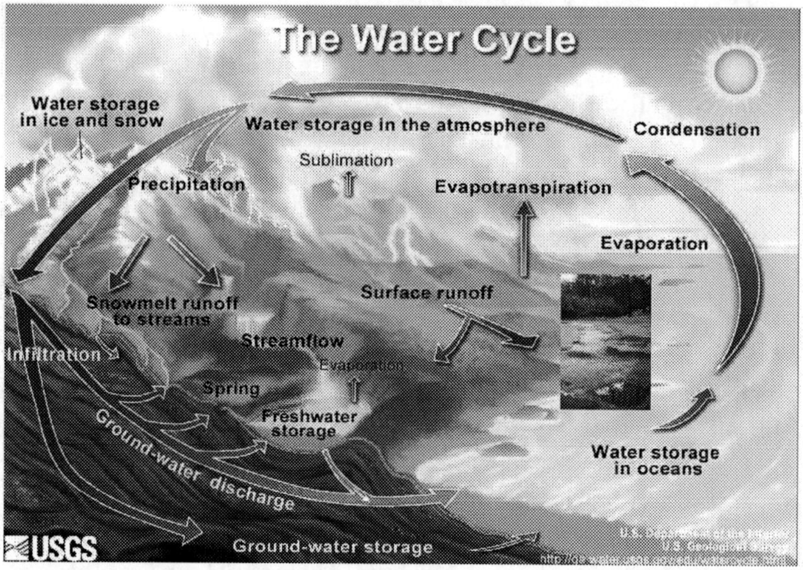

(b)

Figure 1.12. (a) Pressure-versus-temperature diagram of the water phase diagram. (b) An illustration of the water cycle, also known as the hydrologic cycle. It describes the continuous flow of water on, above, and below the surface of the Earth. Water transitions through the liquid, solid (ice), and gas (vapor) phases. [Adapted from USGS for (b)]

rocks being the main types), and in every living thing. Water is solid in glaciers, liquid in oceans, and in a vaporous state in the clouds. Glaciers will eventually melt and will dramatically affect all of life on Earth. Again, the physical states of water are equally important because they are all connected in what is known as the water cycle (also known as the hydrologic cycle).

Let us now discuss the water cycle, depicted in Figure 1.12. The key phenomena in the water cycle to pay the most attention to are (i) evaporation, (ii) condensation, (iii) precipitation, (iv) runoff, (v) sublimation, (vi) transpiration, and (vii) evapotranspiration. Evaporation occurs when the physical state of water is changed from a liquid state to a gaseous state. Condensation is the process by which water vapor changes into a liquid.

Evaporation takes water from oceans and land into the atmosphere, and precipitation takes water vapor from the atmosphere to land and oceans. Precipitation is the process that occurs when any and all forms of water particles reach the ground. Evaporation and precipitation are very important factors in climate studies. Runoff is the flow from a drainage basin or watershed that appears in surface streams. Sublimation refers to the evaporation of snow and ice. Transpiration refers to the expiration of water through the minute pores or stomata of trees. *Evapotranspiration* is the term used by hydrologists to describe the processes of transpiration, sublimation, and evaporation.

Some of the water that precipitates does not runoff into the rivers. It is rather absorbed by the plants and moves deep into the soil. This process is called infiltration. Infiltration is central to the growth of vegetation and soil temperature. High infiltration rates occur in dry soils, with infiltration slowing as the soil becomes wet.

The water cycle is a continuous cycle during which water evaporates, travels into the air, and becomes part of a cloud, falls down to the Earth as precipitation, and then evaporates again. This process is a never-ending cycle. In other words, the water cycle transfers water from one state or reservoir to another. Reservoirs include atmospheric moisture (snow, rain, and clouds), oceans, rivers, lakes, groundwater, subterranean aquifers, polar ice caps, and saturated soil. Solar energy, in the form of heat, light (insolation), and gravity causes the transfer of water from one state to another over periods of time from hours to thousands of years. Most evaporation begins in the oceans and is returned to the Earth as snow or rain. Let us emphasize some of these points by quoting author Marq de Villiers (1999, *Water: The Fate of Our Most Precious Resource*), who says, "On Earth, the total amount of water has almost certainly not changed since geological times: what we had then we still have. Water can be polluted, abused, and misused but it is neither created nor destroyed—it only migrates. There is no evidence that water vapor escapes into space."

Every year the turnover of water on Earth involves about 600,000 km^3 of water. This amount of water represents the total runoff of the Earth's rivers and the direct groundwater runoff to the ocean—

that is, the difference between the amount of water that has fallen as atmospheric precipitation and the amount that represents all evaporation from the Earth's surface. Moreover, this amount of water is the principal source of fresh water needed to support life and man's socioeconomic activities.

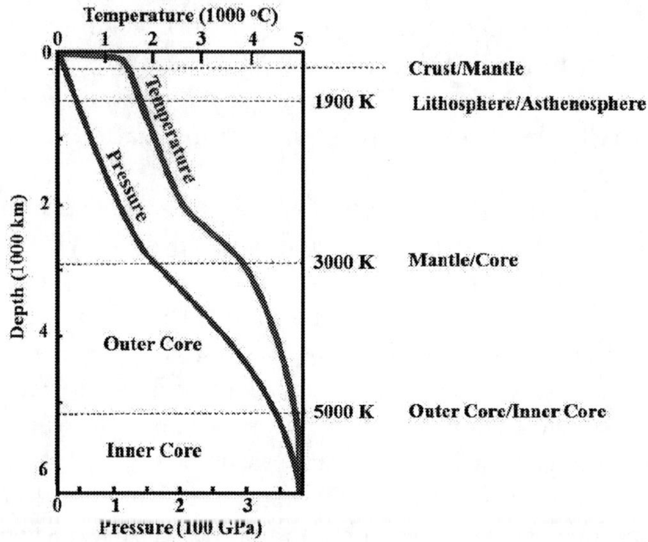

Figure 1.13. An illustration of the geothermal gradient of the Earth. As the depth of the solid Earth increases and away from tectonic-plate boundaries and hotspots, the temperature increases as well. The deeper a rock is, the hotter and denser it is. The temperature increases by about 25 degrees Celsius (77 degrees Fahrenheit or 298 degrees Kelvin), and the pressure increases by about 0.025 GigaPascal (GPa) for every kilometer of depth. Note also that Earth's surface pressure was 2 bars during the Archean period (4.5 to 2.5 billion years ago) and 1 bar since then.

There is a fourth state (gas to plasma) of matter, in addition to the three states, in Figure 1.12a. This state is called plasma.[3] It occurs when temperatures rise to super-high levels, between 10^3 degrees Celsius and 10^9 degrees Celsius. Plasma is not a common state of matter on Earth, but it may be the most common state of

[3]The word *plasma* is attributed to Nobel prize winner Chemist Irvin Langmuir, who was reminded of corpuscles being carried in the blood.

matter in our universe. Plasma is found in stars and lightning. The sun is in a plasma state.

Box 1.2. Temperature Scales and Pressure

Temperature is the measure of the average molecular motion in a system. As heat energy increases, the temperature will increase. Temperature is the degree or intensity of hotness or coldness. A hot body becomes cooler and a cold body becomes warmer as long as heat flows from one body to another. A hot body has a greater ability to give up heat and therefore has a higher temperature than a cold body. There are three temperature scales: Celsius, Fahrenheit, and Kelvin. Temperature scales differ in two ways: the point chosen as the zero degree and the magnitude of incremental units or degrees on the scale. The Celsius scale (C) is used for common temperature measurements. Its zero point, 0 degrees Celsius, is defined as the freezing point of water, with additional degrees defined so that 100 degrees Celsius is the boiling point of water at sea-level atmospheric pressure. The United States commonly employs the Fahrenheit scale (F), on which water freezes at 32 degrees Fahrenheit and boils at 212 degrees Fahrenheit at sea-level atmospheric pressure. Scientists generally use the Kelvin temperature scale (unit symbol K) because it preserves temperature information, even when dealing with temperature to an even power (e.g., T^4). Its zero point, 0 degree Kelvin, coincides with the coldest possible temperature (called absolute zero). The temperature of absolute zero occurs at 0 degree Kelvin = −273.15 degree Celsius = −459.67 degrees Fahrenheit, and the freezing point of water at sea-level atmospheric pressure occurs at 273.15 degrees Kelvin = 0 degree Celsius. The following are useful formulas for moving from one temperature scale to another:

$$\begin{cases} K = C + 273.15 & \text{from the Celsius scale to the Kelvin scale} \\ F = (9/5)C + 32 & \text{from the Celsius scale to the Fahrenheit scale} \\ F = (9/5)K - 459 & \text{from the Kelvin scale to the Fahrenheit scale} \end{cases}$$

Stars are hotter than planets, planets are hotter than icy asteroids, and icy asteroids are hotter than the vacuum of space between astronomical objects.

Pressure. There are two basic types of pressure, hydrostatic and lithospheric. The pressure that a rock is subjected to at a given depth is determined by the weight (hence, the density) of the rock above. This pressure is called lithostatic pressure or sometimes overburdened pressure. For the shallow part of the subsurface, we can mathematically define it as

$$P_{\text{litho}}(z) = \rho_{\text{over}} g z = P_{\text{hydro}}(z) + P_{\text{matrix}}(z) , \tag{1.2}$$

where $P_{\text{litho}}(z)$ is the lithostatic pressure at depth z, ρ_{over} is the mean rock density above z, and g is the gravitational acceleration. The lithostatic pressure at any depth z can alternatively be described as the sum of the overlaying fluid column pressure, $P_{\text{hydro}}(z)$, and the overlaying grain or matrix column pressure, $P_{\text{matrix}}(z)$. The pressure in a fluid, with a connected pathway, occupying a rock formation, is

$$P_{\text{hydro}}(z) = \rho_{\text{fluid}} g z , \tag{1.3}$$

where $P_{\text{hydro}}(z)$ is the hydrostatic pressure at z and ρ_{fluid} is the mean rock-fluid density above z. Because the density of fluids is smaller and about a third of that of rocks, the hydrostatic pressure is smaller than the lithostatic pressure.

1.1.3. Biosphere

Composition. The biosphere includes all life on Earth, including animals and plants on land and those in the sea and air (such as tree roots, flying insects, and birds). The biosphere interacts with and exchanges matter and energy with the other spheres, helping to drive the global recycling of carbon, nitrogen, phosphorus, sulfur, and other elements. In other words, the physical properties of the biosphere in terms of its surface reflectance (albedo), and the exchange of heat and moisture with the atmosphere, are also critical for understanding the worldwide circulation of heat and moisture and therefore, the Earth's climate. Alterations in both the physics (albedo and heat exchange) and the chemistry (carbon dioxide, methane, etc.) of the Earth's systems by the biosphere are fundamental to understanding anthropogenic global warming. Actually, we have just summarized the climate-change chapter of Ikelle (2022c and 2022d). Note that *anthropogenic* refers to the fact that the effects under consideration are of human origin. *Anthropogenic* is an important word in modern climate-change studies.

The deep biosphere. There is life on the ocean floor despite the fact that the pressure there is extremely high and no sunlight penetrates. Deep under the ground and on the seafloor, life is very abundant, albeit mainly in the form of microorganisms that are not visible to the naked eye. In fact, the deep subsurface biosphere represents a substantial fraction of the Earth's total biomass; estimates suggest a contribution of 10 to 30 percent. Biomass is the total weight of the organic material of a particular species or in a particular habitat per unit of area or volume.

Amazingly, life exists deep under the ground. Jules Gabriel Verne's famous book, *A Journey to the Centre of the Earth* (1864), reports how Professor Lidenbrock, his nephew, Axel, and their guide, Hans, find a versatile realm of living organisms during their descent into the interior of our planet. When the novel was first published, the idea of life existing in the deep subsurface, remote from sunlight and air, was genuine science fiction. But toward the end of the twentieth century, when scientific drilling opened the window into environments hundreds of meters deep, it became clear that Jules Verne was in a sense right.

The discovery of the deep biosphere has given rise to fundamental new questions about the evolution and distribution of life. Considering the broad range of temperatures and high pressures in the subsurface (see Figure 1.13) and the lack of fresh organic matter as a source of energy and nutrients, the existence and survival of an extensive deep biosphere are remarkable. Geofuels (i.e., gases that are generated by mineral-water interactions or derived from the deep crust) might be an alternative energy source in the absence of light and photosynthesis. So far, little is known about the individual inhabitants of the deep biosphere, such as their genetic diversity, metabolic capabilities, and survival strategies. Moreover, the relationships between deep subsurface ecosystems and the surface world remain to be explored. Let us now draw your attention to the coastal zone of the Red Sea and the Gulfs of Aden, Aqaba, and Suez (Figure 1.14).

These regions contain some of the world's most unique coastal and marine environments. Among the most notable is the extraordinary system of coral reefs and their associated animals and plants. Surrounded by arid terrestrial environments, which are themselves unique, these environments support rich biological communities and representatives of several endangered species. These natural resources have supported coastal populations for thousands of years and nourished the development of both a maritime and trading culture linking Saudi Arabia and Africa with Europe and Asia.

Lands bordering the Red Sea are subject to extremely hot temperatures, resulting in an evaporation rate of over 80 inches a year from the sea itself. Moreover the region's rainfall is minimal. Yet, almost all water lost to evaporation is automatically and entirely replaced by an ongoing influx of water from the Gulf of Aden driven by winds through the eastern channel of the Babel-Mandeb Strait. It is estimated that the entire waters of the Red Sea are replaced by this process every 20 years. In other words, this unique geography is due to the tectonic plates about 50 million years ago, when the Arabia Plate separated from the Africa Plate. The extremely hot temperatures cause the evaporation of the Red Sea. Fortunately, there is an influx of water from the Gulf of Aden to the Red Sea which ensures that the Red Sea is not going dry. Figure 1.14 shows a very small sample of the rich biodiversity of the Red Sea. It contains spectacular fauna and flora and coral reefs that thrive among the mangroves and sea-grass beds, the salt-pans, and salt marshes.

Quiz: (1) How did we prove that there is life deep in the ground? (2) What is the difference between evaporation and precipitation? (3) Lands and oceans constitute the Earth's surface. What percentage of the Earth's surface is occupied by water? The Earth's surface can be divided into the Northern Hemisphere and the Southern Hemisphere. Which hemisphere has the larger mass of ocean?

1.1.4. Lithosphere

A layer model of solid Earth. Let us now finally turn to the solid Earth beneath our feet. The solid Earth sometimes called the geosphere, has a layered structure in a spherical-shell form. These layers can be defined by their chemical or physical properties, including their rock minerals. Physical properties include deformability—in other words, are we dealing with solid, liquid, or gas?

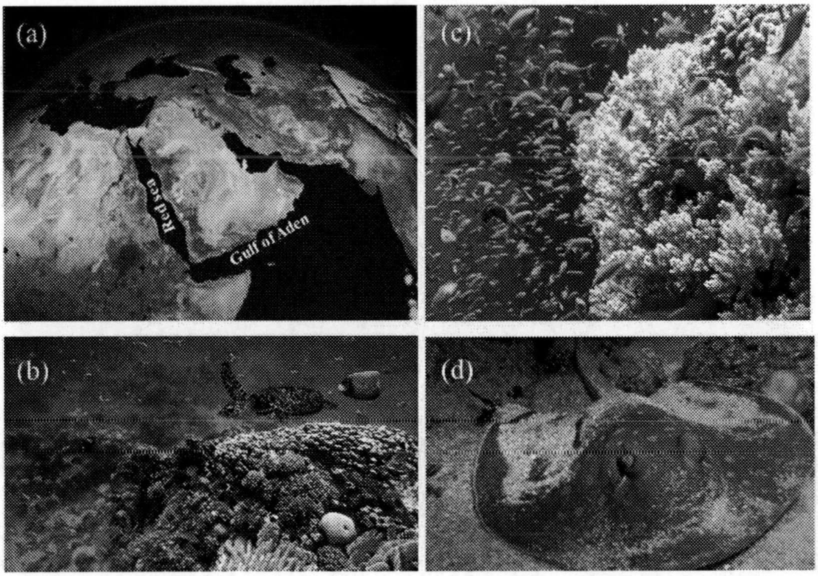

Figure 1.14. (a) The Red Sea and Gulf of Aden. (b),(c), and (d) Three illustrations of the rich biodiversity of the Red Sea. (courtesy Aramco).

Based on their chemical properties, these layers are (see Figure 1.15) the crust, the mantle, and the core. The crust has 1/250 (0.5 percent) of the Earth's mass, the mantle has about two-thirds (67 percent) of the Earth's mass, and the core has one-third (32.5 percent) of the Earth's mass. In other words, the mass of the crust is tiny compared to that of the core and the mantle. Similarly, in terms of volume, the crust is very small. The crust has 1/100 (1 percent) of the Earth's volume, the mantle has about five-sixths (83 percent) of the Earth's volume, and the core has one-sixth (16 percent) of the

Earth's volume. Clearly the mantle is the largest and heaviest layer (2,900 km thick). It is composed of silicate rocks, with abundant iron and magnesium. Its mass density (mass/volume) ranges from 3.2 to 5 g/cc (cc stands for cubic centimeters). The core, which is composed of iron, has an average mass density of 10.8 g/cc.

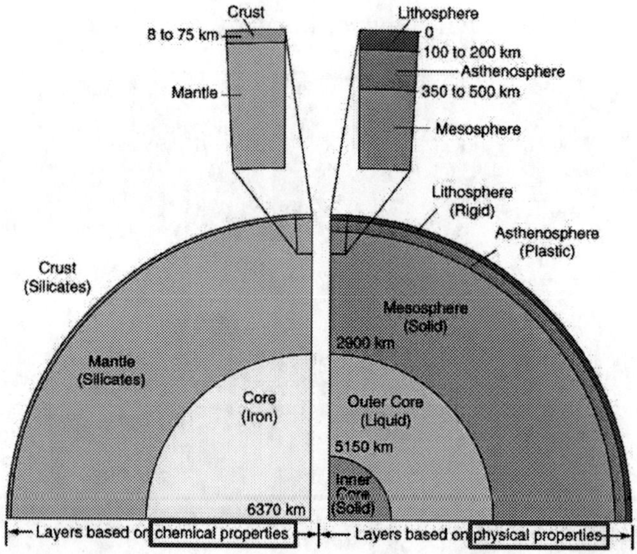

Figure 1.15. The structure of the solid Earth is based on the chemical (left) and physical (right) properties. The crust, the mantle, and the core are layers describing the solid Earth, based on its chemical properties. The lithosphere, the asthenosphere, the mesosphere, the outer core, and the inner core are layers of solid Earth based on its physical properties. The outer core is liquid, whereas the inner core is solid.

As illustrated in Figure 1.15, the crust is the outermost compositional layer. The Earth's crust is made up of several elements: oxygen, 47 percent; silicon, 27 percent; aluminum, 8 percent; iron, 5 percent; calcium, 4 percent; and magnesium, potassium, and sodium, 2 percent each. There are two types of crust: the continental crust (up to 75 km thick in the Himalayas) and the oceanic crust (about 8 km thick). The continental crust is thick (see Figure 1.16), with a mass density of about 2.7 g/cc. It is quite old—maybe billions of years old. Its composition ranges from mafic to felsic. The oceanic crust (which underlies the ocean basins) is thinner, with a mass density of about 3.0 g/cc, and it is comparatively undeformed and much younger, less than 200 million years old. The crust "floats" on a "weak" mantle.

Based on their physical properties, the layers that constitute the solid Earth are the lithosphere, the asthenosphere, the mesospheric mantle, the outer core, and the inner core. In other words, the interior of the Earth is divided into five layers based on its physical properties and into three layers based on its chemical properties. The inner core is solid, and the outer core is liquid iron (along with some nickel and other elements). In fact the outer core is the only internal layer of the Earth that is a true liquid. The core-mantle boundary is the one physical boundary that is also a chemical boundary.

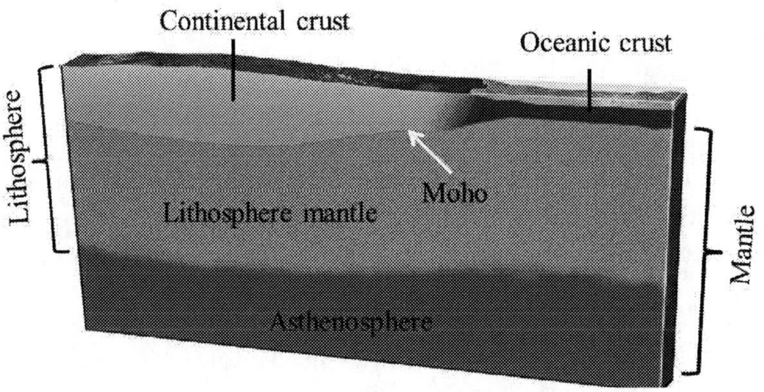

Figure 1.16. The crust is the outermost "skin" of Earth and has variable thicknesses; it is thickest under mountain ranges (70 km) and thinnest under mid-ocean ridges (3 km). The Mohorovicic discontinuity, or Moho, is the boundary separating the crust and the mantle. There are two types of crust—the continental crust and the oceanic crust. The continental crust underlies the continents, and the oceanic crust underlies the ocean.

The lithosphere includes the crust, plus the upper portion of the mantle. It is a rigid solid. Its thickness ranges from 10 km beneath the oceans to 300 km in continental areas. The outer core has 31 percent of the Earth's mass, and the inner core has 1.5 percent of the Earth's mass. In volume, the outer core has 15.5 percent of the Earth's volume, and the inner core has 0.7 percent of the Earth's volume. In other words, at the center of the Earth is the core, which has two parts. The solid inner core of iron has a radius of about 1,220 km. It is surrounded by a liquid outer core composed of a nickel-iron alloy. It is about 2,180 km thick.

Quiz: (1) Define the mass density. (2) What are the differences between continental crust and oceanic crust? (3) What are the differences between the crust and the lithosphere? (4) Is the solid Earth homogeneous? (5) What is the temperature at the base of a 40-km-thick granitic crust with a geothermal gradient of 16 degrees Celsius per kilometer? (6) What is the pressure at the base of a 35-km thick granitic crust with a density of 3000 kg/m^3? (7) What is the radius of the Earth? (8) What are the differences between minerals, glasses, and rocks? (9) List two tools (methods) used to look beneath our feet, inside the solid Earth.

Computational exercise:
Table 1.1 provides Matlab commands for predicting temperature variations with altitude in the atmosphere. The model here assumes that the air is devoid of dust, moisture, and water vapor. (1) Reproduce the results in Figure 1.17. (2) Our description is limited to the troposphere and stratosphere. Determine the temperature in the stratosphere. (3) Temperature linearly decreases with altitude in the troposphere—determine the rate of this variation.

Table 1.1. Matlab commands for predicting pressure variations with depth.

```
>> set(0,'defaultlinelinewidth',2.5);
>> set(0,'defaultaxeslinewidth',2.5);
>> h =1000:100:15000; % altitude in meters
>> [T,a,p,rho]= atmosisa(h); h1=h*0.001;
>> plot(T,h1); grid on
>> axis([210 290 1 15]); print fig17 –dpdf
```

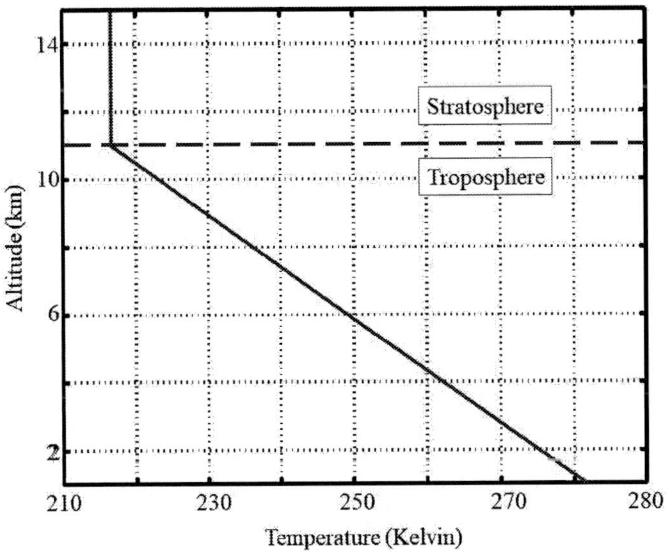

Figure 1.17. Atmosphere temperature variation with altitude.

Computational exercise: We often use latitude and longitude to describe points on the Earth's surface. As illustrated in Figure 1.18, latitude, λ, is parallel to the equator and longitude, ϕ, is parallel to Greenwich meridian; λ can be referenced with a north latitude (e.g., 70N) and south latitude (e.g., 25S)and ϕ, is referenced by a west longitude (e.g., 30W) and an east longitude (e.g., 15E). From λ and ϕ, we can deduce the Cartesian coordinates:

$$\begin{pmatrix} x = R\cos\lambda\,\cos\phi \\ y = R\cos\lambda\,\sin\phi \\ z = R\sin\lambda \end{pmatrix}, \tag{1.4}$$

where x is the projection of the plane of the Earth's equator along line of longitude of 0 degree, y is the projection of the plane of the Earth's equator along line of longitude of 90 degrees East, and the z-component points toward the North Pole. Use the commands in Table 1.2 to estimate (x, y) for these points: (λ, ϕ)=(40N, -90W), (60N, 90W), (25S, 45W), (90N, 35E), and (0N, 15E).

Computational exercise: Mineral exploration is influenced by pressure variations with depth in the solid Earth. At the near Earth's surface (with the assumption that the density is nearly uniform), pressure is given by

$$P = g\rho\left(z - z_{\mathrm{surf}}\right) \ , \tag{1.5}$$

where $g = 9.8\,\mathrm{m/s^2}$ is the acceleration of gravity, ρ is the density in $\mathrm{kg/m^3}$, and z_{surf} and z are the depths in meters at the top and bottom of the region under consideration, respectively. Table 1.3 provides Matlab commands for predicting pressure variations with depth in the atmosphere. Reproduce the results in Figure 1.19.

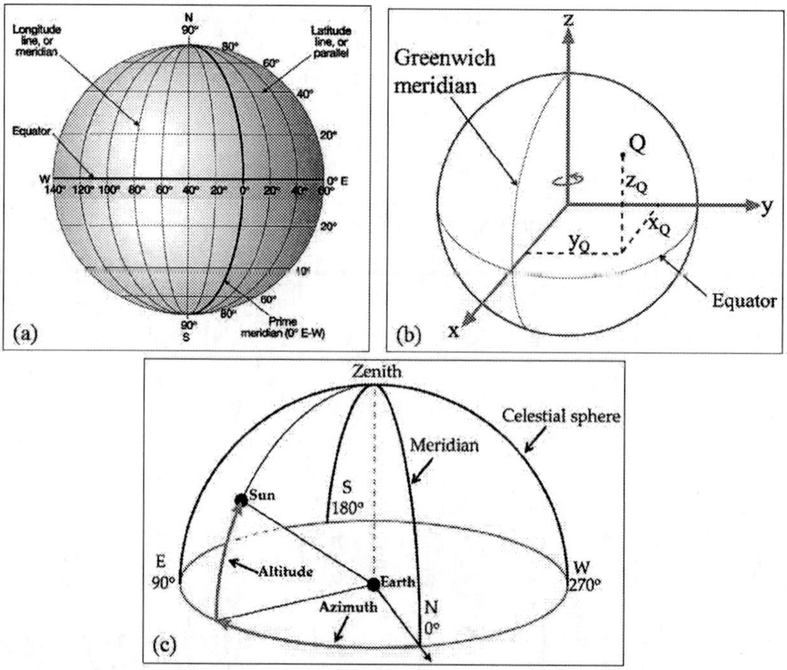

Figure 1.18. (a) Longitude lines are perpendicular to the equator and latitude lines are parallel to the equator. (b) Cartesian coordinates. (c) Definitions of altitude and azimuth for a position (e.g., the Sun) in the atmosphere and/or outer space with respect to an observation point (e.g., solid Earth). Sun motion projected on a sphere concentric to the Earth is known as the celestial sphere.

Table 1.2. Matlab commands for determining Cartesian coordinates of a given point from its latitude and longitude.

```
>> dangle= pi/180; lat = 45*dangle; long=-90*dangle;
>> Re=6370;
>> rr = Re*[cos(lat)*cos(long), cos(lat)*cos(long), sin(lat)]
rr = 1.0e+003*[0.0000 0.0000 4.5043]
```

Table 1.3. Matlab commands for predicting pressure variations with depth.

```
>> g = 9.8;zsurf=0; zb=1000*10;
>> z =zsurf:1:zb; rho = 2500; p = g*rho*z;
>> plot(p*1e-6, z*1e-3); grid; set(gca,'ydir','reverse')
>> ylabel('Depth (km)'); xlabel('Pressure (MPa)')
```

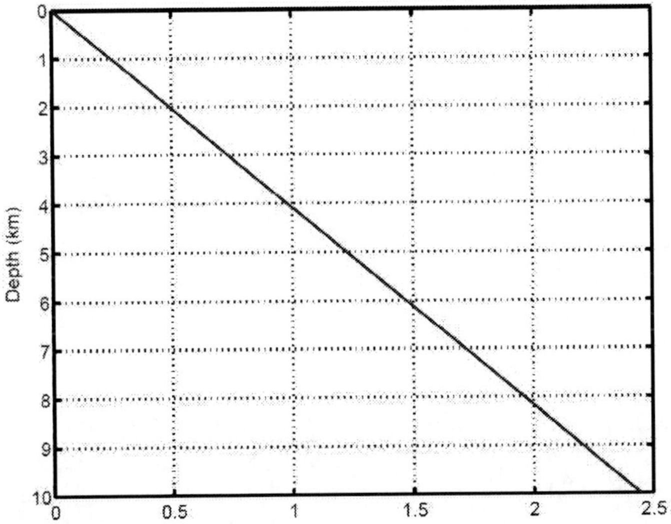

Figure 1.19. Pressure variations with depth.

1.1.5. Glaciers, Ice Ages, Wildfires and the Holocene Period

Let us be more precise about glaciers, ice ages, and wildfires. We will also introduce the Holocene period, which is quite an important period in climate studies.

Figure 1.20. Here is an example of the Greenland ice sheet. Sheets like this one are known as continental glaciers. Ninety-six percent of all glaciers are continental glaciers. (Adapted from http://tvl1.geo.uc.edu/ice/Image/pretty/green.html).

Glaciers and ice ages. Glaciers are very large ice sheets that spread outward in all directions like pancake batter. They are formed from layers of snow that collect in cold areas. More snow is collected than melts each year, so the snow piles up. The snow on the bottom is compressed into ice. Before long an expanding field of ice covers the ground. Glaciers commonly occur in Antarctica and Greenland, but they are also found in some mountain valleys. Rivers of ice are also glaciers. Figure 1.20 shows an example of the Greenland ice sheet. Ice sheets like this one are also continental glaciers. Ninety-six percent of all glaciers are continental glaciers. They are found in Antarctica and Greenland, and covered most of Canada and the northern United States during the last ice age (up until about 18,000 years ago). Figure 1.21 shows an example of a mountain glacier in the Saint Elias Range, Yukon Territory, Canada. As with all mountain or valley glaciers, glaciers begin in the mountains at high altitudes. As snow thickens, it flows downhill, following the shape of the valley.

Figure 1.21. An example of a mountain glacier is in the Saint Elias Range, Yukon Territory, Canada. As with all mountain or valley glaciers, this glacier begins in the mountains at high altitudes. As it thickens, it flows downhill, following the shape of the valley. (Adapted from http://nsidc.org/glaciers/gallery).

An ice age or, more precisely, a glacial age, is a period of long-term reduction in the temperature of the Earth's surface and atmosphere, resulting in the presence or expansion of continental ice sheets, polar ice sheets and alpine glaciers. In other words, successive years of cold weather constitute an ice age. Glaciers are formed during these years of cold weather. In the past million years, the Earth has seen some dramatic climate fluctuations, with five major and 30 minor glaciations. Figure 1.22 shows four examples of ice ages in the last 500,000 years. During a long-term ice age, individual pulses of extra-cold climate are termed "glacial periods," and intermittent warm periods are called "interglacials" (see Figure 1.22). Several ice ages have caused the northern United States to be covered with glaciers. The last ice age was at its maximum approximately 25,000 years ago. Glaciers 1 to 2 kilometers thick covered the mountain peaks of New England (Figure 1.23). By 18,000 years ago, the ice had all melted and retreated, leaving behind a changed landscape.

On the time-scale of the entire history of the Earth, there have been four major glaciations, from about 2.4 to 2.1 billion years ago, from 850 to 630 million years ago, from 460 to 430 million years ago, and from 360 to 260 million years ago. Notice these ice ages last 30 to 300 of millions of years, a long time. The main trigger in initiating ice ages is probably the periodic blockages of the sea water circulations. Such blockages limit the flow of equatorial warm water to the poles. Ice sheets may arise and eventually produce an ice age.

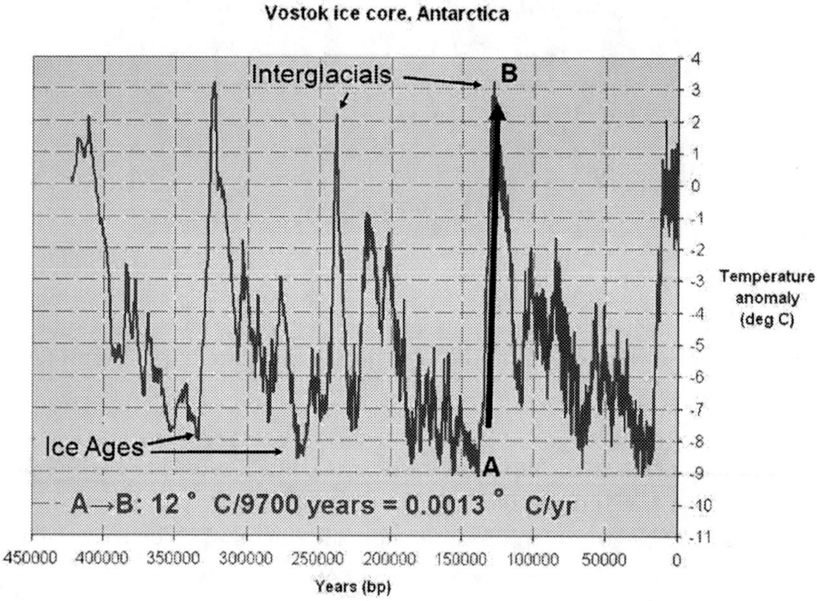

Figure 1.22. Here are four examples of ice ages occurring in the last 500,000 years. We can see individual pulses of very cold climates (glacial periods) and intermittent warm periods (interglacials). (Adapted from Petit et al., 1997 and 1999).

Wildfires. Fire is a rapid oxidation chemical reaction (and not a chemical) that releases light and heat that you can feel and see the flames. A flame itself is a mixture of gases, including oxygen, carbon dioxide, carbon monoxide, and water vapor, and so a flame is matter. The light produced by the flame is energy, not matter. As described in Ikelle (2022a), oxidation, which includes combustion, means chemically of oxygen and other materials (also known as fuels). So the three things required to start a fire are heat (i.e., ignition temperature), oxygen, and fuel (burning materials). In wildfires, the fuels are woody materials, pitch, and other burnable elements found in the forest environment. The atmosphere provides oxygen. Solar heat, volcanic activity, meteorite impacts, and/or lightning can ignite fire if one of these phenomena causes the temperature to be high enough to cause combustion, and if atmospheric oxygen is sufficient to sustain rapid combustion. The atmospheric oxygen concentration must be above 16 percent for an ignited fire to propagate in the forest because fire produces its own heat. At our present atmospheric oxygen level of 21 percent, an ignited fire propagates readily. Above 23 percent, wildfires become more frequent and very destructive—almost anything can ignite wildfires, including a match, which is a common ignition device, and embers from cigarettes. Between 13 and 16 percent of fires would be rare and would only burn very dry plant material. Ecologically, only vegetation grows in environments

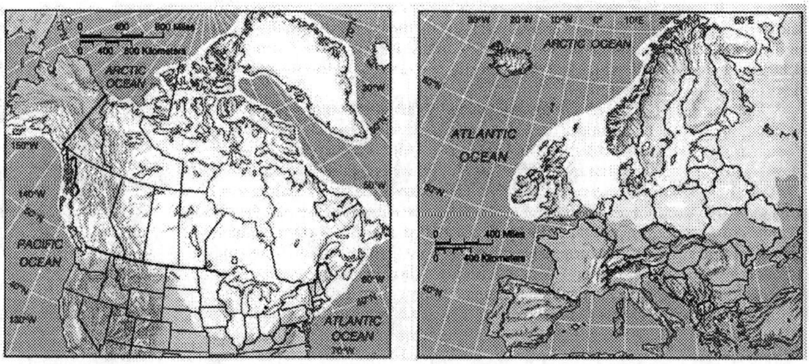

Figure 1.23. The extent of continental glaciers was present 18,000 years ago. Basically, ice sheets covered most of Canada and the northern United States until about 18,000 years ago. The sea level was 100-125 m lower than at present. (Credit: Stockholm Geo Visualization Lab and Microsoft Encarta).

liable to drying would burn. At a level smaller than 13 percent, except under exceptional circumstances, wildfires will not ignite and spread irrespective of moisture content (Scott, 2000; Glasspool et al., 2004 and 2010; and Belcher et al., 2013; Lenton, 2013; and Watson and Lovelock, 2013).

Nine hundred or so wildfires burned a record 2 million acres in

California in 2020 (U.S. Forest Service). Most of them were ignited by an intense series of thousands of lightning strikes. Record-breaking temperatures associated with the heat generated by fires were observed.

For most of the Earth's 4.6 billion years of its history, wildfires were nonexistent; atmospheric oxygen was absent or insufficient until the Phanerozoic period (570 million years ago to the present). This period also marked the arrival of multi-celled living (animals and plants) on Earth and plants had moved from the seas onto land. Conditions were now favorable for wildfires. The wildfire occurred about 420 million years ago according to proxy measurements of charcoal in sedimentary rocks (Scott, 2000). Charcoal, soot, and polyaromatic hydrocarbons are typical evidence of fire, charcoal, usually found in sedimentary rocks, being the most useful in studies of deep time. Charcoal abundance is correlated with the frequency and extent of fire activity: the more charcoal is present in the fossil record, the more fire is inferred to have been on the landscape. These records indicate that, while fire has been a presence throughout most of the past 420 million years, the extent of fire activity has varied considerably over this period. Wildfires increase CO_2 and therefore increase temperature, and O_2 decrease, as more plants are burned. Less wildfires, less CO_2 and therefore more temperature and O_2, as more plants grow. In other words, wildfires as just a consequence of CO2, and O_2 through the Phanerozoic period. Atmospheric oxygen has just fluctuated throughout the Phanerozoic period. A large volcanic eruption, say, VEI =7, and/or simultaneous earthquakes, would break up this small cycle. Maybe wildfires will erase almost all land vegetation.

Again, about 250 million years ago, wildfires reduced atmospheric oxygen content from 30 percent to 13 percent, which corresponds to the current oxygen concentration at the height of 5 km in the atmosphere. In addition, the ashes that were released into the atmosphere caused some cold spells, similar to nuclear winter conditions. The decreasing oxygen level and the quickly alternating climate led to massive extinctions also on land. It is possible that a meteorite impact still added to the destruction.

Holocene period. The Phanerozoic period is geologically divided into Cenozoic (65 million years ago to the present), Mesozoic (246 million to 65 million years ago), and Paleozoic (570 million to 246 million years ago). The cenozoic is further divided into Paleocene, Eocene, Oligocene, Miocene, Pliocene, Pleistocene, and Holocene. The last 15,000 years or so of Earth's history, the time since the end of the last major ice age, is called the Holocene period. Since then, there have been small-scale climate shifts, notably the "Little Ice Age," which occurred between about 1200 and 1700 A.D.[4]—but, in general, the Holocene has been relatively warm.

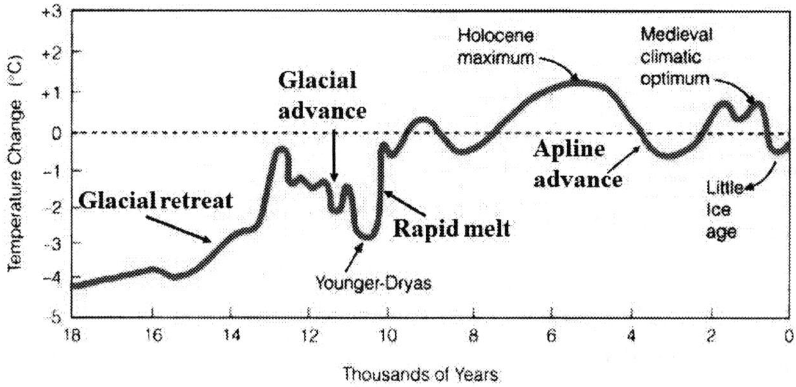

Figure 1.24. Temperatures during the Holocene. Note that there was global warming after the end of the last ice age, as well as cooling periods, but climate changes became more regional at the start of the Younger Dryas. Notice also that the medieval warming period was followed by the Little Ice Age, which was a period of significant cooling. (Adapted http://www.atmo.arizona.edu/holocene.html).

Again, the Earth's climate has been fairly stable over the Holocene period, as we can see in Figure 1.24. However, there is no reason to believe that it represents a permanent end to the current ice age. Ice-core records show that before the Holocene, there was global warming after the end of the last ice age and cooling

[4]Jesus Christ's birth is used as a starting point to count the years preceding (B.C.) and after (A.D.) his birth. For example, the year 1,000 B.C. refers to the time 1000 years before A.D. 1, when Christ is assumed to have been born.

periods, but climate changes became more regional at the start of the Younger Dryas. Compared to glacial conditions, habitable zones expanded northward, reaching their northernmost point during the Holocene-maximum period, also known as the Holocene climate optimum (HCO). Actually, HCO has been known by many other names, including hypsithermal, altithermal, climatic optimum, Holocene optimum, and Holocene thermal maximum and Holocene megathermal. The warm period (HCO) was followed by a gradual decline that lasted until about 2000 years ago. At that time the climate was not unlike today's, but there was a slightly warmer period, from 2,000 to 1,400 years ago, known as the medieval warm period. This period was followed by the Little Ice Age, which lasted from 1,300 years ago to the mid-nineteenth century. This was a period of significant cooling.

1.1.6. Carbon Cycle

Carbon is the fourth most abundant element in our universe. Typical

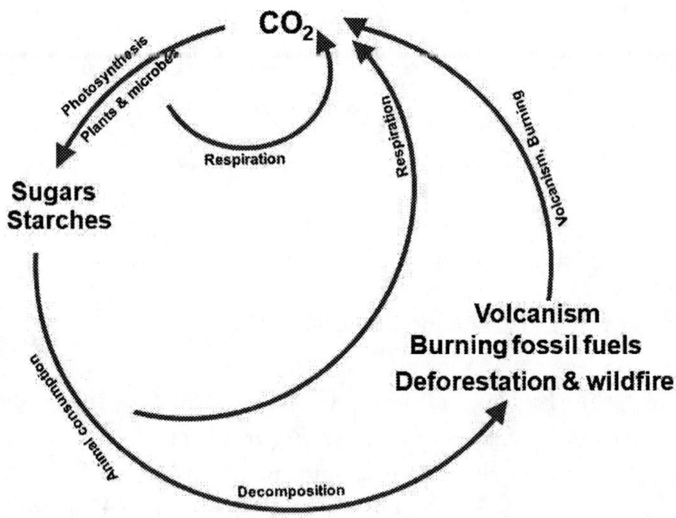

(a)

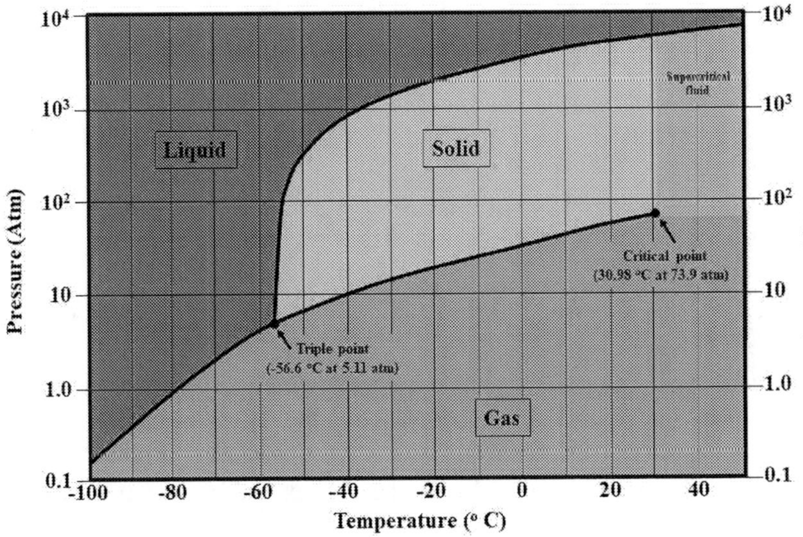

(b)

Figure 1.25. (a) The carbon cycle. CO_2 is absorbed by land plants and aquatic biomass through photosynthesis. We also put CO_2 in the atmosphere through volcanism, the burning of fossil fluids, animal respiration, deforestation, and wildfires. Also atmospheric CO_2 dissolves in rainwater. Rain erodes minerals that flow into the oceans. Minerals combine with carbon to make rocks on the ocean floor. Subduction carries carbon rocks into the mantle. Rocks melt in the mantle and outgas CO_2 back into the atmosphere through volcanoes. (b) Pressure-temperature diagram of CO_2.

carbon compounds are carbon dioxide (CO_2), glucose ($C_6H_{12}O_6$), methane (CH_4), carbon monoxide (CO), chlorofluorocarbon (CFC), hydrofluorocarbon (HFC), and hydrochlorofluorocarbon (HCFC). Water and carbon dioxide cycle through rocks on the Earth, like carbonate rocks such as limestone ($CaCO_3$), and through the life of living things (microbes, animals, and plants). Both cycles have enormous impacts on the properties and composition of rocks, soils, and sediments. They are essential components of the tectonic movements of the lithosphere. Also, as illustrated in Figure 1.25, life on Earth alters the carbon cycle, and through this alternation they have changed the composition of our atmosphere.

Major carbon reservoirs are the lithosphere (about 8,000 trillion tons, including 4 trillion tons of fossil fuels), oceans and ocean sediments (about 40 trillion tons), and the atmosphere (about 2 trillion tons). In other words, most carbon is locked in rocks. Note also that the oceans are a very significant carbon reservoir. Carbon is held in the oceans in dissolved form and in the tissues of ocean-dwelling organisms. Dissolved carbon dioxide in seawater is just like CO_2 dissolved in a can of soda. Some dissolved carbon is released from seawater to the atmosphere just as some CO_2 from a can of soda is released to the air when opening it. This transfer of gas back and forth between a liquid and the atmosphere is an extremely important process in the global carbon cycle since the oceans are such an enormous reservoir with the potential to store and release significant quantities of CO_2. Also this transfer shows the direct interactions of the carbon and water cycles. Furthermore, experiments with isotopes showed that oxygen in the atmosphere comes from water.

So, the natural reservoirs, enumerated above, change over time, in particular, in response to climate change. The connections of these reservoirs, along with the processes that make these connections possible, form the carbon cycle. Figure 1.25 illustrates this cycle for the carbon dioxide. Processes that put carbon into the atmosphere are plant and living things respiration, volcanic eruption, land use change, and fossil-fuel combustion. Processes that take carbon from the atmosphere are plants/vegetation (long lived trees) photosynthesis, burial of organic matter in the ground, and oceans (water, marine sediments, etc.). About 9 billion tons per year of carbon on average is put in the atmosphere and 6.5 billion tons per year of carbon on average is taken from the atmosphere. There is a deficit of about 2.5 billion tons per year of carbon. In other words, the present-day carbon cycle is not in a steady state.

Quiz: (1) Determine the state of water at the following temperatures and pressures: (-10° C and 50 kPa), (25° C and 90 kPa), (80° C and 5 kPa), and (50° C and 0.3 kPa). (2) Determine the state of CO_2 at the following temperatures and pressures: (-30° C and 2000 kPa), (-60° C and 1000 kPa), (-60° C and 100 kPa), and (20° C and 100 kPa).

Figure 1.26. A picture of the Macondo oilfield blowout in the water of the Gulf of Mexico on April 20, 2010, and of the inferno that occurred from the escaping methane. The Deepwater Horizon rig, 65 kilometers off the Louisiana coast, above the 5.5-kilometers-deep Macondo well, was operated by BP. Fireboats pour water onto the rig in the hours with very limited success. The bottom right-corner picture shows Deepwater Horizon before the explosion. Following the explosion and sinking of the oil rig, oil flowed for 87 days until it was capped on July 15, 2010. Eleven people were never found. The blowout is considered the largest accidental marine oil spill in the history of the petroleum industry. The U.S. government (DOE and DOI) estimated the total discharge at 4.9 million barrels. Impacts on wildlife and fisheries were huge with lasting effects. The top left-corner picture shows oiled pelicans near Grand Isle, Louisiana. Photos adapted from the report of the National Commission on the BP Deepwater Horizon Oil Spill and Offshore Drilling, June 11, 2011.

1.1.7. Earth's Sustainability

The Earth's sustainability is sometimes defined as meeting our needs without compromising the ability of future generations to meet their own needs. Meeting our needs includes feeding ourselves, having light to see through darkness, our ability to have indoor and outdoor constructions that can maintain adequate temperatures for human lives, the ability to travel and communicate long distances, and the ability to cure our sicknesses, to combat future pandemics, and to extend healthy human life beyond the current 80 years. While meeting our needs is intrinsic to our survival, and it is our prime duty, our ability of understanding the needs of the next generations, let alone affecting them, may be unrealistic, or even pretentious.

Our key tools today for predicting the needs of future generations are the sciences. Yet experience has shown again and again that modern science has become very effective at understanding the past, but it is less effective at predicting the future. For example, with the dwindling of conventional hydrocarbon resources in the late 1990s and early 2000s, it was acceptable that we were entering an unprecedented period of political, economic, and social instability in the world. Thanks to the human brain and consequently the discoveries of abundant unconventional resources and the exploitation of renewables, this period of instability did not materialize. Moreover, potential energy resources are considered more abundant today than at any other time in modern human history.

In geophysics and in Earth sciences, in general, we are essentially concerned with the processes of withdrawal from and injection into the Earth, especially the solid Earth. The withdrawal process includes oil and gas production, geothermal production, volcanic eruptions, deforestation, freshwater withdrawal, etc. The injection process includes subduction, disposal of municipal and industrial wastes (including nuclear wastes), water injection after and during energy production, etc. Withdrawal from and injection into the Earth affects the equilibrium of the Earth system, including climate, ecology, and agriculture. The causes and effects of these withdrawals and injections still include a number of open questions. Remember, nothing is either created or destroyed on a certain time scale, several hundred times larger than human life. We are dealing with temporary transformations. Yet near simultaneous multiple volcanic eruptions can cool the Earth's atmosphere for a very long time and near simultaneous multiple earthquakes can almost wipe out *Homo sapiens.*

The concept of sustainability is useful in posing the present and future challenges associated with these processes and even in stimulating research funding for addressing these challenges. I believe course corrections and transformational solutions are the answers.

Box 1.3. Withdrawal from Solid Earth and Injection into the Earth's Atmosphere

The process of fossil fuels (oil, gas, and coal) withdrawal from solid Earth and the injection of CO_2 into the Earth's atmosphere resulting from the burning of these fluids and solids is one of the processes of withdrawal from solid Earth and injection into the Earth's atmosphere.

More than half of all the more than 2 trillion barrels of fossil fuel on the planet have been withdrawn from solid Earth mainly by the oil, gas, and coal industries, and burned out in the atmosphere by society at large. Because oil and gas have very low density (less 0.9 g/cm^3g), compared to surrounding rocks (between 2.1 and 2.7 g/cm^3), they have upward motion, and mingle with some lateral deviations, through a process known as oil migration (see Ikelle, 2022b). This upward motion exerts huge pressure on the surrounding rocks. A significant amount of fossil fuels will have reached the Earth's surface without human intervention and eventually burned out in the atmosphere (see Figure 1.26).

So one can argue about the amount of CO_2 related to fossil fuels in the atmosphere and the rate at which it gets there. Actually, the real debate is between oil and gas gushing naturally to the Earth's surface and the human (oil, gas, and coal industries) tapping into some of the resources before they reach the Earth's surface—what is not in doubt that is some increased emission CO_2 in the Earth's atmosphere, due to the burning of fossil fuels will have occurred, irrespective of the mechanism for oil reaching the Earth's surface and burning methods. In other words, the increase and decrease of CO_2 in the Earth's atmosphere are the course of natural planetary evolution, and human forcings are marginal, at best.

The most important process of withdrawal from the solid Earth and injection into the Earth's atmosphere is a volcanic eruption. As described in Chapter 3, a volcano is any opening in the Earth's crust through which magma (i.e., molten rocks below the Earth's surface-lava are molten rocks reaching the Earth's surface) will reach the Earth's surface. Gases (mainly H_2O, CO_2, and SO_2) expand within magma as it nears the Earth's surface due to decreasing pressure. Gas pressure builds as crystallization takes place. The violence of an eruption is related to how easily gases escape from magma. The VEI-6 eruption of Mount Pinatubo, in the Philippines, on June 15, 1991, produced high-speed avalanches of hot ash and gas (more than 10 billion cubic meters of ejecta), a giant mudflow, and a cloud of volcanic ash hundreds of kilometers across. The effects of the eruption were felt worldwide. It ejected millions of tons of SO_2 into the atmosphere and changed the Earth's climate almost instantaneously. Global temperatures dropped by about 0.5 degrees Celsius (0.9 degrees Fahrenheit) in the years 1991-1993, and ozone depletion temporarily increased substantially. The 1815 Tambora (Indonesia) VE-7 eruption injected more than 100 billion cubic meters of ejecta into the Earth's atmosphere in a matter of days. A VEI-8 eruption will produce more than a trillion cubic meters of ejecta and will have cloud-column heights of over 40 km. The last VEI-8 eruption was in New Zealand (Oruanui Ignimbrite), about 27,000 years ago, which means that there have not been any Holocene eruptions with VEI=8.

1.2. A Description of Our Solar System and Its Planets

Our solar system consists of our star, the sun, and its nine orbiting planets (including Earth), along with more than 170 moons, a multitude of comets and asteroids, meteoric rocks, and dust (zodiacal light). It is the gravitational forces (see Ikelle, 2022a) that cause these objects to orbit around the sun and the planets to orbit in the same plane. Note that the moons (also known as planetary satellites) orbit around orbit planets instead of around the sun. So moons are associated with planets.

The moons are considered important in stabilizing planet's axial tilt against chaotic excursions; thus, leading to a relatively stable climate. Also the existence of moons is considered a prerequisite for life on a given planet. Earth's axis is presently tilted at about 23

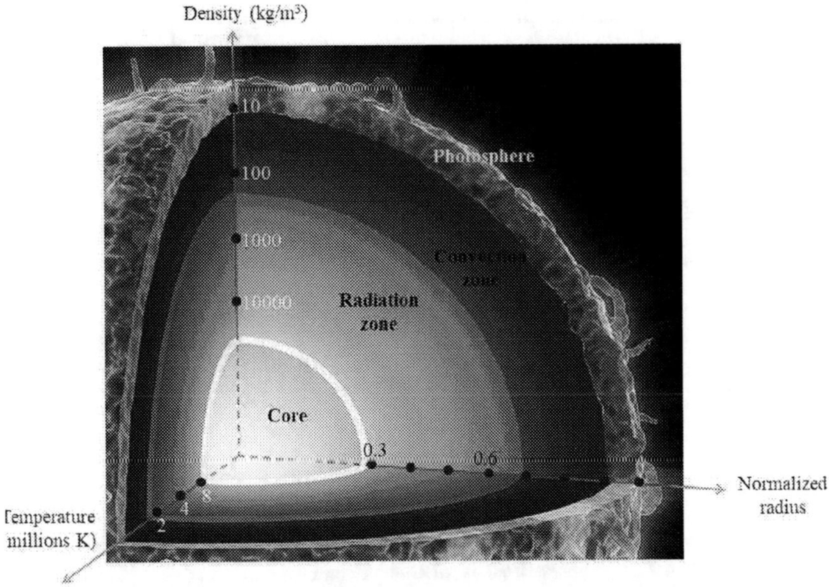
Density (kg/m³)
10
100
1000
10000
Photosphere
Convection zone
Radiation zone
Core
0.3
0.6
Normalized radius
Temperature (millions K)
2
4
8

(a)

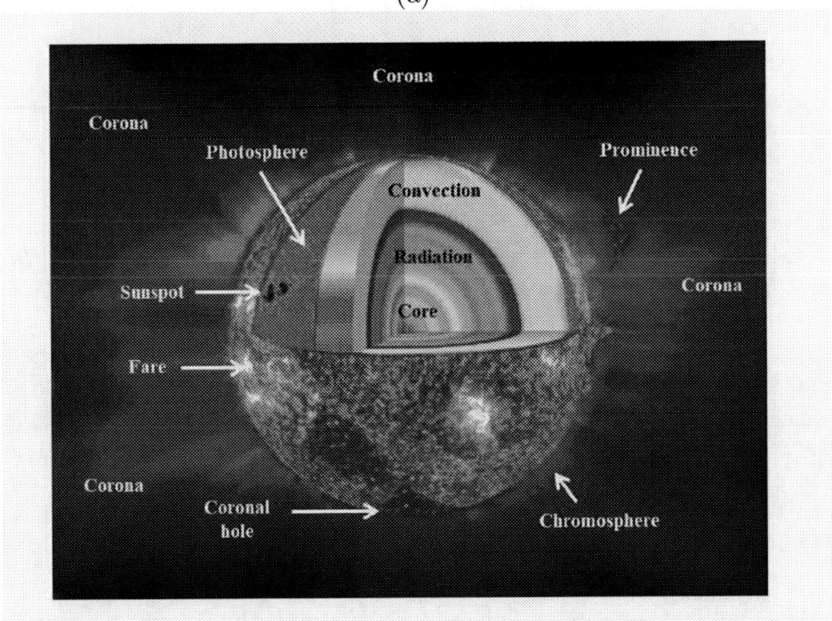
Corona
Corona
Photosphere
Prominence
Convection
Radiation
Sunspot
Core
Corona
Fare
Corona
Coronal hole
Chromosphere

(b)

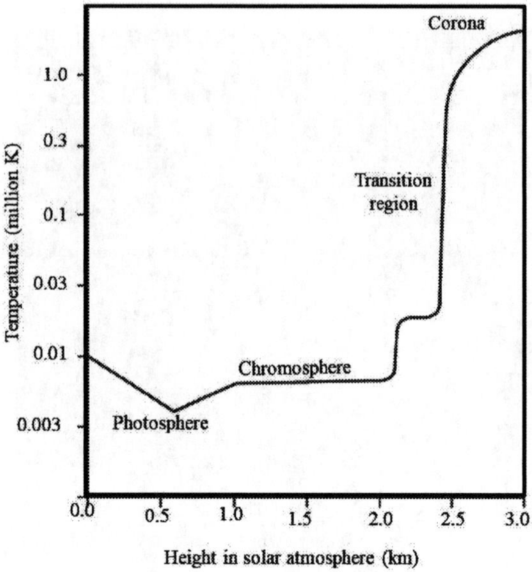

(c)

Figure 1.27. (a) Inner layers of the Sun. The core has temperatures between 15 and 8 million degrees Kelvin and it is located between the Sun center and 0.3 R_{sun}. Radiative zone has temperatures between 8 and 2 million degrees Kelvin and it is located between 0.3 R_{sun} and 0.7 R_{sun}. Convection zone has temperatures between 2 million and 6000 degrees Kelvin and it is located between 0.7 R_{sun} and 0.9 R_{sun}. The photosphere has temperatures between 6500 and 4000 degrees Kelvin and it is located between 0.9 R_{sun} and R_{sun}. (b) Different parts of the Sun, from the hot core where energy is generated through regions where energy is transported outward, first by radiation, then by convection, from the convection zone to the photosphere, and then out through the solar atmosphere. Some typical features of the atmosphere are shown, such as sunspots, fares, coronal holes, and prominences. (c) Temperatures in the solar atmosphere increase upward above the photosphere. Note the very rapid increase in temperature over a very short distance in the transition region between the chromosphere and the Corona. (Adapted from NASA/Goddard images).

degree and stays almost constant. Without the Moon, this angle would vary erratically between 0 and 180 degrees. At 90 degrees, half of Earth will be in constant darkness for six months and the life on Earth would go extinct at some point.

1.2.1. Sun Structure and Solar Wind

The sun is an enormous ball of a mean radius with about 110 times the radius of the Earth (i.e., $R_{\text{sun}} \approx 110\, R_{\text{earth}}$ with $R_{\text{earth}} = 6371$ km) and a mass of $m_{\text{sun}} \approx 333,000\, m_{\text{earth}} = 1.989 \times 10^{30}$ kg with $m_{\text{earth}} \approx 5.972 \times 10^{24}$ kg. It is extremely hot, with surface temperatures between about 4,000 K and 6,500 K. Unlike the Earth, the sun is not solid at any of its parts. The sun is in the plasma state, which is the fourth state of matter, as defined in the previous section, because it is so hot that no matter can survive as a liquid or a solid. In fact, the sun is so hot that many of the atoms in it are ionized–that is, stripped of one or more of their electrons. This removal of electrons from their atoms means that there is a large quantity of free electrons and positively charged ions in the sun, making it an electrically charged environment quite different from the neutral environment on the Earth's surface, and it consists almost entirely of hydrogen and helium.

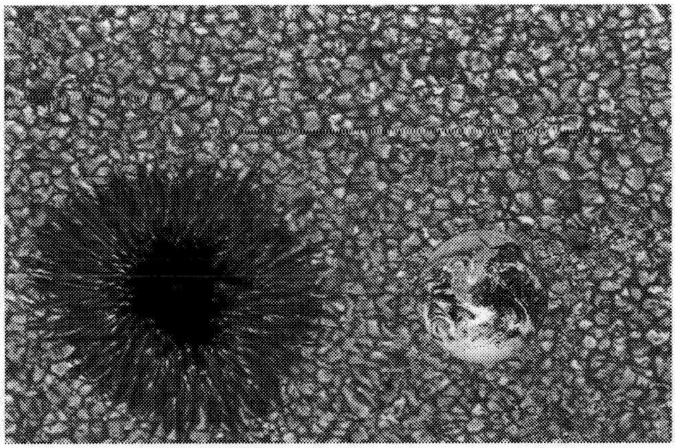

Figure 1.28. A large view of a sunspot. Sunspots appear as dark spots on the surface of the Sun. They are formed over periods of days and weeks and typically last for several days, although very large ones may live for several weeks. Sunspots look small compared to the Sun but the one shown here is as large as solid Earth.

Despite the fact that the sun is not solid, it does have a definite internal structure, as illustrated in Figure 1.27a. The layers of the sun are (1) the core, which occupies the innermost quarter of the sun's radius ($R_{\text{sun}}^{\text{core}} \approx 0.3 R_{\text{sun}} = 33\, R_{\text{earth}}$; space is big!), with an average temperature of approximately 12 million degrees Kelvin, (2) the radiative zone (approximately 280,000 kilometers thick), with an average temperature of about 5 million degrees Kelvin, (3) the convective zone (approximately 210,000 kilometers thick), with an average temperature of 1 million degrees Kelvin, and (4) the photosphere, which is the visible surface, again, with an average temperature of 5,000 K. The photosphere looks sharp only from a distance. If you were falling into the sun, you would not feel any surface but would just sense a gradual increase in the density of the gas surrounding you, assuming that you would not vaporize long before reaching the photosphere. It is much the same as falling through a cloud in the Earth's atmosphere. From far away, the cloud looks as if it has a sharp surface, but you do not feel a surface as you fall into it.

Let us continue with the description of the layers of the sun by looking at its atmosphere (Figures 1.27b and 1.27c). The chromosphere is a layer in the sun's atmosphere between about 400 km and 2,100 km above the photosphere. The temperature in the chromosphere varies between about 4,000 K at the bottom and 8,000 K at the top. It gets hotter if you go further away from the photosphere, unlike in the lower layers, where it gets hotter if you go closer to the center of the sun. Above the chromosphere, we have the Corona, with an average temperature of nearly a million degrees Kelvin. The region of rapid temperature rise occurs between the chromosphere and the Corona and is called the transition region. It is only a few tens of kilometers thick.

Photosphere, chromosphere, transition region, and corona describe the big picture of the sun's atmosphere. However, the sun's atmosphere also has some hot and cool regions intermixed in these layers and numerous features with lifetimes of seconds to months. For example, sunspots, which are dark blotches on the sun's surface, in Figures 1.27b, reveal the presence of powerful disturbances of the sun's magnetic field. These disturbances generate solar flares and coronal mass ejections, which are actually two types of immense explosions associated with the release of magnetic energy stored in the solar corona. There are also solar prominences, filaments, and coronal loops, which are structures formed from plasma suspended by magnetic fields in the sun's atmosphere. Voids called coronal holes associated with high-velocity solar wind sometimes appear in the solar atmosphere, allowing fast streams of solar wind to flow freely outward into space.

The sun's core produces a lot of energy that travels extremely slowly through the radiative zone to the convection zone and by convection it reaches the sun's surface (see Ikelle, 2022a and 2022c). Anyway, once the energy reaches the sun's surface, it is mainly transmitted by charged particles to the rest of the solar system and in all directions at an average speed of about 400 km/s (almost 1.5 million kilometers per hour), but it can range from 200 km/s to as high as 1,000 km/s. These streams of charged particles are called the *solar wind* or plasma (Kivelson et al., 1995). The solar wind is invisible to the naked eye.

From time to time, the solar wind through the sun's atmosphere generates violent outbursts of material and energy leading to what is now called *space weather*. Coronal mass ejections and solar flares are examples of such violent outbursts. Space weather can affect space activities, including human satellites and spacecrafts, communications of all kinds on Earth, from telephone to television, navigational and geographic positioning systems, even the operation and integrity of electric power grids, and the safety of astronauts, space visitors, and space immigrants. For example, a solar flare (Figure 1.28) observed by the GOES-13 spacecraft on December 5, 2006, was so intense that it damaged the camera taking the picture and disrupted satellite navigation systems on Earth. On March 13, 1989, a 600 nanoTesla magnetic storm interrupted the entire power grid in Quebec, Canada. As our use of space technology increases, we become more exposed to the risks of space weather and therefore it is becoming more and more important to be able to predict when space weather may get worse, and in what way.

1.2.2. Terrestrial and Jovian Planets

As illustrated in Figure 1.29 and Tables 1.4 and 1.5, our solar system has two types of planets: terrestrial (or "rocky") planets (Mercury,

Venus, Earth, and Mars), and gas giant (Jovian) planets (Jupiter, Saturn, Uranus, and Neptune). The surfaces of terrestrial planets are made of materials that have densities between 3.3 and 5.5 g/cm. Only solid rocks and metals have densities in this window; therefore, the surfaces of terrestrial planets are made of solid rocks. A standard spacecraft can land on their surfaces.

The density of Jovian planets varies between 0.7 and 1.6 g/cm^3—that is, we are dealing with soft materials. Jovian planets are made primarily of gas, mostly hydrogen, and do not have solid surfaces. On top of that, they have thick atmospheres of hydrogen and helium gases. A spacecraft cannot land on their surfaces but can crash through their atmospheres—you cannot stand on the surface of a Jovian planet. As we can see in Table 1.5, Jovian planets are very large, farther from the sun, and therefore cooler than terrestrial planets. They have rings, while terrestrial planets do not (see Figure 1.30). The ring systems on Uranus and Neptune are much less substantial than those of Saturn and Jupiter. The atmospheres of these planets are diverse but largely made of the same six elements—hydrogen (H), helium (He), oxygen (O), carbon (C), nitrogen (N), and sulfur (S). These are the most abundant elements in the sun and stars. With the exception of helium, they are also the main elements in living things.

Basically, distances from the sun to planets are fundamental parameters of our solar system. Too close to the sun, too much water enters the atmosphere, and we end up with a runaway CO_2 in the atmosphere. Too far from the sun, too much water freezes, and we end up with runaway glaciations—planetary rotation (or planet day) is another fundamental parameter. It affects the planet's day and night temperature variation and the possibility of magnetic field generation, as addressed in Ikelle (2022a and 2022c). Unless a terrestrial planet has a thick carbon-dioxide atmosphere, slower rotation will result in larger day-night temperature differences. Mercury has almost no atmosphere and a slow planetary rotation (once every 59 Earth days versus 24 hours for Earth). The average daytime temperature is near 400 degrees Celsius (too hot) and the average night temperature is -200 degrees Celsius (deep freeze). This temperature range is the most extreme temperature range in the whole solar system. Venus also has a slow planetary rotation. But its runaway CO_2 atmosphere keeps day–night temperature differences quite small. Its thick atmo-

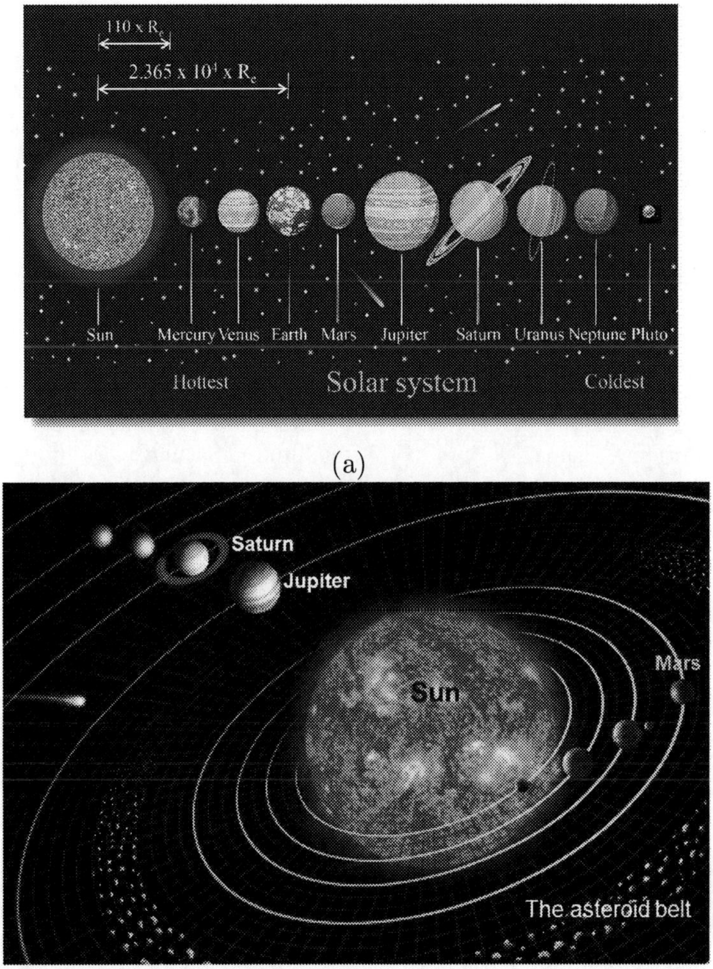

(a)

(b)

Figure 1.29. (a) Nine known planets of the solar system. Terrestrial planets are Mercury, Venus, Earth, and Mars. Jupiter, Saturn, Uranus, and Neptus are gas giant (jovian) planets, which are made primarily of gas, mostly hydrogen. (b) Artist's conception of Earth's solar system (not to scale). (Image credit: NASA/JPL).

sphere of CO_2 also prevents direct visual observation of the planet's surface and is largely responsible for its large surface temperatures of more than 400 degrees Celsius.

There is a planet in our solar system, included in Table 1.5 and in Figure 1.29, that does not fit the category of terrestrial or Jovian planets—it is known as Pluto. Pluto is small (less than one-sixth of the size of the Earth), like terrestrial planets, but lies far away from the sun and has low density (about 2 g/cm^3) just like Jovian planets. In fact, some studies consider it to be a comet rather than a planet. Pluto consists of a solid core with a thick outer layer of ice. It lacks a thick atmosphere. It takes 248 years for Pluto to orbit the sun. All the planets (except Pluto) have the same orbital plane.

Table 1.4. Characteristics of terrestrial and Jovian planets

Terrestrial planets: Mercury, Venus, Earth, Mars	Gas giant (Jovian) planets: Jupiter, Saturn, Uranus, and Neptune
Are made of solids like rocks, metal	Are made primarily of gas, mostly hydrogen
Have solid surfaces	Do not have solid surfaces
A spaceship can land on it	A spaceship cannot land on it
Are smaller than jovian planets	Are larger than rocky planets
Are closer to the Sun	Are farther from the Sun
Are warmer than jovian planets	Are cooler than rocky planets
Have few moons and do not have rings	Have rings and many moons

Asteroids are possibly fragments of broken planets. Their diameters vary from 1,000 km to a few meters. An extensive belt of a thousand asteroids occurs between the orbits of Mars and Jupiter (see Figure 1.29b). This belt starts about 2.0 AU from the sun and ends around 3.2 AU, a distance of 1.2 AU (or 180 million kilometers). It is also used as the region that separates terrestrial and Jovian planets. Some asteroids have collided with the Earth in the past. Also, when asteroids collide, they produce small fragments that occasionally fall on the Earth. These rocks are called meteorites and provide valuable information about the primordial solar nebula. Most of these fragments are the size of sand grains. They burn up in the

Earth's atmosphere, causing them to glow like meteors (or shooting stars). Comets can be described as dirty snowballs of frozen gases. Some comets contain organic materials. The composition of comets (abundance of the light elements, carbon, nitrogen, and oxygen) is very important because it is very likely one of the sources, along with asteroids, of water in terrestrial planets. The delivery is through collisions of comets with planets. A typical comet can contain about 78.5 percent H_2O (frozen water), 4.5 percent H_2CO_2 (carbonous acid), 4.0 percent H_2CO (formaldehyde), 2.6 percent N_2 (nitrogen gas), 1.5 percent CO (carbon monoxide), 1.5 percent C_2H_4 (ethylene), 1.0 percent HCN (formonitrile), 0.8 percent NH_3 (ammonia), and 0.8 percent N_2H_4 (hydrazine). Meteoroids, asteroids, and comets also follow orbits around the sun. There are millions of comets that orbit the sun beyond Pluto.

Each of the planets in the solar system has its own moons, as depicted in Figure 1.30, except Mercury and Venus. Terrestrial planets have a few moons. Jovian planets have many moons. For example, Earth has one moon, Mars has two moons (Phobos and Deimos), Saturn has 62 moons, and Jupiter may have 79 moons. Titan is the largest of Saturn's moons. Titan has an icy solid surface (see Figure 1.31); therefore, it is more similar to terrestrial planets than Jovian planets, in this respect. Like many moons, Titan is not a true sphere. Its shape is distorted by the gravitational pull of Saturn, making the moon sort of oblong along its equator and a little flatter at the poles. Jupiter's moon, Ganymede, is the largest moon in our solar system, bigger than the planet Mercury, and NASA's Hubble-Space-Telescope images suggest a possibility for a very large underground saltwater ocean on Ganymede. This ocean is estimated to be more than 100 kilometers thick—about 10 times deeper than the Earth's ocean—and is thought to be buried under a 150-kilometer-thick crust of mostly ice. Moreover, NASA's Galileo spacecraft discovered in 1996 that Ganymede has its own magnetic field (see Ikelle, 2022d), the only moon known to have its own magnetic field. It may even have a thin oxygen atmosphere on Ganymede. Most of the moons revolve around the planets in the same direction as the planets revolve around the sun. Pluto has one moon, Charon, which is 1,300 km in diameter.

How are we certain of the above description and the data summaries in Tables 1.4 and 1.5? The answer is that the terrestrial and even Jovian planets have been visited by several spacecrafts, and some have stayed in orbit for years. Here are some examples:

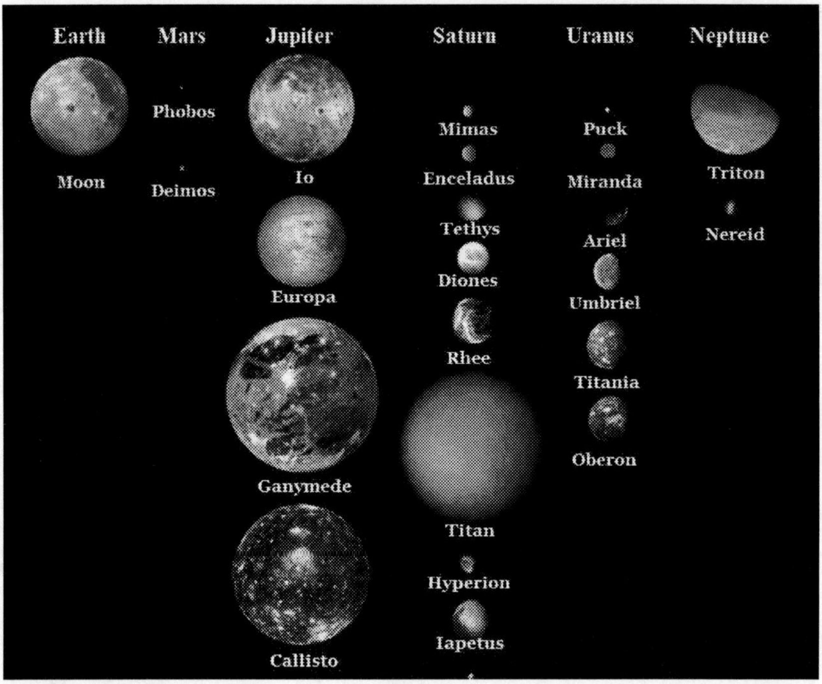

Figure 1.30. Moons of Earth and Mars, and selected moons of major Jovian planets. Actually, Jupiter has 67 moons, Saturn has 62, Uranus has 27, and Neptune has 13. Only the medium and large moons of the Jovian planets are shown here. The diameters of Ganymede, Titan, Callisto, Io, our Moon, Europa, and Triton are 5262 km, 5150 km, 4806 km, 3642 km, 3476 km, 3138 km, and 2706 km, respectively. Io is very volcanic, probably the most volcanic object in the solar system; Europa's crust is fractures; Ganymede has a mixture of old and new terrains, and hints of tectonic activity; and Callisto has a heavily cratered surface.

- *MESSENGER (Mercury Surface, Space Environment, Geochemistry, and Ranging)*, a NASA spacecraft, left Earth in August 2004 and reached Mercury in March 2011, with a successful Venus flyby.

- *Pioneer Venus Orbit*, a NASA spacecraft, orbited Venus from 1978 to 1992. It entered the Venus atmosphere on October 8, 1992, and burned up by the entry heat.

- *Viking 1* and *Viking 2*, NASA spacecrafts, visited Mars for six years, from 1976 to 1982. They were the first U.S. missions to land on Mars. They took photographs, collected data on the Martian surface, and performed biological experiments.

- *Pioneer 10* and *Pioneer 11*, NASA spacecrafts, visited Jupiter and Saturn from 1972 to 2003. They made direct observations of Jupiter and Saturn and obtained close-up images of these planets, including the magnetospheres of Jupiter and Saturn (see Ikelle, 2022d). They discovered that Jupiter and Saturn are exclusively fluid planets.

- *Voyager 1* and *Voyager 2*, NASA spacecrafts, visited all four Jovian planets and their moons. They were launched in 1977 and are still sending signals daily to Earth.

- *Ulysses*, NASA-ESA (European Space Agency) spacecrafts, flyby of Jupiter, visited their moons. They determined that the Sun's magnetic field "reverses" every 11 years.

- *Galileo*, a NASA spacecraft, orbited Jupiter and probed Titan, 1995-2003. Galileo's demise was in an impact with Jupiter in September 2003.

- *Cassini*, NASA-ESA spacecraft, flyby of Jupiter, Saturn orbit since 2004, and probed Titan.

- *New Horizons*, NASA spacecraft, flyby of Jupiter and the asteroid Ultima Thule, en route to Pluto.

- *Juno*, NASA spacecraft was launched on August 5, 2011 and arrived at Jupiter on July 4, 2016. It is currently in orbit around Jupiter. At the time, Juno was flying near Ganymede, within 1,038 kilometers of the moon's surface.

Each of these missions has provided information about the planets and their moons, and rings. Future missions in our solar system to the moons of Jupiter and Saturn, including Europa, Io, Enceladus, and Titan, where some life forms may exist, are being planned.

Table 1.5. Some key properties of the solar system's planets, including their surface temperatures (T), their distances to the sun in AU with AU $= 1.5 \times 10^8$ km ($d_{\rm p-s}$), their radii ($R_{\rm planet}$), the compositions of their atmospheres, their orbital periods around the sun ($\tau_{\rm orbit}$), the rotation period (or planetary rotation or planet day) is the time that a planet takes to complete one revolution around its axis of rotation relative to the background stars ($\tau_{\rm rot}$) in Earth days, and their surface pressures (P). We added Antarctica and coal here as a reference in terms of temperatures. Note also that Mercury is colder than Antarctica in some parts and hotter than coal in others.

Planet	$R_{\rm planet}$ ($R_{\rm earth}$)	$d_{\rm p-s}$ (AU)	$\tau_{\rm orbit}$ (years)	$\tau_{\rm rot}$ (days)	T(^{o}C)	P (bar)
Mercury	0.38	0.387	0.241	58.65	-202 to 396	10^{-13}
Venus	0.95	0.723	0.616	243.01	426 to 482	92
Earth	1	1.0	1	1	-4 to 28	1.013
Mars	0.53	1.524	1.88	1.03	-124 to -23	0.006
Jupiter	11.19	5.20	12	0.41	-145 to -108	10
Saturn	9.46	9.54	29	0.44	-185 to -122	1
Uranus	3.98	10.2	84	0.61	-213	$\gg 10^3$
Neptune	3.81	30.1	165	0.76	-225	$\gg 10^3$
Pluto	0.1667	39.5	248	6.39	-226 to -240	10.0^{-5}
Antarc.					-60 to -10	
Coal					35 to 245	

Planet	**Atmosphere composition**
Mercury	N/A
Venus	96.5% CO_2 and 3.5% N_2
Earth	78% N_2, 21% O_2, and 1% Ar, CO_2, and H_2O
Mars	95% CO_2, 2.6% N_2, 1.9% Ar, and 0.16% O_2
Jupiter	90% H_2 and 10% He
Saturn	96.3% H_2, 3.25% He, and 0.45% CH_4
Uranus	82.5% H_2, 15.2% He, and 2.3% CH_4
Neptune	85% H_2, 13% He, and 2% CH_4
Pluto	97% N_2 and 2.5% CH_4

Figure 1.31. Evidence of the rocky red surface of Saturn's moon, known as Titan. Titan has a radius of about 2,575 kilometers and about 1.2 million kilometers from Saturn. Analogous to Earth, features on Titan include volcanoes, sand dunes, lakes, rivers, and a nitrogen (N_2)-rich atmosphere. The Cassini mission discovered that Titan's lakes are filled with liquid hydrocarbons (e.g., CH_4 and C_2H_6), the only body in the solar system beyond Earth known to have liquid on its surface. Through gravity measurements, it also revealed that Titan may have an ocean of liquid water about 55 to 80 kilometers below the icy ground and could potentially contain habitable environments, even biological life as we know it. (Credit: NASA's Scientific Visualization Studio).

Quiz: (1) What are the terrestrial planets? What do they have in common? (2) Why does the Earth have lots of oxygen compared to, say, Venus? (3) Why is Mercury's surface similar to that of the Moon's? What feature do they share that causes this? (4) Which of the Jovian planets have solid surfaces? (i) Jupiter and Saturn, (ii) Uranus and Neptune, (iii) all four planets, and (iv) None of them. Select one answer.

Quiz: The rover, Curiosity, landed on Mars and began operations on August 6, 2012. When did the rover mark its one Mars year? (2) List two planets in our solar system that have both volcanoes and craters and water features on their surfaces.

1.2.3. Layer Models of the Planets of Our Solar System

The four terrestrial planets have roughly the same structure: a central metallic core, mostly iron, with a surrounding silicate mantle underlying a lithosphere layer (Rivoldini et al., 2011). Three of the four solar terrestrial planets (Venus, Earth, and Mars) have substantial atmospheres, as illustrated in Figure 1.8. Yet there are differences and similarities in the surfaces and interiors of these four terrestrial planets. Study of these differences and similarities helps us to better understand the Earth's dynamics and how unique our current state of our planet is. We here briefly describe the surfaces and interiors of Mars, Mercury, and Venus.

Mars is often referred to as the *Red Planet* because the iron oxide (hematite) prevalent on its surface gives it a rust-red appearance. The other surface colors, shown in Figure 1.9—golden, brown, tan, and greenish—depend on minerals (Peplow, 2007; Karttunen et al., 2017). The rotational period of Mars is similar to that of Earth (see Table 1.5). Its average temperature is 220 degrees Kelvin (-63 degrees Fahrenheit or -53 degrees Celsius), much colder than Earth's. Remember that the Earth's maximum surface temperature is 293 degrees Kelvin, or about 68 degrees Fahrenheit (20 degrees Celsius). Like the Earth, Mars has a dense metallic core of about 1794 ± 65 kilometers in radius, consisting primarily of iron and nickel (Rivoldini et al., 2011), surrounded by a silicate mantle that formed many of the tectonic and volcanic mountains on the planet's surface. These mountains appear to be dormant (Jacqué, 2006; and Webster et al., 2014). The average thickness of the planet's crust is about 50 km, with a maximum thickness of 125 km. In contrast, the Earth's crust average thickness is 40 km. Also, in contrast to Venus, Mars's atmosphere is thin with compressed gases. The dominant gas is the condensed (ice) carbon dioxide (CO_2). In summary, Mars has a thick lithosphere, compared to that of the Earth.

Notice that Antarctica on Earth, with very cool surface temperatures of between -10 degrees and -60 degrees Celsius, is not too dissimilar from those of Mars and Titan. The fact that it never had indigenous life tells us that our physiology is unlikely to be repeated elsewhere in our solar system. However, penguins have lived and are living in Antarctica.

The Earth is the third planet from the sun. It has liquid water. Unlike the Earth, the atmospheres of its two neighboring planets, Venus and Mars, consist mainly of carbon dioxide. In other words, the planet's atmosphere conditions do not readily relate to the planet's distance from the sun. We further discuss to this point later in Ikelle (2022a).

Venus is the most similar planet to the Earth in size, mass, density, and distance from the sun (see Table 1.5 and Figure 1.29). However, conditions on the Venusian surface differ radically from those on the Earth. With a mean surface temperature of 735 degrees Kelvin (863 degrees Fahrenheit and 462 degrees Celsius), Venus is by far the hottest planet in the solar system, even though Mercury is closer to the sun. The surface of Venus is relatively flat compared to the Earth's surface. The total distance from the lowest point to the highest point on the entire surface is about 13 kilometers, while on the Earth the distance from the basins to the Himalayas is about 20 kilometers. Much of the Venusian surface appears to have been shaped by volcanic activity. Venus has several times as many volcanoes as the Earth, and it possesses 167 large volcanoes that are over 100 km across (Nimmo and McKenzie, 1998; Faure and Mensing, 2007). The only volcanic complex of this size on Earth is the Big Island of Hawaii.

The interior of Venus is still poorly known, and that is why we did not include it in Figure 1.32. However, the similarities in size and density between Venus and the Earth suggest they share a similar internal structure: a core, mantle, and crust (Nimmo and McKenzie, 1998; Faure and Mensing, 2007). However, Venus has a thin lithosphere compared to that of Earth. Its thick atmosphere is caused by an extreme greenhouse effect (CO_2 effect). Also, the surface of Venus cannot be seen directly yet from Earth because of this dense cloud covering its atmosphere.

Mercury is about 50 million kilometers from the sun, with a mean radius of 2,439 km. It is the smallest terrestrial planet in the solar system and the closest planet to the sun (Figure 1.29), but its day is very long. Surface temperatures of Mercury can reach over 800 degrees Fahrenheit (700 degrees Kelvin, or 427 degrees Celsius) because Mercury is so close to the sun. Its closeness to the sun means that some parts are cold because of its long nights. In fact, Mercury is simultaneously one of the hottest and one of the coldest places in our entire solar system. It might even have some ice on its poles.

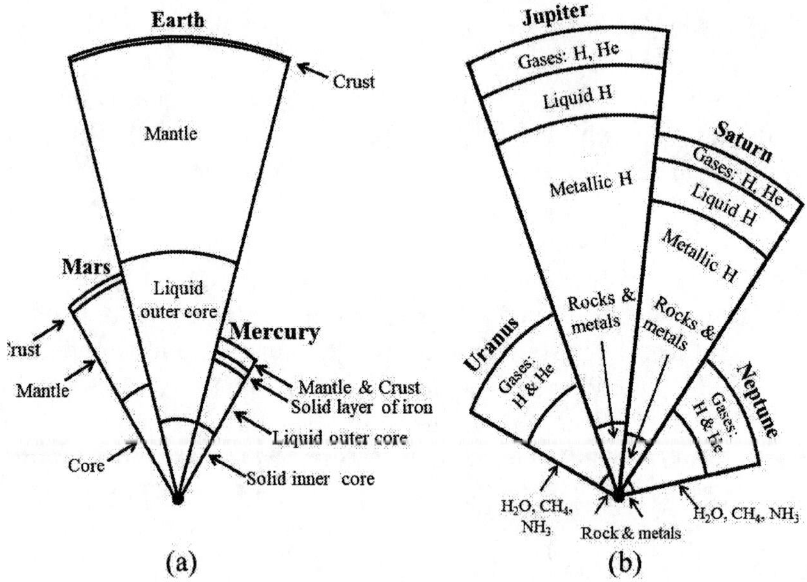

Figure 1.32. (a) The internal structures and sizes of Mercury, Earth, and Mars. (b) The internal structures and sizes of Jupiter, Saturn, Uranus, and Neptune.

Mercury has no atmosphere and therefore it has little protection from meteors, unlike the Earth, which is somewhat more protected by its atmosphere. So Mercury's surface shows impact craters, intercrater plains, basins, and smooth plains. Impact craters on Mercury are progressively degraded by subsequent impacts. The interior structure of Mercury includes a metallic core, an intermediate rocky layer, and a thin crust.

Jovian planets are mostly made of hydrogen and helium. Jupiter is the largest planet in our solar system. The fact that Jupiter gives off almost twice as much energy as it receives from the sun suggests that its planetary formation may still be ongoing. Saturn is the second-largest planet in the solar system and has the largest ring system. Uranus is far from the sun; its core is rich in ice and has a thin atmosphere. Neptune is the farthest planet from the sun in our solar system. It contains winds that exceed 1,000 km/hour. Like Uranus, its core is essentially composed of ice and it has a thin atmosphere. As we can see in Figure 1.32b, all Jovian planets have small icy rocky cores surrounded by thick accumulations of gases.

Quiz: Anywhere inside a planet (or star), the pressure is the weight per unit area of the material above that point. On Earth, a column of air 2.54 cm on a side, extending from sea-level to the top of the atmosphere, weighs 6.6678 kg. Show that the corresponding pressure is 1 bar.

Computational exercise: Table 1.6 provides Matlab commands for simulating the rotations of the planets of our solar system with respect to the sun in two dimensions (2D). (1) Reproduce the results in Figure 1.33.
(2) Propose an explanation of why Venus spins in the opposite direction from most other planets, including Earth.

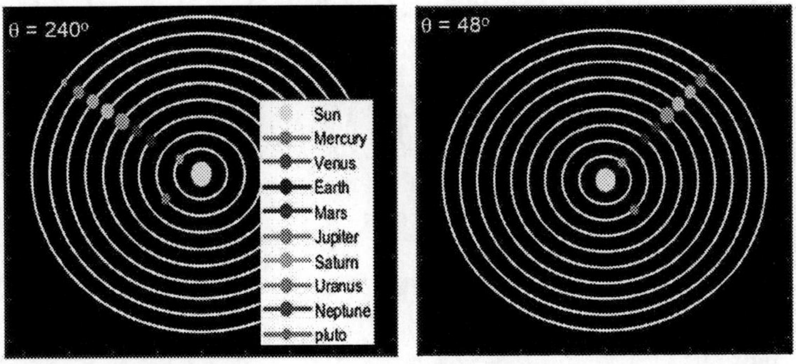

Figure 1.33. Two snapshots of our Solar System and its planets in a 2D motion.

Table 1.6. Matlab code for simulating the rotation of our Solar system.

```
clc; clear all
rm = 3; rv=5; re=7; rma = 9; rj=11; rs=13;
ru=15; rn=17; rp=19; Cc='Color'; Mm='MarkerSize';
theta=-2*pi:0.01:2*pi; xc=cos(theta); xs=sin(theta);
plot(rm*xc,rm*xs); hold on; plot(rv*xc,rv*xs);
plot(re*xc,re*xs); plot(rma*xc,rma*xs);plot(rj*xc,rj*xs);
plot(rs*xc,rs*xs); plot(ru*xc,ru*xs); plot(rn*xc,rn*xs);
plot(rp*xc,rp*xs); sun=plot(0,0,'.y',Mm,60);
cme=[0.95,0.45,0.75]; cve=[0.5,0.5,0]; cea=[0,0,1];
cma=[0.8,0,0]; cju=[1,0.5,0]; csa=[1,0.7,0];
cur=[0.2,0.8,0.8]; cne=[0,0.5,1]; cpl=[0,0.7,1];
Mk = 'marker'; xS='Sun'; xM='Mercury'; xV='Venus';
xE='Earth'; xMa='Mars'; xJ='Jupiter'; xSa='Saturn';
xU='Uranus'; xN='Neptune'; xP= 'Pluto';

for i=1:5
  for t=1:18
    th=2*(pi*2)/t; x1 = cos(th); x2 = sin(th);
    me=plot(rm*x1,rm*x2,Mk,'.',Mm,25,Cc,cme);
    ve=plot(rv*x1,-rv*x2,Mk,'.',Mm,28,Cc,cve);
    ea=plot(re*x1,re*x2,Mk,'.',Mm,30,Cc,cea);
    mar=plot(rma*x1,rma*x2,Mk,'.',Mm,27,Cc,cma);
    ju=plot(rj*x1,rj*x2,Mk,'.',Mm,42,Cc,cju);
    sa=plot(rs*x1,rs*x2,Mk,'.',Mm,38,Cc,csa);
    ur=plot(ru*x1,ru*x2,Mk,'.',Mm,36,Cc,cur);
    ne=plot(rn*x1,rn*x2,Mk,'.',Mm,34,Cc,cne);
    pl=plot(rp*x1,rp*x2,Mk,'.',Mm,20,Cc,cpl);
    legend([sun me ve ea mar ju sa ur ne pl],xS, ...
    xM,xV,xE,xMa, xJ, xSa, xU, xN, xP);pause(0.001)
    delete(me); delete(ve); delete(ea); delete(mar);
    delete(ju); delete(sa);delete(ur);
    delete(ne);delete(pl)
    set(gca,Cc,'k'); axis([-22 22 -22 22])
  end
end
```

1.2.4. Hypervelocity Impact Craters

Hypervelocity impact craters are the most abundant morphologic features on rocky planetary bodies (e.g., Mercury, Mars, and Venus) in the solar system, except on Earth. When large bodies traveling through space intersect the Earth, their speed is hardly affected. They crash into the Earth at speeds of tens of kilometers per second (on average 20 kilometers per second). These hypervelocity impacts often leave oval or circular features that can be detected by changes in topography or by geophysical methods. So a simple impact crater typically comprises a circular bowl-shaped depression. When fresh, it has an uplifted rim covered in impact ejecta. On Earth, impact craters have mostly been erased by plate tectonics and erosion, are buried under sediments, and/or altered by post-impact processes such as alteration. Convergent plate tectonic processes of subduction and crustal collision have probably obliterated discernible topographic structures of a great number of impact structures. So some impact craters can only be discovered on the Earth by using ground-truth geophysics, such as magnetic and gravity surveys, and

(a)

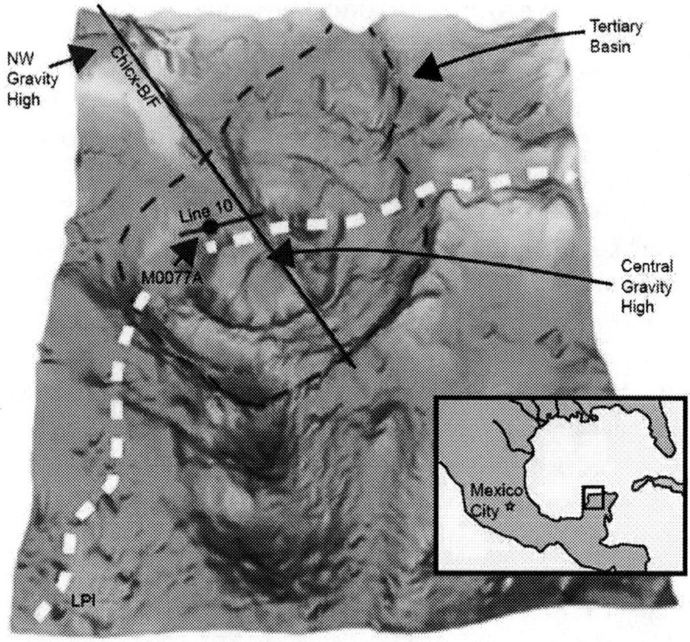

(b)

Figure 1.34. (a) Barringer Meteorite Crater (1.6-km wide, and up to 174 meters deep), Arizona is one of the best known impact craters on Earth. (b) Gravity anomaly map of Chicxulub (Mexico) crater created by the impact of a 10-km-diameter cosmic object over 66 million years ago. Red and yellow are gravity highs; and green and blue are gravity lows. This impact is thought to have caused the extinction of dinosaurs on Earth.

geological studies (Ikelle, 2022a and 2022b). As we can see in Figure 1.34, hypervelocity impacts led to the Barringer crater in Arizona and the Chicxulub crater in the Gulf of Mexico, which still show clear circular shapes of anomalies induced by old impact events. Barringer Meteorite Crater (Figure 1.34a), near Winslow, Arizona is one of the best known impact craters on planet Earth. It is 1.6-km wide, and up to 174 meters deep and was struck about 50,000 years by a 40-meter-diameter cosmic object traveling over 42,000 kilometers per hour. Chicxulub (Mexico) crater in Figure 1.34b was created by the impact of a 10-kilometer-diameter cosmic object over 66 million years ago. This impact is thought to have caused the extinction of the dinosaurs on Earth, along with about 70 percent of the species living on Earth. The colliding object (asteroid) was 10 to 20 kilometers across and ejected dust into the atmosphere. Months of darkness and cold, death of plants, starvation of herbivores, and starvation of carnivores followed. Not to mention the tsunamis, forest fires, and acid rain.

Hypervelocity impacts of meteorites and comets generate very large amounts of energy at the impacts. The impact at Chicxulub is believed to have delivered an estimated energy of about 21 billion Hiroshima atomic bombs—the atomic bomb that destroyed Hiroshima in Japan in 1945, close to the end of the Second World War, was roughly equivalent to the explosive power of 15,000 tons of TNT explosives. They are capable of modifying the composition of rocks by the intense heat they generate in a manner of seconds to minutes. Unlike geological processes, hypervelocity impacts affect large volumes of the Earth's crust (as deep as 100 kilometers) in a very short time interval (minutes). Shock waves generated radiate from the point of impact at speeds that can exceed 10 km/s, although they attenuate very rapidly, and peak shock pressures can reach several hundred GigaPascals, while normal rock deformations occur around one GigaPascal (French, 1998). In fact, one of the definitive proofs of hypervelocity impacts is the presence of irreversible rock deformations that are only possible at these very high pressures. Consequently, meteorite impacts have tremendous effects on the Earth, and its materials, including its mineral resources and biological niches, and life at its surface. Maybe the rare carbonados with porous polycrystalline diamond aggregates and with highly polished ceramic-like surfaces, which are found in the Central African Republic, are due to a large meteorite impact in Bangui (the capital of the Central African Republic in Africa) about 540 million years ago (e.g., Ravat et al., 2002; and Ikelle, 2020). A recent geomagnetic anomaly map, shown in Figure 1.35, obtained using data from the CHAMP and Swarm satellite missions, including magnetic-gradiometer data, increases the possibility that the Bangui anomaly is the result of a large meteorite impact because it has the tendency of circularity as at some other known impact sites. Its differences in shape with Kursk anomalies, the other strong magnetic anomaly, due to the effects of

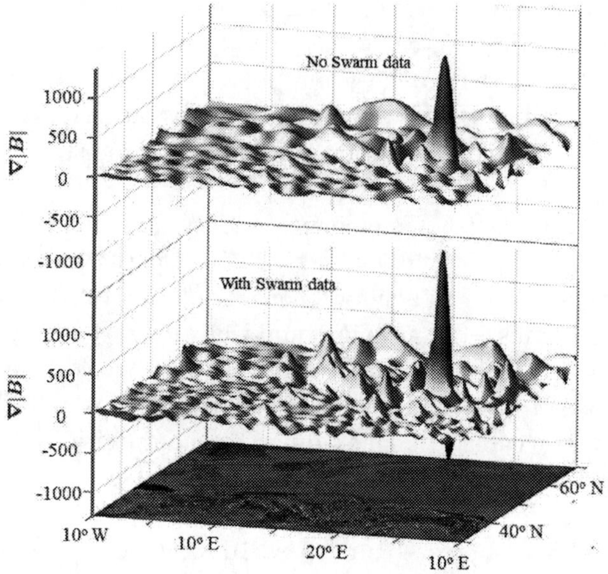
No Swarm data
With Swarm data
1000
500
0
-500
-1000
1000
500
0
-500
-1000
∇|B|
∇|B|
10° W
10° E
20° E
10° E
60° N
40° N

(a)

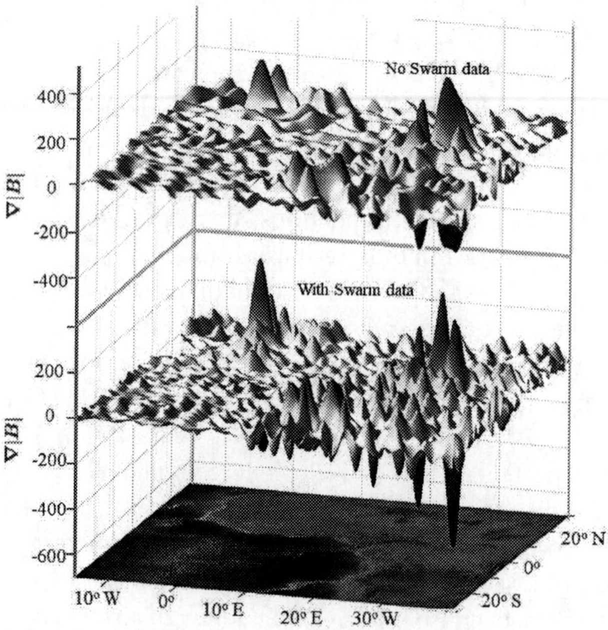
No Swarm data
With Swarm data
400
200
0
-200
-400
200
0
-200
-400
-600
∇|B|
∇|B|
10° W
0°
10° E
20° E
30° W
20° N
0°
20° S

(b)

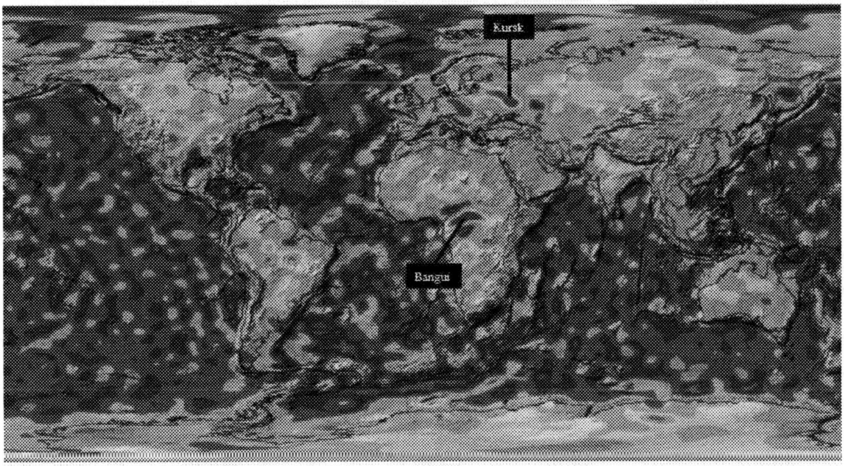

(c)

Figure 1.35. (a) Reconstruction of the Kursk magnetic anomaly with and without Swarm gradiometer data (see Ikelle, 2022d). It generates the strongest signal in satellite altitude that has been associated with the iron ore with the same name in Russia. $|B|$ is the intensity of the magnetic field. Swarm is a satellite mission equipped with magnetometers for measuring the full magnetic field vector. (b) Reconstructions of the Bangui anomaly produced in Central Africa with and without Swarm gradiometer data. (Adapted from Olsen et al., 2015). (c) Magnetic intensity from satellite data. The Bangui anomaly is the high-intensity (red) anomaly in central Africa, while the Kursk anomaly is the one at the north. (Source: USGS).

the anomalously thickened crust (Taylor and Frawley, 1987), rather than meteorite impact, also suggests the Bangui anomaly is the result of a large meteorite impact.

More than 250 hypervelocity meteorite and comet impact structures have been identified on Earth (Pesonen et al., 1992a, 1992b, and 2005). Compared to other terrestrial planets and moons, this number is a relatively small number on Earth due to the thick atmosphere. There are probably still more to discover. The active geological processes that continuously reshape the Earth's surface make these discoveries very challenging. For example, no impact craters on the oceanic crust have been found so far, albeit hypervelocity impacts and/or gravitational pull may have created oceans and some seas in the first place. Anyway, the studies of meteorite impacts can eventually lead to a unified geological description of all terrestrial planets.

Impact craters, large and small, are common on all solar system bodies with a solid surface. One of the most striking craters on Mars is Hellas Planitia. As shown by the NASA image in Figure 1.36, the Hellas impact structure spans more than 2,000 kilometers across in the Southern Hemisphere. Its lowest point is about 8.2 kilometers below the Martian "sea level." The formation of this impact structure is believed to have taken place between 3.9 and 4.6 billion years ago. On Mercury, hypervelocity impact structures are everywhere, as we can see in Figure 1.37, partially because it has almost no atmosphere. Cosmic objects reach Mercury's surface without any deceleration.

Figure 1.9 shows Venus, where violet indicates low elevations and red indicates the highest elevations. Venus is presumed to have emerged about 4.6 billion years ago and its crust was heavily bombarded and cratered. Today, there is a low number of visible impact craters on Venus. Most of the early impact craters have been destroyed by lava flows because there is a very large number of volcanoes on Venus and the resulting lava filled impact craters. Note also that the surface of Earth's moon is heavily cratered.

The largest known crater is the Caloris Basin on Mercury, with a diameter of 1,550 km. About 15 other impact basins have been identified on Mercury (Figure 1.37). Again, these crater plains appear to have obliterated many earlier craters. On July 16, 1994, comet fragments struck Jupiter with energies equivalent to an object of 2 km striking the Earth (Solem, 1995; de Pater and Lissauer, 2010)—this collision of our solar-system objects was the first direct observation of such a collision by scientists and the interested public worldwide. It received a large amount of coverage in the public press. The results of the collision were firestorms and a year-long blackened region of Jupiter's atmosphere as large as the entire Earth. This event was also a wake-up call that big collisions still occur in our solar system. The prediction was that similar impacts are known to occur every several hundred thousand years (Harrington et al. 2004). However, 15 years later, on July 19, 2009, a second large impact occurred involving Jupiter and an icy body (probably an undiscovered asteroid), with a size of about 1 km. Despite these collisions, craters are not seen on the Jovian planets because they do not have a solid surface.

Box 1.4. Parametrization of Earth's Atmosphere

In modern life, the science of the Earth's atmosphere has become a cross-disciplinary subject because it is central to climate studies. In physics, we describe the Earth's atmosphere by temperature and pressure profiles at altitudes, gravity, and magnetic fields, sources of solar energy, and the Earth's surface emissions. In chemistry, we describe the atmosphere by its concentrations of oxygen gas, nitrogen gas, ozone, greenhouse gases (e.g., water vapor, carbon dioxide, methane, nitrous oxide, sulfur hexafluoride, and carbon monoxide), halo-compounds (methyl chloroform, carbon tetra-chloride, methyl chloride, chlorofluorocarbon-11, chlorofluorocarbon-12, chlorofluorocarbon-113, halon-1211, carbonyl sulfide, methyl bromide, ethane, nitrogen tri-fluoride, propane, sulfuryl fluoride, and tetrafluoromethane), and aerosols (e.g., smoke, fog, sea spray, and pollution particles from heat engines). Any attempt to develop a model of the Earth's atmosphere that explicitly captures all these parameters is very likely to be very inefficient. These parameters are really not independent as far as the Earth's atmosphere. For example, in the first computational problem of this chapter, we established that a parametrization of the Earth's troposphere by temperature is equivalent to describing it by the ratio of O_2/N_2. In other words, temperature variations at altitudes tell us directly how O_2/N_2 the Earth's troposphere, is varying. More such relations will be established throughout the book.

(a)

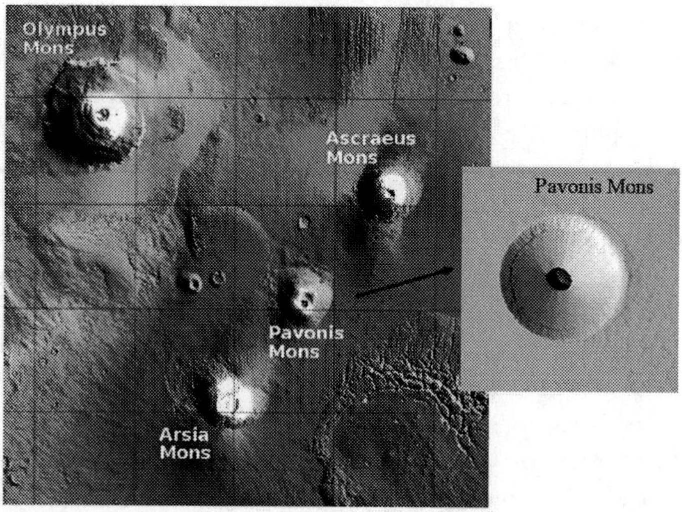

(b)

Figure 1.36. (a) Topography of Mars in kilometers showing highlands (red and orange) dominating the southern hemisphere of Mars, lowlands (blue) the northern. Tall volcanoes are in white. Because Mars has no oceans and hence no "sea level", zero altitude is defined as the height at which there is 610.5 Pa (6.105 mbar) of atmospheric pressure. (Credit: NASA's Scientific Visualization Studio). (b) Pavonis Mons contains a natural tunnel within a solidified lava flow, formerly occupied by flowing molten lava, known as lava tubes. Kazumura Cave in Hawaii is the world's most extensive lava tube.

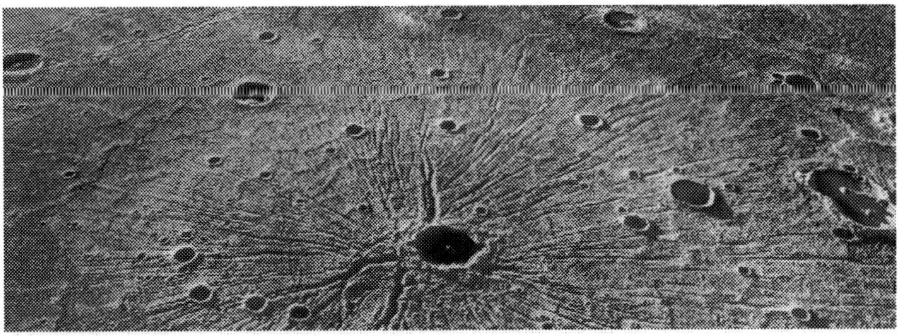

Figure 1.37. The NASA MESSENGER spacecraft returned this unprecedented view of the Caloris Basin of the planet Mercury (greens and blues are low, whites and red are high). Caloris Basin on Mercury is the largest known crater, with a diameter of 1,550 km. About 15 impact basins have been identified on Mercury, with more to be revealed.

1.3. The Formation of Our Solar System and Its Planets

Our goal in this section is to understand the composition of the system—that is, the different sizes, compositions, and orbital and rotational motions of planets (Terrestrial, Jovian) and minor objects (e.g., asteroids, comets).

1.3.1. The Formation of the Solar System: Kant-Laplace Nebular Hypothesis

The two main stages of star system formation are the formation of the star nebula and the accretion disc (Figure 1.38), as described by Kant (1755) and Laplace (1799). Let us elaborate further on these two astronomical objects for our solar system.

Solar nebula: The solar system (the sun, planets, moons, comets, meteorites, and asteroids) formed about 4.6 billion years ago, within 50 to 100 million years, when the attractive forces of gravity pulled together a low-density cloud of interstellar gas and dust (hydrogen 70 percent; helium, 28 percent; hydrogen compounds like methane

(CH_4), ammonia (NH_3), water (H_2O), 1.4 percent; silicon-based minerals, 0.4 percent; and iron, nickel, aluminum, 0.2 percent). This cloud of interstellar dust and gas is known as *solar nebula* (*nebula* is the Latin word for "cloud"). The nebula was already thicker than the average interstellar region—the initial cloud was spread very large, several light years across, and most of the motions of the cloud particles were random and therefore diffuse. Yet the nebula had a net rotation, albeit imperceptibly slow.

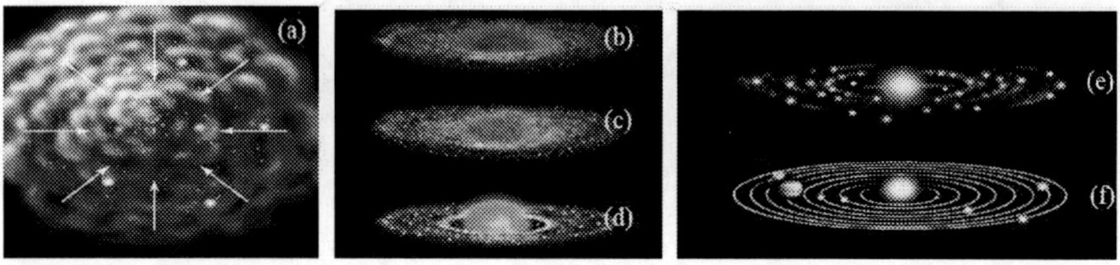

Figure 1.38. (a) The starting point of Star System formation is the formation of nebulae. It is a cloud of interstellar gas and dust is formed. The nebula is thicker than the average interstellar region. A small overdensity in the cloud caused contraction to begin and the overdensity to grow, thus producing a faster contraction. The cloud starts collapsing under its own gravity. (b) Accretion disk: The matter around the center spins up and flattens into a disk, while heat vaporizes the dust. The flattening produces the accretion disc. (c) Protostar forms in the center, when the core becomes opaque; later it will become the Sun. (d) Condensation: The disk radiates away its energy and cools off; some gas condenses into tiny dust grains of metal, rock and, far enough from the forming star, outside the frost line. (e) Planetesimals: Dust grains stick to each other (ice helps) and sweep their paths, forming larger particles; this accretion goes on until the particles become the size of boulders or small asteroids, which attract matter with their gravity. (f) Protoplanets: The larger particles' growth accelerates, and they accumulate all of the solid matter close to their own orbit. In 100,000 to 20,000,000 yr, the protoplanets' size is large asteroid/lunar size in the inner solar system, and several times the Earth's size in the outer solar system (lower temperature). Fragmentation: In 10 to 100 million years, while larger planetesimals become more massive, smaller ones break into smaller pieces when they collide, end up as meteorites on larger objects, and/or their orbits are altered.

Accretion disc: The cloud began to contract under its own gravity due its small portion of over-dense materials. In other words, the cloud began to collapse because its lower internal pressures, which are smaller than gravity, trigger collapse—remember that when the internal pressure of the nebula is greater than the gravitational force, particles oscillate; when the internal pressure of the nebula is smaller than the gravitational force, particles collapse; and when the gravitational force balances the internal pressure, we have hydrostatic equilibrium, because gravity pulls material in and gas pressure opposes (Figure 1.39). Any stationary object in space, like the sun and Earth, the force of gravity must be balanced by an outward pressure if the object is to retain its shape and size. As the collapse proceeded, the rotation speed of the cloud was gradually increasing due to the conservation of angular momentum (i.e., d^2/τ, where d is the nebula diameter and τ is its period, which is constant) and the cloud was continuing to contract (shrink). For example, when an initial nebula with a diameter of 10,000 AU and a rotation period of one million years shrunk to 200 AU (twice Pluto's orbit), its period becomes 400 years only, thus, the nebula moves faster.

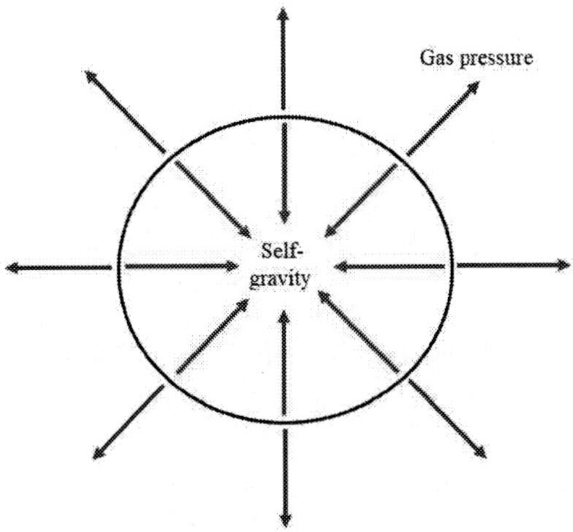

Figure 1.39. Hydrostatic equilibrium: the downward force of gravity is balanced by force of pressure.

Also, conservation of energy ensures that the cloud heats up as it collapses. This gravitational collapse causes material to fall toward the center, where things become more and more concentrated and hot because the increasing temperatures in the shrinking nebula vaporized most of the solid material that was originally present; over 100,000 years, it shrinks down to 100 AU, heats up (thermal energy), and compresses in the center. The fast rotating cloud produces collisions between particles in the cloud that cause the cloud to finally collapse into a wide thin disc, with a diameter of about 100 AU (0.00159 light years)—collisions between gas particles in a cloud gradually reduce random motions and these collisions also reduce up-and-down motions. This disk-shaped rotating nebula, also known as *protoplanetary disc*, explains the primary motions in the solar system (Figure 1.38). The solar nebula became hotter (over 10 million Kelvin) near the center, where much of the mass was collected to form the *protosun* (the cloud of gas that became the sun). The collisions among the atoms were so violent that nuclear reactions began (see Ikelle, 2022a), at which point the sun was born as a star and it began to shine.

One may ask, what prevented further collapse? As the temperature and density increased toward the center, so did the pressure causing a net force pointing outward. The sun reached a balance between the gravitational force and the internal pressure, also known as hydrostatic equilibrium, after about 50 million years.

A large amount of evidence exists today that supports the above description of the formation of the solar system. They include the physical observations of young star formation and stars near the end of their life, the discovery of other star systems with planets in orbit around stars, the observations of discs of gas and microscopic solid particles around young stars are now common, the physical properties of the planets of the solar system, studies of meteorites, and numerical simulations that confirm observations (e.g., Wetherill, 1996; Raymond et al., 2006; and Barnes and Raymond, 2004). These simulations also show how accretion works by the progressive gathering together of smaller particles into large bodies and confirm the differences between the heavier terrestrial planets close to the sun and the lighter, more gaseous planets at greater distances.

So the solar nebula model fits the facts that terrestrial planets are composed of rocky substances, Jovian planets are composed primarily of hydrogen and helium, terrestrial planets are smaller than Jovian planets, all of the planets orbit the sun and are nearly in the same plane [all the planets (except Pluto) have orbital planes that are inclined by less than 6 degrees with respect to each other], terrestrial planets orbit close to the sun, and the Jovian planets orbit further away. The dating of Earth and moon rocks provides most of the dates mentioned above (see also Ikelle, 2022a). The oldest objects in our solar system are 4.57 billion years old, mm-sized grains found in some meteorites. Some of the meteorites even give us evidence that a star exploded in our galaxy around the time the solar system formed. Close-up of the Orion-Nebula images, obtained with the Hubble Space Telescope in 1992, show discs of dust and gas surrounding a newly formed star. This protoplanetary disc spans about 0.14 light years. In other words, we can observe protoplanetary systems that are being formed right now using space telescopes. In 2017, the European Space Agency launched a satellite (CHEOPS), which is currently studying exoplanets ranging in sizes from Earth to Neptune.

1.3.2. The Formation of the Planets

The protoplanetary disc contained only 0.2 percent of the mass of the solar nebula with particles moving in circular orbits. The rotation of the disc prevented further collapse of the disc. As the disc radiated away, its internal heat temperature dropped and the heaviest molecules began to form tiny solid or liquid droplets, a process called *condensation.* Near the sun, where the temperatures were high, only the heaviest compounds condensed, forming heavy solid grains, including compounds of aluminum, titanium, iron, nickel, and, at somewhat cooler temperatures, silicates. In the outskirts of the disc, the temperatures were low enough that hydrogen-rich molecules condensed into lighter ices, including water ice, frozen methane, and frozen ammonia. At other words, depending on the temperature, different materials can condense out of the nebula and form solids (see Table 1.7). The initial temperature of the nebula, before disk formation, was close to 50 degrees Kelvin. Formation of the protosun caused the temperature in the center of the nebula to rise to about 2,000 degrees Kelvin, so all elements could condense. However, H and He remained in gaseous form at all stages of solar formation because their condensation temperatures, 22 degrees Kelvin and 4 degrees Kelvin, respectively, are lower than the initial temperature of the nebula. The great temperature differences between the hot inner regions and the cool outer regions of the disc determine condensates available for planet formation. The frost line, in Figure 1.40, is the distance in the solar nebula from the center of the protoplanetary disc, where ices like hydrogen-rich compounds can condense, while only metals and rocks can condense inside this line. This line marks a transition from the warm inner regions where terrestrial planets form to the cool outer regions where Jovian planets form.

Table 1.7. Relation between the temperature and the mass of the particles that become solid during planet formation—near the sun, where the temperature was higher, only the heaviest compounds condensed, forming heavy solid grains, including compounds of aluminum, titanium, iron, and nickel. At the outskirts of the disc, where the temperature was low enough, hydrogen-rich molecules condensed into water ice, frozen methane, and frozen ammonia.

	Condensation temperature (K)	Portion of the disk
Metals: iron, nickel, aluminum	1,300 – 1,600	0.2%
Rocks: silicon-based minerals	500 – 1,300	0.4%
Ices: methane, ammonia, water	100 – 300	1.4%
Light gases: hydrogen and helium	never condensed in the disk	98%

Solid particles orbited the sun in nearly circular orbits right next to each other, as the gas from which they condensed. Gently collisions allowed the flakes to stick together and make larger particles, which, in turn, attracted more solid particles. This process is called *accretion*. The objects formed by accretion are called planetesimals ("small planets"). They are between 1 and 10 km in diameter. Gravity draws *planetesimals* together to form planets. For the construct of terrestrial planets, planetesimals are made of grains of rock and metals from inside the frost line and they are made of ices outside the frost line for the construct of Jovian planets.

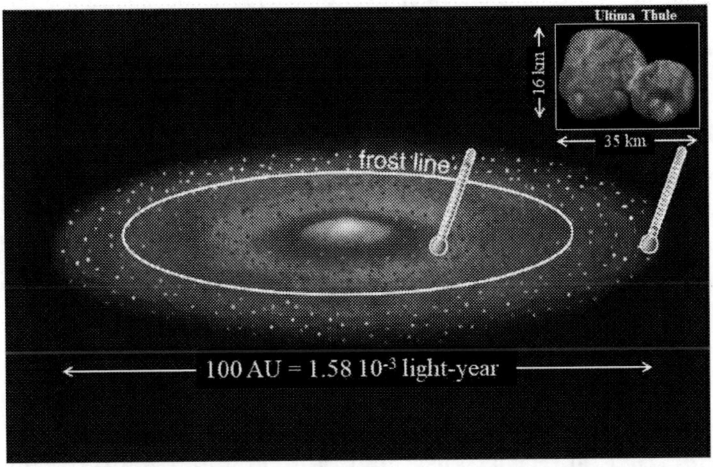

Figure 1.40. The frost line, also known as the ice line, separates the Jovian and terrestrial planets. Within the frost line, rock and metals condense, hydrogen compounds stay gaseous. It is too hot for hydrogen compounds to form ices inside the frost line. Beyond the frost line, hydrogen compounds, rocks, and metals condense. It is cold enough for ice to form outside the frost line. We also appended a picture of asteroid, Ultima Thule, here. This asteroid is also known as Arrokoth. It is located beyond the orbit of Neptune and within the Kuiper Belt. It is one of the small remnants left over from the formation of the solar system. It is considered to be the most distant and primitive object ever to visited by a man-made space probe.

Jovian planets are larger than terrestrial planets because the protoplanetary disc contains several times more gases and ices (see Table 1.7) used in the formation of Jovian planets than metals and rocks that are used in the formation of terrestrial planets. Also, the stronger gravitational pull of Jovian planets (see Ikelle, 2022b) allows them to attract larger quantities of hydrogen and helium than terrestrial planets, which have weaker gravity. Note also that Jupiter's massive gravity may explain Mars's smaller mass, with the larger planet consuming more material. Moreover, Jupiter's gravity accelerates the movement of nearby materials, generating destructive collisions rather than constructively gluing material together. Similar explanations hold for Mercury due to its closeness to the sun and its massive gravity.

The newly formed planets (protoplanets) were undifferentiated (i.e., homogenous). Driven by density differences, differentiation occurred within about 100 million years—that is, Mercury, Venus, Earth, Mars, Jupiter, Saturn, Neptune, and Uranus moved from homogenous planets into heterogeneous ones. Let us start with the terrestrial planets. The early planets were characterized by volcanism, asteroid impact, and partially molten surfaces with all elements evenly distributed through its volume. Gravitational compression (see Ikelle, 2022b), radioactive decay (see Ikelle, 2022a), and asteroid impacts made the young planets very hot. As the planets cooled and solidified from their molten state, minerals started to crystallize and settle, resulting in a separation of minerals, based on density, and the creation of the crust, mantle, and core occurred. The less-dense minerals floated to the surfaces of the planets to form the crusts, while the denser materials sank to form the mantles, and the highest-density iron and nickel sank into the cores.

Jovian planets began as terrestrial planets, and then acquired envelopes of H and He—that is, in a nutshell, the differentiation of Jovian planets. Basically, in the first stage, Jovian planets were formed by initial accretion of planetesimals containing ice into larger rocky cores. In the second stage, gas atoms moving at slower speeds in the cooler outer nebula were captured by gravity to produce gas envelopes of Jovian planets. This stage of accretion was completed when the mass of the core and gas became equal. In the final stage, the envelopes pulled in all the surrounding gases, dramatically increasing the planet's mass and diameter. The rapid growth would continue until all available gas in the nebula was used up, producing extremely thick hydrogen-rich envelopes surrounding the rocky cores.

1.3.3. The Formation of Comets, Asteroids, Meteorites and Moons

Comets may be leftover icy planetesimals that could not accrete the surrounding gas, so they remained like small dirty snowballs. In

> **Computational exercise:** Table 1.8 is Matlab code for simulating collisions between gas particles in the cloud and collapse at the center of the cloud. (1) Reproduce Figure 1.41; rand: uniformly distribution and rand: normally distribution. (2) Why some particles will never gravitationally collapse at the center of the cloud.

Table 1.8. Code for simulating the rotation of our Solar system.

```
clc; clear all; dt=0.0001; tm=0.2; t=0:dt:tm;G =1;
d=5*(G*m*dt*dt/4*pi*pi)∧(1/3); b=1; bb=[-b b]; idm=1/d;
hfl=['history_' num2str(now*1e6,'%16.0f') '.mat'];
r=randn(1,n,2); v=0.005*randn(1,n,2); r1=shiftdim(r);
dd2=sum(r1.∧2,2); dd=sqrt(dd2); ddm=mean(dd);
f=0.1*sqrt(G*mm/(ddm*ddm*ddm))/(2*pi);
v=bsxfun(@minus,v,(mean(v,2)));
r=bsxfun(@minus,r,(mean(r,2)));w=2*pi*f; wv=[0;0;w];
for nc=1:n
  rx=r(1,nc,1); ry=r(1,nc,2); rv=[rx;ry;0]; vv=cross(wv,rv);
  v(1,nc,1)=v(1,nc,1)+vv(1); v(1,nc,2)=v(1,nc,2)+vv(2);
end
hp=plot(r(1,:,1),r(1,:,2),'k.', 'MarkerSize', 15);
ht=title(' '); axis equal; dlc=5; lc=1;
for tt=t
  dr=bsxfun(@minus,permute(r,[2 1 3]),r);id2=1./sum(dr.∧2,3);
  id2(1:n+1:end)=0; id=sqrt(id2); bi=(id<idm); id2e=id2.*bi;
  nn=bsxfun(@times,dr,id); F=sum(bsxfun(@times,G*nn,id2e),1);
  dv=bsxfun(@minus,permute(v,[2 1 3]),v);
  dve=bsxfun(@times,dv,~bi);
  F=F+al*sum(dve,1); v=v+(F/m)*dt; r=r+v*dt;
  if mod(lc,10)==0
    set(hp,'Xdata',r(1,:,1),'Ydata',r(1,:,2));
    mr=(mean(r,2)); bb1=bb+mr(1,1,1); bb2=bb+mr(1,1,2);
    set(ht,'string',['lc=' num2str(lc)]); drawnow;
  end
  lc=lc+1;
end; grid on
```

fact, beyond Neptune and even beyond Pluto, in the coldest regions of the nebula, there are surviving icy planetesimals—some were originally near the orbits of the big planets (Jupiter, Saturn, Uranus, and Neptune) and may have gotten kicked out into the far distance by those planets' gravity. Beyond the frost line, lots of icy planetesimals remain.

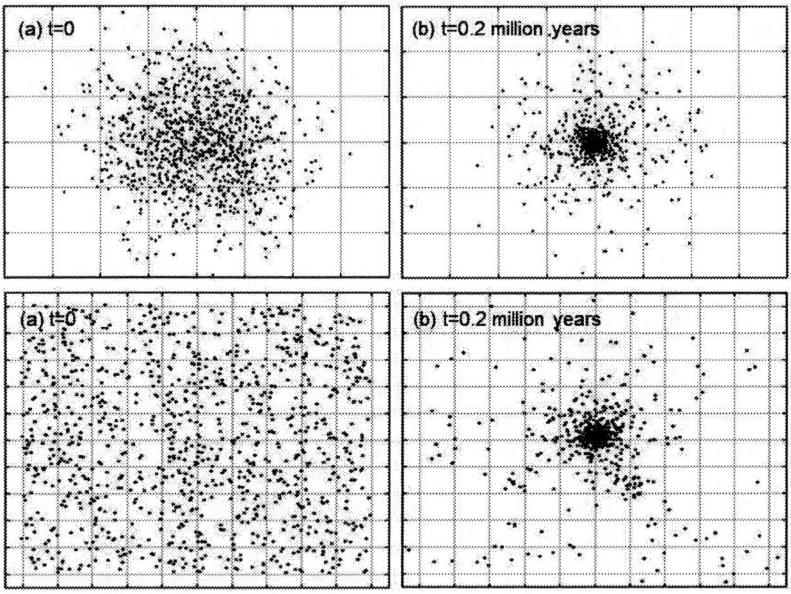

Figure 1.41. A simulation of star formation from rotating small particles representing clouds of gas and dust. Top row: (a) The cloud is formed by a Gaussian distribution. (b) A result of the Gaussian cloud contraction due to self-gravity. Bottom row: (a) The cloud is formed by a uniform distribution. (b) A result of a uniform cloud contraction due to self-gravity.

Asteroid belt is located between Mars and Jupiter. It contains thousands of rocky planetesimals from 1,000 km to a few meters across. These are also thought to be debris from the formation of the solar system that could not form a planet due to Jupiter's gravity. When asteroids collide, they produce small fragments that occasionally fall on Earth. These rocks are called meteorites and provide valuable information about the solar nebula. Again, most of these fragments are the size of sand grains. They burn up in the Earth's atmosphere, causing them to glow like meteors (or shooting stars).

Jupiter's gravity is so strong (see Ikelle, 2022c) that it prevents asteroids from ever coalescing into a planet. It prevents a single planetary body from forming in that region. It protects Mercury, Venus, Earth, and Mars from repeated asteroid collisions. In other words, without Jupiter exerting its gravitational force on the asteroids in the belt, the inner planets would still be constantly bombarded by large asteroids today.

Note that, as mentioned, NASA's New-Horizons Spacecraft returns the image of Ultima Thule, an asteroid of Kuiper Belt in 2019, which was unknown before Horizons launched. Ultima Thule is still the most distant and primitive object ever to visited by a man-made space probe. It consists of two objects stuck together (see Figure 1.40). It is a Kuiper Belt object left over from the birth of the solar system 4.6 billion years ago. Therefore, Ultima Thule may have information on the origin of our solar system.

Moons: There are numerous moons orbiting the Jovian planets. They are generally characterized as small moons when their diameters are less than 300 km, medium-size moons when their diameters are between 300 and 1,500 km, and large moons when their diameters are greater than 1,500 km. Some of the largest moons are Europa, Io, Callisto, and Ganymede, with diameters between 3,000 and 5,500 km. They are four of the moons of Jupiter and are called the Galilean moons.

Many of these have changed very little in the past 4.5 billion years, but others show extensive signs of recent and even ongoing geological activity. Medium and large moons of the Jovian planets are most closely related to the terrestrial planets in their compositions, which include a considerable amount of ice in addition to rock and metal. Titan, a moon of Saturn (in Figure 1.31) is a typical example. However, the origin of the Jovian moons remains a mystery. The current working model (Canup and Ward, 2009; Alibert et al., 2005; and Chown, 2009) is that these moons are formed in a process similar to that of solar nebulae, consisting of contraction, spinning, flattening, and heating, but with smaller discs of material around these planets. In other words, condensation and accretion took place within the Jovian nebulae, creating a miniature solar system around each Jovian planet. So when small amounts of silicon-based minerals, iron, nickel, and/or aluminum were available during the moon formations, the resulting moons can have some similarities with terrestrial planets.

Also, there is no consensus on the formation of the moons of terrestrial planets. The two Martian moons—Phobos (23 km in diameter) and Deimos (12 km in diameter)—look more like asteroids. Some studies suggest these moons of Mars were formed elsewhere and then captured by the gravitational pull of Mars (e.g., Landis, 2001). After all, Mars orbits just one step inward from the asteroid belt, as shown in Figure 1.29.

Even the origin of our moon has still to be proven, despite our large geological knowledge of this satellite. The current model of the origin of our moon is that it is the result of a giant impact of a large body on a young Earth (Hartmann et al., 1986; Hartmann and Miller, 1991; Hartmann, 1997; and Canup and Asphaug, 2001). A fraction of the debris from the impact went into orbit around the Earth and aggregated into the moon. This model is supported by the fact that our moon has exactly the same oxygen isotope composition as the Earth, whereas Mars' rocks and meteorites from other parts of the solar system have different oxygen isotope compositions. Although not conclusive, this observation shows that our moon was formed from materials in the Earth's neighborhood.

Recent numerical modeling suggests that Mars suffered a large collision very early in its history, similar to the idea that the Earth's moon is the result of an impact, and that Phobos and Deimos may have been ejected from Mars (Canup and Salmon, 2018). So the bulk composition of these moons should be similar to that of Mars. The Japan Aerospace Exploration Agency's (JAXA) Mars-moons mission in 2024 will shed more light on this idea. If it holds, we will have a unique explanation for the formation of moons associated with terrestrial planets. Again, Mercury and Venus have no known moons.

Interstellar object: Oumuamua: The first known interstellar object to visit our solar system was discovered on October 19, 2017 by the University of Hawaii's Pan-STARRS1 telescope and confirmed by other telescopes around the world, including ESO (European Southern Observatory)'s Very Large Telescope in Chile (Meech et al., 2017; Jewitt et al., 2017; Bannister et al., 2017; and Knight et al., 2017). This interstellar object has been named *Oumuamua.* It appears to be rocky, about 400 meters long and about 60 meters wide. It varies in brightness by a factor of 10 as it spins on its axis every 7.3 hours. No known asteroid or comet in our solar system varies so widely in brightness, with such a large ratio between length and width. The most elongated objects we have seen to date are no more than three times longer than they are wide (Meech et al., 2017; Jewitt et al., 2017; Bannister et al., 2017; Bolin et al., 2018; and Knight et al., 2017). The present observations suggest this unusual object had been wandering through the Milky Way, unattached to any star system, for hundreds of millions of years before its chance encounter with our star system. Almost 5 years later (2022), Oumuamua is still orbiting our solar system. However, we are still not able to trace back its original star system.

Note that most of the objects in our solar system, including the Sun, planets, and asteroids, rotate counterclockwise. This is due to the initial conditions in the cloud of gas and dust from which our solar system formed. These initial conditions may include an explanation for Venus' rotation. Oumuamua spins in the same direction as Venus rotates.

1.3.4. Recycle between Star Systems

The old stars shed materials from their outer layers (see Ikelle, 2022a). The ejected matter contains heavy chemical elements such as carbon, oxygen, nitrogen, phosphorus, iron, and nickel. In other words, as stars die, they enrich the interstellar medium by casting a lot of heavy matter back into space. The formation of new stars is fueled by these heavy elements, in addition to helium and hydrogen in the interstellar medium. These heavy elements are also used to form planets, asteroids, moons, and comets. So our solar system contains recycled material that was passed on to it from older stars. The heavy elements that make up the unique composition of the Earth were recycled from older stars. Note that the dominance of helium and hydrogen is not limited to our solar system; it is everywhere in our Milky Way and in all galaxies. The rarity of the heavy elements partly explains the limited size of terrestrial planets.

1.4. Our Universe

1.4.1. Our Galaxy, the Milky Way

The universe is all of space and time and their contents, including planets, moons, stars, and galaxies (i.e., a huge collection of gas, dust, and billions of stars and their solar systems, all bound together by gravity), and all other forms of matter and energy (Clayton, 1983; Hooft, 1997). When you look up at stars in the night sky, you are seeing our sun (a star) and hundreds of billions of other stars that form our galaxy, known as the Milky Way. If it is really dark, far away from the lights of cities and houses like in some West African villages, you can even see the dusty bands of the Milky Way stretch across the sky (see Figure 1.42). In fact, our galaxy is called the Milky Way because it appears as a milky band of light in the sky when you see it in a really dark area. Note that the sun is not among the brightest or the faintest stars in our galaxy, the Milky Way. It is not the most massive star; even though it is more massive than about 96 percent of the stars in the Milky Way, there are billions of stars more massive than the sun. More than half the stars found in the Milky Way are older than the sun.

Quiz: (1) Why are the inner and outer planets of our solar system so different? (2) Suppose that the mass of the sun is made up of hydrogen (73 %), helium (25 %), and heavy elements (which include everything else, 2 %). Compare the heavy elements in the solar system that are found in the sun to those in the planets.

The Milky Way is a large barred spiral galaxy. Our solar system is about 25,000 light-years [5] away from the center of the Milky Way

[5] One light-year is about 9 trillion kilometers or 9×10^{12} kilometers. It is the distance light travels in one Earth year.

(a)

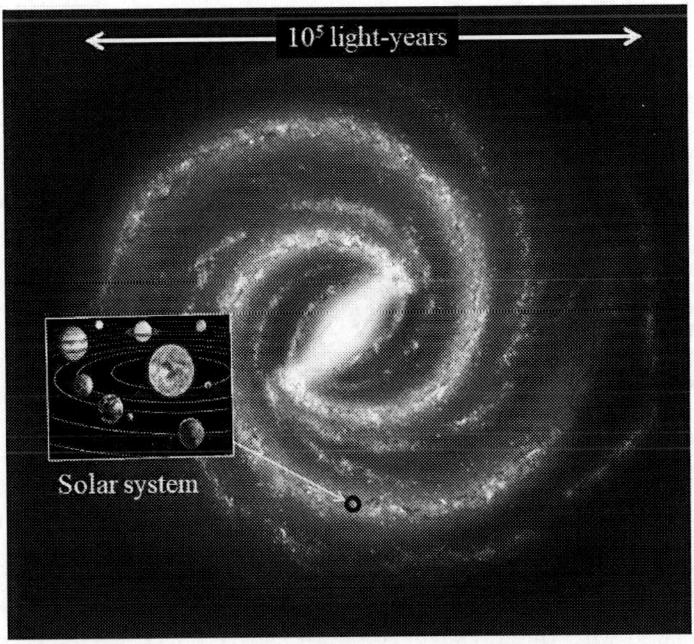

(b)

Figure 1.42. (a) A picture of the Milky Way from inside the galaxy, which means we do not have an image of the Milky Way as a whole just as we cannot take a whole picture of our house from inside. (b) A simulated plane view of the Milky Way.

galaxy—we live in the suburbs of our galaxy. Just as the Earth goes around the sun, the sun goes around the center of the Milky Way. It takes 250 million years for our sun, and therefore our solar system, to go all the way around the center of the Milky Way. Many of the stars in our galaxy have planets. NASA's Spitzer Space Telescope has revealed the first-known system of seven Earth-size planets around a single star (see Figure 1.43). Planets outside our solar system are known as exoplanets. This system of exoplanets is about 40 light-years (360 trillion kilometers) from Earth. The densities of these exoplanets suggest that they are rocky planets. This example confirms that within our galaxy we have other star systems and consequently even more planets. Alpha Centauri AB, (a binary pair) and Proxima Centauri (Alpha Centauri C) are the three stars closest to the sun. They are about 4.21 light-years from us. A planet about the same mass as the sun orbits Alpha Centauri B. Other stars within 20

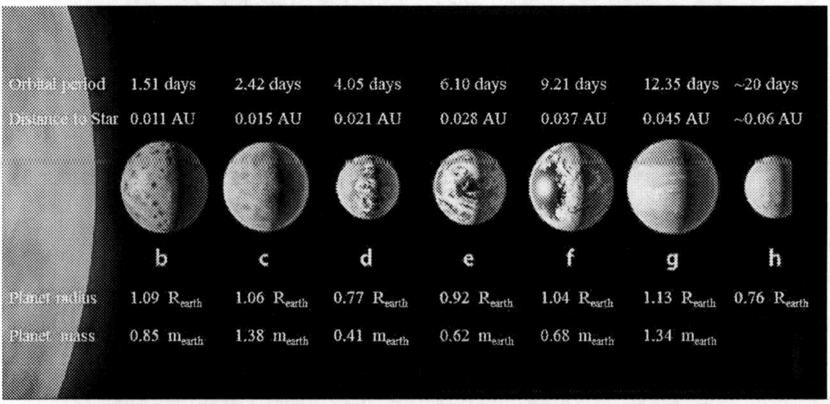

Figure 1.43. NASA's Spitzer Space Telescope revealed the first known system of seven Earth-size planets around a single star. The star system associated with these planets is known as the TRAPPIST-1 system. It is located about 39.5 light-years away from Earth.

light-years of the Earth include Sirius A and B (8.6 light-years away from us), Epsilon Eridani (10.4 light-years away from us), Procyon A and B (11.4 light-years away from us), Tau Ceti (11.7 light-years away from us), and 82 G. Eridani (20 light years away from us).

There are also many galaxies besides ours (Figure 1.44). There are so many, we cannot even count them all yet! Some studies suggest there could be as many as two trillion galaxies in our universe (Siegel and Bang, 2018), meaning an infinite number of galaxies. One of the largest visible galaxies, Andromeda, may have about 300 billion stars and the Milk Way may have 200 billion stars. One of the smallest stars, M33, may have 25 billion stars. Some galaxies are spiral-shaped like ours. They have curved arms that make them look like a pinwheel. Other galaxies are smooth and oval-shaped. They are called elliptical galaxies. And there are also galaxies that are not spirals or ovals. They have irregular shapes and look like blobs. The light that we see from each of these galaxies comes from the stars inside them. At the largest scale, galaxies are distributed uniformly in all directions (isotropic), meaning that our Universe has neither an edge nor a center. At smaller scales, galaxies are distributed in clusters and superclusters which form immense filaments and voids in space, creating a vast foam-like structure. Sometimes galaxies get too close and smash into each other. Our Milky Way galaxy will someday bump into Andromeda, our closest galactic neighbor. But it will not happen for about five billion years, and even if it happens tomorrow, you might not notice. Galaxies are so big and spread out that even galaxies bump into each other, the planets and solar systems often not affected.

With two or so trillions of galaxies (i.e., several quadrillion stars) distributed almost uniformly without a center, our universe is, for all practical purposes, spatially infinite. Moreover, a large number of galaxies are spiral-shaped like our galaxy and even have almost the same size as our galaxy. With billions of stars associated with each of these galaxies and multiple planets associated with a large number of these stars, it is mathematically and statistically impossible that only our star, and only our galaxy, has one planet with our biosphere and hydrosphere. Physical and chemical properties differentiating stars are less than 10—in permutation terms, that means there are less than 400,000 types of stars in a set of several quadrillion stars. Surely there are replicas of our planet in other galaxies and probably in other star systems than in our Milky Way galaxy. One of the obstacles to real evidence of such similar planets is the large distances between stars. For example, multiple trillion kilometers separate us from the nearby stars of the Milky Way other than our sun. The real issue is to understand how our biosphere and hydrosphere vary in the solar system with time. In other words, is the evolution of the Earth, in the last 4.5 billion years, repeatable in our solar system, in the next 6 or so billions of years the sun is estimated to last? As we learn more in the coming chapters, at least partial answers to this question will emerge.

1.4.2. Ordinary and Dark Matter

By studying the movement of galaxies, it has been discovered that our universe contains much more matter than is accounted for by visible objects: stars, galaxies, nebulas, and interstellar gas. This invisible matter accross the entire electromagnetic spectrum is known as dark matter (Zwicky, 1933; Babcock, 1939; Weinberg, 2011). It neither emits nor absorbs light or any other electromagnetic radiation at any noticeable level. Yet there is a wide range of strong indirect evidence that it exists, but we have not yet figured out how to detect it directly. The existence and properties of dark matter are inferred from its gravitational effects on visible matter, radiation, and the large-scale structure of the universe (see Ikelle, 2022a and 2022b).

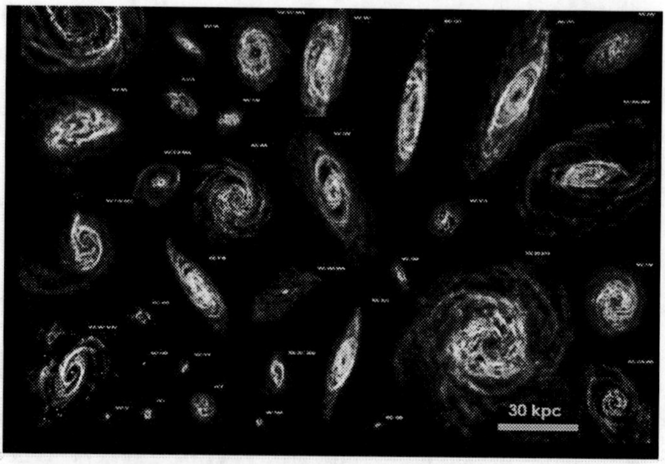

(a)

(b)

Figure 1.44. (a) An illustration of galaxy-mass distributions; pc stands for parsec and one parsec equals to 30.86 trillion kilometers. [Adapted from Walter et al. (2008)]. (b) A picture of thousands of galaxies taken by the Hubble Space Telescope. Even the tiny dots are whole galaxies. Some galaxies are spiral-shaped like ours. Others are elliptical or just have irregular shapes.

In other words, we cannot see dark matter, but we can feel its effects. Our own Milky Way galaxy has an enormous amount of dark matter—it has a total mass of more than 5×10^{11} m_{sun} and more than 80 percent of this mass is dark matter. Dark matter is estimated to constitute 84.5 percent (a huge number!) of the total matter in our universe (www.cam.ac.uk/research/news/planck-captures- portrait-of-the-young-universe-revealing-earliest-light). Exactly what dark matter is made of is unknown despite many suggestions.

The first detection of dark matter is attributed to Zwicky (1933), who found that the velocities of some galaxies far exceed those that could be attributed to the luminous matter in the galaxies themselves. Babcock (1939) used optical spectroscopy to measure the rotation of the Andromeda galaxy (M31), the nearest of the large galaxies in the vicinity of the Milky Way, and found the rotational velocity at large distances from the M13 center to be too large to be easily attributed to the luminous components. From the velocities of the Milky Way and M31 toward each other, Kahn and Woltjer (1959) estimated the mass of the local group of galaxies (in which the Milky Way and M31 are the dominant members). By comparing this estimate to that expected from the luminous matter in these two objects, they concluded that most of the mass of the local group must be dark. In the 1970s, dark matter became a well-recognized concept. The rotational evidence for dark matter in M31 grew stronger (e.g., Rubin and Ford, 1970; Roberts and Whitehurst, 1975) and kinematic investigations of other large disk galaxies (Einasto et al. 1974; Ostriker et al. 1974) gave similar results. These observations indicated that dark matter was a common feature among galaxies, but did not constrain its spatial distribution. Ostriker and Peebles (1973) showed that galactic disks by themselves would be unstable and suggested that they may be surrounded by massive spherical halos. Hence, the important concept of dark halos, today believed to be common to all galaxies, was born.

Quiz: (1) What are the differences and similarities between stars and giant (gaseous) planets?
(2) Venus spins backwards compared to most of the other planets in our solar system. Why?

1.4.3. James Webb Telescope in Space: Expectations

NASA's James Webb Space Telescope (JWST) was launched into space on December 25, 2021 at 6:20 AM CST from Kourou, French Guiana, and arrived at the Sun–Earth L2 Lagrange point (about 1.6 million kilometers from Earth) on January 2022. The James Webb Space Telescope is today the most complex and powerful telescope ever built by man and the largest and most powerful space telescope ever launched, as of 2022. It carries the largest mirror (6.5-meter-diameter mirror) ever flown into space and a suite of cameras and spectrometers. Webb aims to address fundamental questions in science, including how did the Big Bang make us, humans, possible? Where are the other intelligent life-forms in our universe? And many other questions we haven't even thought of asking yet.

Webb can capture both the sunlight it reflects and the heat that it emits. These reflections and emissions allow it to look back in time and to hunt for the first galaxies, the first stars, the first supernovae, and the first black holes. It will measure the effects of dark matter and dark energy, even though it cannot see them. We will learn how galaxies grow by comparing young and mature galaxies. We will watch new stars being born.

Figure 1.45. The first publicly released science-quality image from NASA's James Webb Space Telescope, revealed on July 11, 2022, is the deepest infrared view of the universe to date. (Image credit: NASA, ESA, CSA, and STScI).

After JWST arrived at L2, it unfolded its mirror and engineers and scientists on Earth spent about four months making sure that its optics were aligned to near-nanometer precision. Finally, the first image from NASA's JWST was released on July 11, 2022 (see Figure 1.45). The Webb telescope's infrared eyes have pulled normally faint and distant objects into sharp focus, providing a view of some never-before-seen galaxies and star clusters. The oldest light from some of these objects dates back roughly 13 billion years ago, in the early days of the universe. It shows the galaxy cluster SMACS 0723 as it appeared 4.6 billion years ago. The mass of the galaxy cluster is magnifying and distorting far more distant objects behind it—a phenomenon known as gravitational lensing.

The characterization of exoplanets is one of the key expectations from observations of the James Webb Space Telescope. We hope that JWST will eventually see some exoplanets in a format similar to those of the planets of our solar system in Figures 1.9 and 1.29. It will also see moons of exoplanets and surfaces of rocky exoplanets and moons. It will even see large objects like multiple-kilometer-long aircraft with kilometer-wide cross-sections. However, it may not be able to detect biological activity directly. Moreover, most biological life on exoplanets is very likely located underground just like in ancient Mesopotamia (present-day Iraq and Kuwait and parts of present-day Iran, Afghanistan, Syria, and Turkey), Ethiopia before Europeans unsuccessfully attempted to colonize it, and huge underground military bases like the Kunia tunnel, Hawaii. Göbekli Tepe, in Turkey, is an example of an underground city built about 12,000 years ago and there are more than 50 kilometers of tunnels and ancient bib- lical technologies in underground Iraq, including possibly the ark of covenant that contains the two stone tablets of the Ten Commandments. Anyway, it is very unlikely that JWST will locate cities, streets, airports, or skyscrapers on the surfaces of exoplanets. However, JWST will still provide spectroscopic observations of the hallmarks of biological life through the compositions of the exoplanet's atmospheres. The major challenge in this case would be the determination of the type of biological humanoids associated with the compositions of the exoplanet's atmospheres.

Box 1.5. Signals and Systems

Through this chapter, we have repeatedly used the word *system* and we will do so in the coming chapters. So let us briefly recall the notion of system and that of signal, which is directly related to it. A signal is defined as any physical quantity that varies with time, space, or any other independent variable. A temperature variation at one location is an example of a one—variable signal, to it. A signal is defined as any physical quantity that varies with time, space, or any other independent variable. A temperature variation at one location is an example of a one—variable signal, $T(x_0, y_0, t)$ (where (x_0, y_0) is a fixed point and t is the variable), and a global temperature variations is an example of a three-variable signal, $T(x, y, t)$, where x, y, and t are variables. In addition to representing signals as functions of time, they can also be represented as a function of frequency and/or wavenumber by using the Fourier transforms.

A system can be defined as a physical device that performs an operation on a signal. For example, a computer filtering program used to reduce noise interferences on desired signals is a system. In this case, the filtering program has the effect of reducing the noise interferences from the desired signals. In a more general sense, software realizations of operations on a signal are examples of systems. The definition of a system is not limited to software programs or physical devices; systems are far broader than that. The Earth, for instance, can be treated as a system because it responds to a source (e.g., large weight-drop and nuclear explosion) by producing ground motions that are recorded as signals. The source here represents the means by which signals are generated. Moreover, the Earth is a dynamical system because its state changes with time.

The two key properties are commonly used to differentiate systems: passive and linear. A passive system is one that does not generate an output signal if there is no input signal. The Earth, in the context of earthquake studies, can be regarded as a non-passive system, as it can generate signals (ground motion) in the event of an earthquake without an external input signal. We treat the Earth as a passive seismic system when it generates a ground motion (i.e., an output signal) only if there is an input signal (e.g., nuclear explosion).

A system is linear, if upon scaling the input signal by a certain constant value, the system produces an output that is scaled by the same value. Consequently, a superposition of the input signals leads to a superposition of the outputs; this superposition property does not hold for nonlinear systems.

Chapter 2

Ordinary Matter: Atoms, Molecules, and Organic Compounds

Matter is anything that has mass and occupies some space. In our universe, there are two types of matter: ordinary matter and dark matter. The ordinary matter means forms of matter in stars and planets, while dark matter is a hypothetical form of matter that can yet be explained or described using forms of ordinary matter. We here focus on ordinary matter.

Ordinary matter (also simply know as matter) can be solid (e.g., wood, rocks, and people) liquid (e.g., water, gasoline, and vinegar), gas (e.g., air, oxygen, and smoke), and plasma. The plasma state of matter is found in stars, aurora, lightning (i.e., we see plasma every time we see lightning), flames, neon, and fluorescent light tubes (see Figure 2.1). Plasma occurs when temperatures rise to super-high levels, more than 1,000 degrees Celsius. Plasma is not a common state of matter on Earth, but is the most common state of matter in our universe, and is found in stars, including our sun.

A solid has definite shape and volume and a mass. A liquid has an indefinite shape, a definite volume, and a mass. Gases and plasmas have indefinite shapes, indefinite volumes, and masses. We sometimes forget that gases such as air have mass and occupy space. If air did not occupy space, a balloon would not inflate when it was filled with air. Note, however, that kinetic and potential energy, vacuum, light, and sound are not matter because they are mass-less. Remember, a vacuum (also known as free space) is by definition entirely devoid of matter. Outer space is considered a good approximation of a vacuum because it has less matter than anything mankind can reproduce.

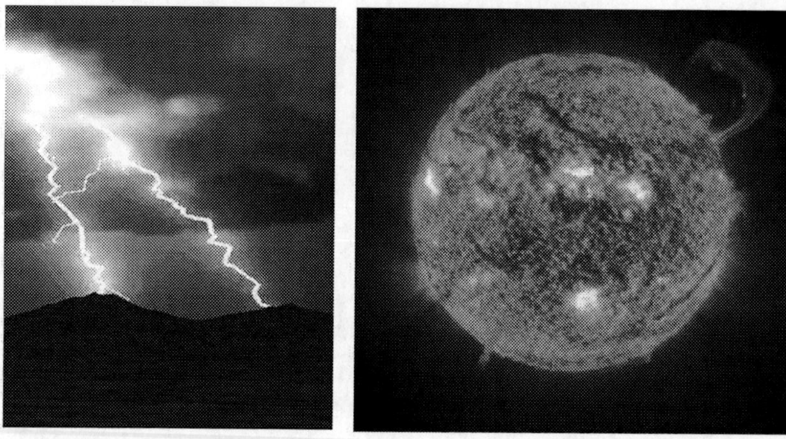

Figure 2.1. Examples of plasma: (left) lightning and (right) the sun, in a plasma state—the surface of the sun has a temperature of about 5,000 degrees Celsius.

Matter is all around us and inside us; therefore, the physical and chemical properties of matter are central to all aspects of the sciences and to all the challenges of our time. Chemical properties include flammability and acidity and chemical changes involve the production of matter (e.g., grape juice ferments, charcoal burns in the air, and iron rusts). Physical properties of matter include shape, volume, mass, and temperature, and physical changes include melting and boiling (the temperature at which a solid substance changes from solid to liquid is called its melting point and the temperature at which it changes from liquid to gas is called the boiling point). They do not involve a change in the composition of matter. Our discussion in this chapter is about the physical description of matter as it exists. As we will see in the later chapters, this description is the building blocks of many applications, including electricity generation, nuclear medicine, and subsurface fluid exploration. Moreover, the description of matter is fundamental to all fields of the sciences.

Note that forces and energy are fundamental concepts of matter that we develop through this book series. The forces are nuclear, gravity, and electromagnetic. Molecules and atoms are extremely small, and the distances between them are very large compared to the size of atoms. At the atomic level, there is a lot of empty space. Our perception of solids and fluids is created by the fast motion of atoms caused by electromagnetic forces. Sending a projectile into matter is likely to hit atoms rather than pass through them since atoms move quickly. A large number of representations in this chapter capture these empty spaces between atoms.

Key sentences: *All chemical elements introduced in this chapter were already present in the first years of the formation of our solar system. They came from the interstellar medium. Some of these elements were put in the interstellar medium (the space between stars) by dying stars and some are the results of nuclear fusion in the stars. Only some isotopes of these elements are man-made, mostly for medical purposes, by nuclear fission. All major living things are predominantly made of hydrogen, oxygen, carbon, nitrogen, and sulfur.*

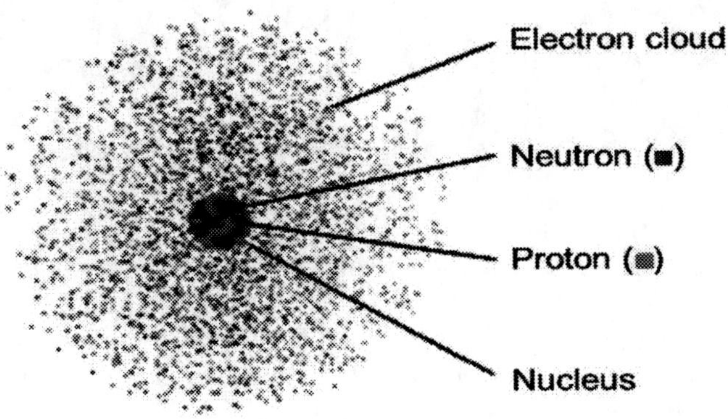

(a)

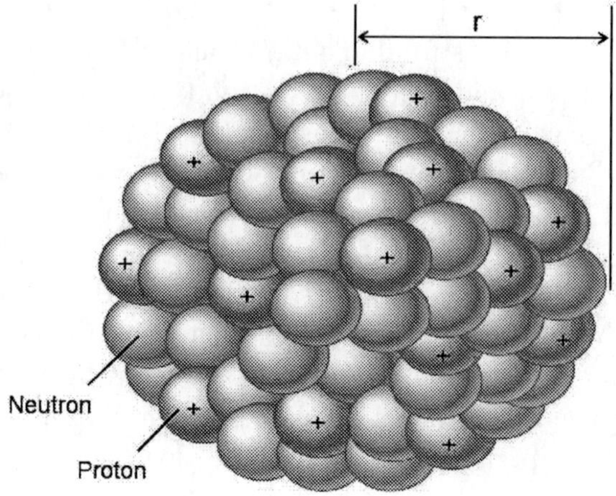

(b)

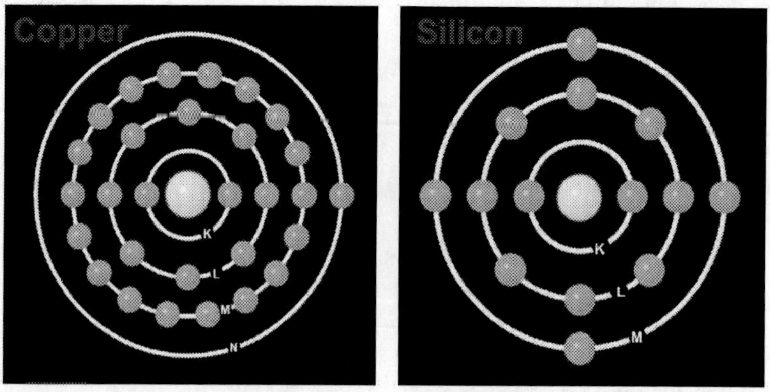

(c)

Figure 2.2. (a) An illustration of atomic composition—an atom has a nucleus (the center of an atom) that consists of protons and neutrons. The electrons are negatively charged particles that revolve around the nucleus. (b) A detailed view of the nucleus. It is approximately a spherical bob in which the positive protons and neutral neutrons are clustered together. The radius of the nucleus is $r \approx (1.2 \times 10^{-15})A^{1/3}$ meter, where A is the atomic mass number (also known as atomic weight). (c) Atomic structures of copper-63 ($^{63}_{29}$Cu) and silicon-28 ($^{28}_{14}$Si). Notice the electron shells. The closest shell to the nucleus is called K-shell"), followed by the L-shell, then the M-shell, and so on further and farther from the nucleus. The K-, L-, M-shell labels were introduced by Charles Glover Barkla (1877-1944).

2.1. Atoms and Ions

We here describe atoms and ions, along with their compositions and their properties. They are the building blocks of our lives and everything around us. The understanding of atomic and ionic behavior, along with molecular behavior, described in the next section, is the foundation of science and its derivatives include better healthcare and more comfortable lives on Earth.

Box 2.1. Definition of Energy

Energy is defined as the capacity for action or the performance of work. According to the first law of thermodynamics (the energy conservation law), energy can neither be created nor destroyed, but it can be transformed into different states or forms. Matter, a special form of energy, is included in the energy-conservation law. Energy that is waiting to do work—that is, energy that has the potential for work—is called potential energy. Potential energy is stored energy that can be converted to kinetic energy. Examples of energy include the potential energy stored in matter (e.g., the gravitational field at a certain elevation), the kinetic energy of a moving object, the energy associated with the chemical bonds that hold atoms and molecules together, and the nuclear energy created by transforming matter into energy. One form of energy that often does not make this list is food. In fact, food, the main source of energy that fuels our bodies, is one of our most important sources of energy. Electrical energy is used to power lights, computers, and industries. Fossil fuels power our cars, boats, and airplanes. Moreover, virtually every task performed by living things requires energy, including heavy labor, exercise, and even during sleep. The kinetic energy (i.e., the energy of a body in motion) of an object of mass (m) and speed (v) is given by: $\mathcal{U}_K = mv^2$. Expressing mass in kilograms and speed in meters per second will yield energy values in units of joules ($\mathrm{J} = N\,m = \mathrm{kg\,m^2\,s^{-2}} = 0.239$ calorie $= 6.242 \times 10^{12}$ MeV (Mega-electronVolt)—notice that kinetic energy increases with the mass and velocity of the body.

For example, a vehicle traveling at a high speed has more kinetic energy than the same vehicle traveling at a low speed. In terms of mass, a sedan car has less kinetic energy than a lorry traveling at the same speed. Notice also that the amount of energy, 1.6×10^{-19} J, is an electron volt (eV). The electron-volt unit is convenient when dealing with energy at the atomic level. Again, energy exists in various forms. Additional forms of energy include thermal (heat), solar, chemical, electromagnetic, gravitational, mechanical, and nuclear energies. Thermal energy is the energy that is associated with the motion of molecules in a substance. Solar energy is radiant energy from the sun that is transferred through electromagnetic waves (see Ikelle, 2022a). Other sources of radiant energy include stars and light bulbs. Chemical energy is energy stored within the chemical bonds in matter, as described in this chapter and in Ikelle (2022a and 2022b).

Chemical energy can be released, for example, in batteries or sugar/food, when these substances react to form new substances. Electromagnetic energy is the energy contained in electromagnetic waves (see Ikelle, 2022a). Gravitational energy arises from gravitational collapse and dissipation during the accretion of stellar structures (see Ikelle, 2022a). Mechanical energy is the energy due to the motion (kinetic) and position (potential) of an object. When objects are set in motion or are in a position where they can be set in motion, they have mechanical energy (see Ikelle, 2022a). Nuclear energy is generated by binding energy per nucleon from strong nuclear forces, introduced in the next section, and it is the basis of fusion and fission reactions, described in Ikelle (2022a).

Our aim is to have you become more familiar with the notions of various chemical elements, their atomic numbers and mass numbers, the notion of isotopes, and nuclear stable and unstable elements. Nuclear medicine, nuclear power plants, and nuclear weapons are constructed from nuclear unstable elements. Rock, archaeological, and meteorite dating are most based on the isotopes.

2.1.1. Atoms, Ions, Electrons, Protons and Neutrons

All ordinary matter is composed of extremely small particles called atoms. An atom is the smallest piece of a chemical element that maintains the identity of that element. Consider, for example, the chemical element known as lead (Pb). By repeatedly cutting a lead nugget in half, we will eventually reach a point where the pieces will no longer be lead. The pieces of lead just before this point are lead atoms. A billion lead atoms weigh about 3×10^{-13} grams, a mass that is far too light to be weighed on even the world's most sensitive balances. Atoms are seen only indirectly by using very complex technology such as tunneling microscopy (see Chen, 1993; Wiesendanger, 1994; and Meyer et al., 2004).

Each atom includes the elementary particles in nature: protons, neutrons, and electrons (see Figure 2.2). An atom also has a dense central nucleus surrounded by a cloud of electrons. This nucleus contains a mix of protons and neutrons. Protons and neutrons are collectively called *nucleons*. Neutrons, protons, and electrons are also called subatomic particles. These subatomic particles have a property called electric charge (the physical property that causes the subatomic particle to experience a force when placed in an electromagnetic field). The electric charge q of a particle can be either positive (e.g., $q > 0$ for a proton), zero (e.g., $q = 0$ for a neutron), or negative (e.g., $q < 0$ for electrons). Note that electrons and protons have an equal and opposite charge (see also Table 2.1). Experiments show that electric charges that occur in nature are quantized-only integer multiples of the elementary charge $e = 1.6021773 \times 10^{-9}$ C. We measure charge in coulombs (C), and 1 C has a very large number of elementary charges. Also, that electric charges cannot be reduced further than $\pm e$, and the electron is the smallest material particle with an indivisible negative charge. An object may have a total charge of 0, $\pm e$, $\pm 2e$, $\pm 3e$, etc. Since all stable matter includes protons, neutrons, and electrons, the net charge q of a material is $q = (Z - N_e)e$, where Z is the integer number of protons in the material and N_e is the integer number of electrons. Thus, q is quantized in multiples of e. If $Z = N_e$, the material is neutrally charged, $Z > N_e$ is positively charged, and $Z < N_e$ is negatively charged.

Table 2.1 also provides nuclear masses. These masses are sometimes measured in terms of the unified atomic mass unit, u $= 1.66 \times 10^{-27}$ kilograms [sometimes denoted as *amu* and sometimes called a Dalton (Da)]. Electrons are much lighter than protons and neutrons; protons and neutrons have a similar mass.

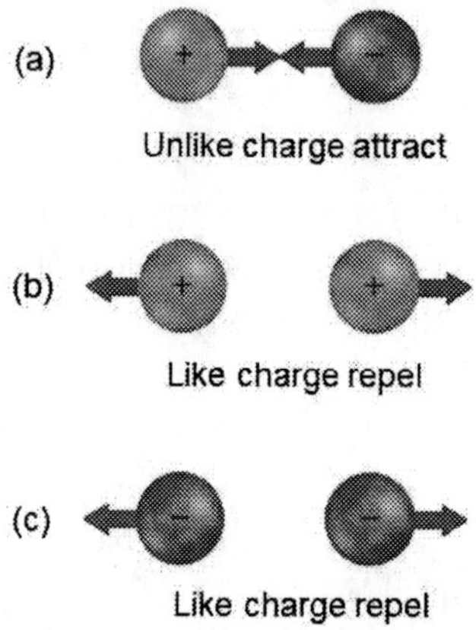

Figure 2.3. Illustrations of electrically charged objects exerting forces on each other—these forces can be attractive or repulsive. If two opposite charges are placed near each other, they form an electric dipole, characterized by an electric dipole moment. (a) If one object is positively charged and the other is negatively charged, they attract each other. (b) If two objects are positively charged, they will repel each other. (c) If two objects are negatively charged, they will repel each other.

Two objects that are electrically charged exert forces on each other (Figure 2.3). These forces can be attractive or repulsive, as shown here. If two objects are positively charged, they repel each other. If two objects are negatively charged, they also repel each other. If one object is positively charged and the other is negatively charged, they attract each other. In other words, like charges repel and unlike charges attract. So for a given atom, electrons are attracted to the positively charged nucleus. Hence, it takes work to keep electrons in orbit—that is, electrons have potential energy, which increases with increasing distance between the nucleus and electrons. Note, however, that electrons do not orbit the nucleus in the manner of a planet orbiting the sun, but instead remain fixed while others between them vibrate with maximum amplitude (i.e., electrons exist as standing waves. Thus the lowest possible energy an electron can take is similar to the fundamental frequency of a wave on a string.

Table 2.1. The electric charge and the mass of elementary particles—protons are much larger than electrons (one proton is approximately 1,836 times larger than the mass of an electron), but they have an equal and opposite charge. The mass of a neutron is approximately equal to that of a proton. The positron is the electron's antiparticle. It has the same mass as the electron, but the opposite electric charge. The neutrino and the antineutrino are particles associated with some nuclear decay, which we will describe later. They were postulated by Pauli (1934, 1961a, 1961b, and 1964) to explain an otherwise unaccountable loss of energy and momentum in some nuclear decay processes, which we will introduce in Chapters 5 and 6 (see also Enz, 1981)

Particle & matter	Charge	Mass (kg)	Mass (u)
Proton	$+e$	1.673×10^{-27}	1.00728
Neutron	0	1.675×10^{-27}	1.00866
Electron	$-e$	9.11×10^{-31}	0.00055
positron	$+e$	9.11×10^{-31}	0.00055
Neutrino	0	very small	
Positively charged matter	$(Z - N_e)e > 0$		
Negatively charged matter	$(Z - N_e)e < 0$		
Neutrally charged matter	$(Z - N_e)e = 0$		

An atom is classified according to the number of nucleons (protons and neutrons) in its nucleus. The number of protons in the nucleus is called the *atomic number* and given the symbol Z. The importance of atomic number comes from the fact that all atoms with the same atomic number have nearly identical chemical properties. Moreover, this number dictates the number of electrons orbiting the nucleus that are available for chemical activity. Each atomic number corresponds to a different chemical element and there are now approximately 111 known chemical elements that correspond to nuclei containing from one to 111 protons. In other words, there are 111 naturally occurring elements, each with a different number of protons and a different arrangement of electrons. Some chemical elements are associated with multiple atoms, as we can see in Tables 2.2 and 2.3.

Quiz: What is the number of protons in a 70-kg body under the assumption that protons and neutrons are nearly equally numerous in our bodies?

The number of neutrons in the nucleus is given the symbol N. The total number of nucleons (protons and neutrons) in the nucleus is called the *atomic mass number* (also known as atomic weight) and is given the symbol A with $A = Z + N$. The atomic masses range from one to more than 273. The physical properties of atoms, such as boiling and freezing points, depend on mass, A, and not on the atomic number, Z, whereas chemical properties of atoms depend on the atomic number, as mentioned above. The atomic symbol (chemical and nuclear shorthand notation) is

$$^{A}_{Z}\mathrm{X}\,,$$

where X is the chemical element symbol, which is a one- or two-letter abbreviation of the name of the element. Table A.1, in Appendix A, lists common elements and their symbols. By convention, the first letter of an element's symbol is always capitalized, while the second letter (if present) is lowercase (e.g., the symbol for hydrogen is H, the symbol for sodium is Na, and the symbol for lead is Pb). Alternatively, we sometimes write out the chemical element name with the mass number to indicate an atom. For example, we will sometimes write hydrogen-1 for $^{1}_{1}\mathrm{H}$, carbon-12 for $^{12}_{6}\mathrm{C}$, and uranium-235 for $^{235}_{92}\mathrm{U}$.

Again, all atoms of the same neutral chemical element have the same number of protons and electrons (that is, the number of protons is equal to the number of electrons, $Z = N_e$, where N_e is the integer number of electrons; the number of positive charges equals the number of negative charges), but the number of neutrons can differ (e.g., $^{1}_{1}\mathrm{H}$, $^{2}_{1}\mathrm{H}$, and $^{3}_{1}\mathrm{H}$). Atoms of the same chemical element but with different neutrons are called isotopes; it is the variation in the number of neutrons in the nuclei of atoms that gives rise to isotopes (see also Table 2.2). Consider, for example, three isotopes of hydrogen with zero neutrons, one neutron, and two neutrons. The atomic symbols of these isotopes are $^{1}_{1}\mathrm{H}$ (protium with zero neutrons), $^{2}_{1}\mathrm{H}$ (deuterium with one neutron), and $^3_1$H (tritium with two neutrons). By the way $^1_1$H is the only nucleus of an atom with zero neutrons, as all other nuclei contain one or more neutrons. The isotope symbol is ^AX, so hydrogen isotopes can be written as ^{1}H (protium), ^{2}H (deuterium), and ^{3}H (tritium). Every chemical element has one or more isotopes.

Table 2.2. Isotopes, isobars, isotones and isomers

Atoms	Characteristics
Isotopes	atoms with the same Z but different A (chemically identical); e.g., ^{12}C and ^{14}C
Isobars	atoms with the same A but different Z (different chemical elements); e.g., ^{11}C and ^{11}B
Isotones	atoms with the same number of neutrons but different A
Isomers	atoms with the same Z and A but with different energy levels

Two other important examples of isotopes are carbon and uranium. Carbon atoms exist naturally with 6, 7, and 8 neutrons. As each atom of carbon has 6 protons, its isotopes have atomic mass numbers of 12, 13, and 14 (as atomic mass number = protons + neutrons). These isotopes are called carbon-12, carbon-13, and carbon-14 which can symbolically be written as ^{12}C, ^{13}C, and ^{14}C, respectively. More than 99 percent of the carbon found in nature exists as carbon-12, ^{12}C. Most of the rest of the naturally occurring carbon is carbon-13, ^{13}C. The rarest carbon isotope is carbon-14, ^{14}C, and is produced in the Earth's atmosphere by cosmic-ray bombardment of nitrogen-14. Unlike the other two isotopes, carbon-14 is unstable: its nucleus tends to break up into elements with lower atomic numbers. As we will discuss in the next subsection, this nuclear breakup, which emits a significant amount of energy, is called radioactive decay, and isotopes that decay in this fashion are known as radioactive isotopes. Uranium also has three naturally occurring isotopes: uranium-234, uranium-235, and uranium-238. These isotopes can symbolically be written as ^{234}U, ^{235}U, and ^{238}U, and as each atom of uranium has 92 protons, they have 142, 143, and 146 neutrons, respectively. All the three isotopes are unstable and isotope $^{235}_{92}$U is widely used for nuclear power generation and nuclear weapon fabrication.

The term *nuclide* (also known as *nuclear species*) is also used to describe particular isotopes, ${}^{A}_{Z}\mathrm{X}$, notably in cases where the nuclear properties rather than the chemical properties of an atom are to be emphasized while the isotope concept (grouping of all atoms of a chemical element) emphasizes chemical over nuclear.

In our universe, hydrogen (${}^{1}_{1}\mathrm{H}$, 77 percent) and helium (${}^{4}_{2}\mathrm{He}$, 21 percent) are overwhelmingly abundant and the other elements combined amount to only 2 percent of the universe's mass. Elements heavier than iron are rare because of the energy required to make them. Similarly, our sun is composed 75 percent of hydrogen and 23 percent helium. The crust of the Earth (i.e., the rock below our feet) contains 88 naturally occurring elements but only eight chemical elements [46.6 percent from oxygen (O), 27.7 percent from silicon (Si), 8.1 percent from aluminum (Al), 5 percent from iron (Fe), 3.6 percent from calcium (Ca), 2.8 percent from sodium (Na), 2.6 percent from potassium (K), 2.1 percent from magnesium (Mg), and 1.5 percent from all others] make up over 98.5 percent of the crust mass. Among the other 1.5 percent that make up the Earth's crust mass, the most important chemical element in terms of mass is the metal titanium (Ti). Because titanium is light, highly resistant to corrosion, and can withstand extreme temperatures, it is used in the aerospace industry, medical implants, and portable computer housings. Some of the Earth's crust elements are essential for human health, especially calcium and sodium. These elements are naturally extracted from rocks as minerals break down at the Earth's surface to form soil. They are then absorbed by crops growing in the soil and find their way into our foods. Also calcium is added to foods such as milk and orange juice to prevent bone disease. Not all elements promote good health. Mercury (Hg) poisoning affects the human nervous system. In fact the most abundant element in the Earth's oceans is hydrogen just like in our universe. The oceans are basically composed of water with dissolved salts. In Earth's mantle, the most abundant element is silicon (Si); and in Earth's core, it is iron (Fe).

The atomic composition of human bodies is: oxygen (O) 65 percent, carbon (C) 18.5 percent, hydrogen (H) 9.5 percent, nitrogen (N) 3.6 percent, phosphorous (P) 0.2 percent, and sulfur (S) 0.1 percent. The body also needs trace amounts of other elements such as calcium (Ca), chloride (Cl), potassium (K), sodium (Na), magnesium (Mg), manganese (Mn), and selenide (S) for proper functioning of muscles, nerves, etc. Inspection of Table 2.3 shows that these elements have atomic numbers less than 21 and, thus, have low atomic masses and the distribution of elements in living systems is by no means accidental. The most common elements inside organisms are not the elements that are most abundant in the Earth's crust. For example, silicon, aluminum, and iron constitute 40.8 percent of the Earth's crust, but they exist in trace amounts in the human body.

2.1.2. Ions, Anions, and Cations

So far, we have described atoms that are electrically neutral—that is, a given atom has one electron for each proton. However, this is not always the case. Sometimes other forces overcome the attractive forces that maintain electrons in atom's orbits and the atom loses one or more electrons. In other cases, atoms may gain additional electrons. Atoms in which the number of electrons does not equal the number of protons are known as ions, and they carry a net electrical charge. They are denoted as follows

$${}^{A}_{Z}\mathrm{X}^{n\pm} ,$$

where $n\pm$ is the signed charge. Atoms with overall positive charges are called cations, while atoms with overall negative charges are called anions. So ions are formed only when electrons move from one atom to another. Typical examples of ions are sodium (Na^+), which has one positive charge, chloride (Cl^-), which has one negative charge, potassium (K^+), which has one positive charge, oxygen (O^{2-}), which has one positive charge, helium (He^{2+}), which has two positive charges, and hydrogen (H^+), which has one positive charge; H^+ is the proton. The dominant ions in stars, including our sun, are H^+, He^{2+}, and O^{2-}.

Table 2.4 provides predictions about whether an atom will form an anion or a cation. This table does not include all elements. You can expand it by using the tables in Appendix A. Atoms of many main-group me-

Table 2.3. Here are some of the stable chemical elements, including their binding energy per nucleon, $\mathcal{U} = \mathcal{U}_B/A$ (MeV), that you will encounter in this book [sources: Audi and Wapstra (1995) and Rosman and Taylor (1998)]. Other stable chemical elements include Chromium-52, Chromium-53, Chromium-54, Copper-63, Copper-65, nickel-60, Nickel-61, nickel-62, nickel-64, Zinc-64, Zinc-66, Zinc-68, Titanium-50, and Vanadium-51. Note that no stable chemical elements exist with a mass number of $A = 5$ or 8

Element		Z	A	$\mathcal{U}$	Element		Z	A	$\mathcal{U}$
Hydrogen	H	1	1	0	calcium	Ca	20	40	8.55
			2	1.11				42	8.61
Helium	He	2	3	2.57				43	8.60
			4	7.07				44	8.66
Lithium	Li	3	6	5.33				44	8.67
			7	5.61	Manganese	Mn	25	55	8.76
Beryllium	Be	4	9	6.46	Iron	Fe	26	54	8.74
Boron	B	5	10	6.48				56	8.79
			11	6.93				57	8.77
Carbon	C	6	12	7.68				58	8.79
			13	7.46	Cobalt	Co	27	59	8.77
Nitrogen	N	7	14	7.48	Cadmium	Cd	48	106	8.54
			15	7.70				108	8.55
Oxygen	O	8	16	7.98				110	8.55
			17	7.75				111	8.54
			18	7.77				112	8.54
Sodium	Na	11	23	8.11				114	8.53
Silicon	Si	14	28	8.45	Cesium	Cs	55	133	8.41
			29	8.45	Lead	Pb	82	206	7.88
			30	8.52				207	7.87
Chlorine	Cl	17	35	8.52				208	7.87
			37	8.57	Bismuth	Bi	83	209	7.85

tals lose enough electrons to leave them with the same number of electrons as an atom of the preceding noble gas. For example, a neutral calcium atom, with 20 protons and 20 electrons, readily loses two electrons to produce a cation with 20 protons, 18 electrons, and a 2+ charge, the same number of electrons as atoms of the preceding noble gas, argon. It is then symbolized as Ca^{2+}. Also, the neutral bromine atom, with 35 protons and 35 electrons, can gain one electron to provide it with 36 electrons to produce an anion with 35 protons, 36 electrons, and a 1- charge (Br^-) which has the same number of electrons as atoms of the next noble gas, krypton. However, some metals often exhibit variable charges that are not predictable by their location in the table. For example, iron can form ions with a 2+ or 3+ charge.

The cloud of electrons, in Figure 2.2a, is actually organized into a set of discrete amounts of energy, as illustrated in Figure 2.2c. Basically, every atom exhibits a ladder of potential energy values, rather than a continuous spectrum. Because the amount of energy an electron possesses is related to its distance from the nucleus, electrons that are the same distance from the nucleus have the same energy. Such electrons are said to occupy the same energy level. In a schematic diagram of an atom (see Figure 2.2c), the electron energy levels are drawn as concentric rings, with the energy level increasing with distance from the nucleus. These rings are sometimes called shells or simply energy levels. Different shells hold different numbers of electrons. The two innermost shells hold two and eight electrons, respectively. The Outer shells can hold more electrons and are considered stable when they contain eight electrons. The electrons in the outer shell are called valence electrons. Atoms may lose or gain electrons to reach a configuration with a stable outer shell. Again, this process results in the formation of negatively or positively charged atoms, known as ions.

For most of the atoms important to life, an outer energy level can contain no more than eight electrons. Elements possessing all eight electrons at their outer energy level are inert or nonreactive. In contrast, elements with seven electrons (one fewer than the maximum number of eight) in their outer energy level, such as chlorine (Cl) are highly reactive. They tend to gain the extra electron needed to fill their outer energy level. Elements with only one electron in their outer energy level such as sodium (Na) and potassium (K), are also very reactive; they tend to lose the single electron in their outer level.

Quiz: Write the nuclide notation, including charge, if applicable, for atoms with the following characteristics: (a) 20 protons, 20 neutrons, and 18 electrons, (b) 19 protons, 20 neutrons, and 18 electrons, (c) 26 protons, 28 neutrons, and 23 electrons; and (d) 90 protons, 142 neutrons, and 90 electrons.

Table 2.4. Symbols and charges for some monoatomic ions—writing in so-called half equations. Atoms may lose or gain electrons to reach a configuration with a stable outer shell. This process results in the formation of ions—that is, negatively or positively charged atoms. The ion names are in brackets. We also included the potential activities of these ions. Elements like magnesium have a greater tendency to lose electrons while elements like zinc have a tendency to gain electrons. Some of them form as X^{n-}, stem + -ide (e.g., S^{2-}; sulfide).

Ion		**Name**	**Ion**		**Name**
H^+;	$H \rightarrow H^+ + e^-$	(proton)	H^-;	$H + e^- \rightarrow H^-$	(hydride)
Li^+;	$Li \rightarrow Li^+ + e^-$	lithium	F^-;	$F + e^- \rightarrow F^-$	(fluoride)
Na^+;	$Na \rightarrow Na^+ + e^-$	sodium	Cl^-;	$Cl + e^- \rightarrow Cl^-$	(chloride)
K^+;	$K \rightarrow K^+ + e^-$	potassium	Br^-;	$Br + e^- \rightarrow Br$	(bromide)
Rb^+;	$Rb \rightarrow Rb^+ + e^-$	rubidium	I^-;	$I + e^- \rightarrow I^-$	(iodide)
Cs^+;	$Cs \rightarrow Cs^+ + e^-$	cesium	O^{2-};	$O + 2e^- \rightarrow O^{2-}$	(oxide)
Be^{2+};	$Be \rightarrow Be^{2+} + 2e^-$	beryllium	S^{2-};	$S + 2e^- \rightarrow S^{2-}$	(sulfide)
Mg^{2+};	$Mg \rightarrow Mg^{2+} + 2e^-$	magnesium	Se^{2-};	$Se + 2e^- \rightarrow Se^{2-}$	(selenide)
Ca^{2+};	$Ca \rightarrow Ca^{2+} + 2e^-$	calcium	N^{3-};	$N + 3e^- \rightarrow N^{3-}$	(nitride)
Sr^{2+};	$Sr \rightarrow Sr^{2+} + 2e^-$	strontium	P^{3-};	$P + 3e^- \rightarrow P^{3-}$	(phosphide)
Ba^{2+};	$Ba \rightarrow Ba^{2+} + 2e^-$	barium	Al^{3+};	$Al \rightarrow Al^{3+} + 3e^-$	aluminum
Ag^+;	$Ag \rightarrow Ag^+ + e^-$	silver	Ga^{3+};	$Ga \rightarrow Ga^{3+} + 3e^-$	gallium
Zn^{2+};	$Zn \rightarrow Zn^{2+} + 2e^-$	zinc	In^{3+};	$In \rightarrow In^{3+} + 3e^-$	indium
Cd^{2+};	$Cd \rightarrow Cd^{2+} + 2e^-$	cadmium	Au^{3+};	$Au \rightarrow Au^{3+} + 3e^-$	(auric)
Fe^{2+};	$Fe \rightarrow Fe^{2+} + 2e^-$	(ferrous)	Fe^{3+};	$Fe \rightarrow Fe^{3+} + 3e^-$	(ferric)
Mn^{2+};	$Mn \rightarrow Mn^{2+} + 2e^-$	(manganous)	Mn^{3+};	$Mn \rightarrow Mn^{3+} + 3e$	(manganic)
C^{4-};	$C + 4e^- \rightarrow C^{4-}$	(carbide)	Ni^{2+};	$Ni \rightarrow Ni^{2+} + 2e^-$	(nickelous)
Bi^{3+};	$Bi \rightarrow Bi^{3+} + 3e^-$	bismuth(III)	Bi^{5+};	$Bi \rightarrow Bi^{5+} + 5e^-$	bismuth(V)
Si^{4-};	$Si + 4e^- \rightarrow Si^{4-}$	(silicide)	Se^{2-};	$Se + 2e^- \rightarrow Se^{2-}$	(selenide)

2.1.3. Stable and Unstable Isotopes and Nuclear Half-Life

Henri Becquerel (1852-1908), Marie Curie (1867-1934) and Pierre Curie (1858-1906), and Ernest Rutherford (1871-1937) were the first scientists to discover that some chemical elements spontaneously emit different kinds of particles and/or energy waves. They described these elements as radioactive and their spontaneous emission property is now called *radioactive decay.* As scientists studied radioactivity more closely, they discovered that radioactive atoms are naturally unstable and unstable atoms emit particles and/or energy waves in order to become stable. They can even transform into other chemical elements (e.g., uranium can decay into thorium). Moreover, the major sources and causes of huge amounts of heat deep in the ground (8×10^{30} Joules) are mostly due to radioactive decay of potassium, thorium, and uranium. As we will see in the next chapters, applications of the radioactive decay are widespread in our daily life. They include nuclear power plants, nuclear bombs, smoke detectors, medicine, the sterilization of food, rock dating, and the analysis of ancient artifacts. Radioactivity is actually the basis for diagnostic and therapeutic methods in medicine, such as positron emission tomography (PET) described in Ikelle (2022a). PET scans can show muscle activity, brain scans, lymph node metastasis, etc. On the other hand, emitted energetic subatomic particles by radioactive substances have the potential to damage living cells, and, at high doses, cell death. Consequently, exposure to radiation is carefully controlled and regulated nowadays.

What is a stable nucleus and therefore a stable isotope? A nucleus is stable if it does not spontaneously emit any kind of particles or energy waves. On the other hand, if the nucleus is unstable, it has the tendency of emitting some kind of particles or energy waves which can include electrons, neutrons, protons, positrons, and helium-4 in order to become stable or less unstable. These spontaneous emissions are collectivity called radioactivity. Therefore, radioactivity is associated with unstable nuclei (also known as radioactive nuclei, radionuclides, or radioactive nuclides). No stable chemical elements exist with a mass number of $A = 5$ or 8.

What makes the nucleus a stable one or an unstable one? The answer to this question comes down to explaining how and why protons and neutrons are held together in the nucleus. Remember that an atom is electrically neutral because it contains an equal number of positively charged protons and negatively charged electrons and their charges balance. The nucleus of an atom, however, contains only positively charged protons, which are closely packed together in a very small volume (remember neutrons have no charge), as illustrated in Figure 2.2b. From the laws of electromagnetic physics, described in Ikelle, 2022a (see also Figure 2.3), one would expect that the protons, being of the same charge and so close together, would exert strong repulsive forces on each other and pull apart in the absence of a compensating attractive force. The combined gravitational force from the protons and neutrons in a nucleus is insignificant as an attractive force to compensate for repulsive electric forces because gravitational forces depend on the masses of protons and neutrons which are tiny (see Table 2.1).

Table 2.5. Here are some of the unstable chemical elements, including their binding energy, $\mathcal{U} = \mathcal{U}_B/A$ (MeV), you will encounter in this book. Note that an isolated proton, a hydrogen nucleus with or without an electron, does not decay. Decay modes are described in Ikelle (2022a) [sources: Audi and Wapstra (1995) and Rosman and Taylor (1998)]

Element	Symbol	Z	A	$\mathcal{U}$	Half-life time	1st Decay
Carbon	C	6	14	7.52	5730 years	beta
Oxygen	O	8	20	7.57	13.51 seconds	beta
Aluminum	Al	13	25	8.02	7.183 seconds	beta
			26	8.15	730000 years	beta
			28	8.31	2.2414 minutes	beta
Potassium	K	19	40	8.54	1.25×10^9 years	beta
			42	8.55	12.36 hours	beta
			43	8.58	22.3 hours	beta
Calcium	Ca	20	45	8.63	162.61 days	beta
			48	8.67	6×10^{18} years	beta
			49	8.59	8.718 minutes	beta
Iron	Fe	26	60	8.76	1.5×10^6 years	beta
			61	8.70	5.98 minutes	beta
Lead	Pb	82	200	7.88	21.5 hours	beta
			204	7.88	1.4×10^{17} years	alpha
			205	7.87	15.3×10^6 years	beta
			209	7.85	3.253 hours	beta
Thorium	Th	90	230	7.63	75380 years	alpha
			231	7.62	25.52 hours	beta
			232	7.62	14.05×10^9 years	alpha
			233	7.60	22.3 minutes	beta
Uranium	U	92	234	7.60	245500 years	alpha
			235	7.59	0.703×10^9 years	alpha
			237	7.58	6.75 days	beta
			238	7.57	4.468×10^9 years	alpha
			239	7.56	23.45 minutes	beta

This observation implies that there must be an additional attractive force similar in size to the electric repulsive force that overcomes the repulsion of the protons and that therefore holds the nucleus together. This additional force is a strong nuclear force. The actions of the strong nuclear force are attractive over short distances, but this is not all. If the force was only attractive, then it would pull the protons and neutrons together into an ever-decreasing small volume of space. But we know that this is not the case because each nucleus contains a distinct number of protons and neutrons, which retain their identity. The more protons and neutrons, the bigger the nucleus, and thus, the separation between the protons and neutrons in a nucleus must be a fixed distance. This means the strong nuclear force must be repulsive at a very small range as well as being attractive at a slightly larger distance. This is the force that holds the nucleus together. The energy associated with this force is called binding energy.

To recapitulate the nuclei of atoms are affected by three forces:

- The strong nuclear force, which acts between protons, between neutrons, and between protons and neutrons and causes protons and neutrons to bind together. It is a short range (less than 10^{-15} meter) attractive force and repulsive at even smaller distances. It has no effect on electrons. Over larger distances and outside the nucleus, the strong nuclear force is essentially nonexistent.

- The electric force, which is manifested by the repulsion of protons and tends to rip the nucleus apart. Only protons feel the repulsive electric force. This force is responsible for atom bombs and nuclear power.

- The weak nuclear force, which causes neutrons to change into protons, and vice versa.

How tightly a nucleus is bound together, and therefore how stable or unstable a nucleus is, is mostly the interplay between the nuclear strong force and the repulsive electric force. The binding energy is the net energy that is the result of the balance between the strong force and the repulsive force. By using Einstein's famous mass-energy formula, $\mathcal{U} = mc_0^2$, where $\mathcal{U}$ is energy, m is the mass of the matter being converted, and c_0 is the speed of light in a vacuum, we can mathematically define the binding energy as the difference between the energy of nucleons before and after they get bound to each other (i.e., before the nucleus is formed and after the nucleus is formed), i.e.,

$$\mathcal{U}_B(A, Z) = [Zm_\mathrm{p} + (A - Z)m_\mathrm{n} - m_\mathrm{nucleus}]\, c_0^2 \, , \tag{2.1}$$

where $\mathcal{U}_B(A, Z)$ is the nuclear binding energy, m_p is the mass of a proton, m_n is the mass of a neutron, m_nucleus is the mass of the nucleus (i.e., the mass of the nucleons after they get bound to each other after the nucleus is formed), and c_0 is again the speed of light in a vacuum. So the binding energy is zero when the energy of a nucleus is equal to the energy of its individual nucleons or equivalently when the mass of the nucleus is equal to the sum of the masses of its individual nucleons. However, the binding energy is nonzero, because when a system of two or more particles gets bound to each other, the energy (and hence, the mass) of that system decreases. Moreover, experiments have shown that the sum of the masses of protons and neutrons is always greater than the nuclear mass, m_nucleus [i.e., $(Zm_\mathrm{p} + Nm_\mathrm{n}) > m_\mathrm{nucleus}$]. The sum of the masses of protons and neutrons can be determined directly from the values in Tables 2.1 and 2.3, and the nuclear mass is determined experimentally. The binding energy is a quantitative measure of the nuclear stability. In some atoms, the binding energy is great enough to hold the nucleus together. The nucleus of this kind of atom is said to be stable. Radioactive isotopes (also known as unstable isotopes, radionuclide, or radioactive nuclide) are atoms that have excess nuclear energy, which makes them unstable.

To express nuclear binding energy in terms of experimentally accessible quantities, we first write the nuclear mass in terms of the atomic mass as

$$m_\mathrm{nucleus} c_0^2 = (m_\mathrm{atom} - Zm_\mathrm{e})\, c_0^2 + \mathcal{U}_b \tag{2.2}$$

where m_atom is the measured atomic mass of the nucleus, m_e is the mass of an electron, and $\mathcal{U}_b$ is the electronic binding energy. We

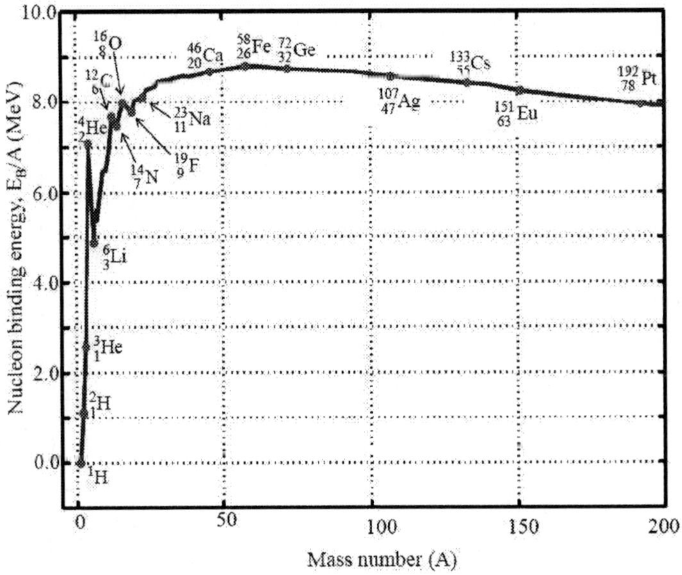

(a)

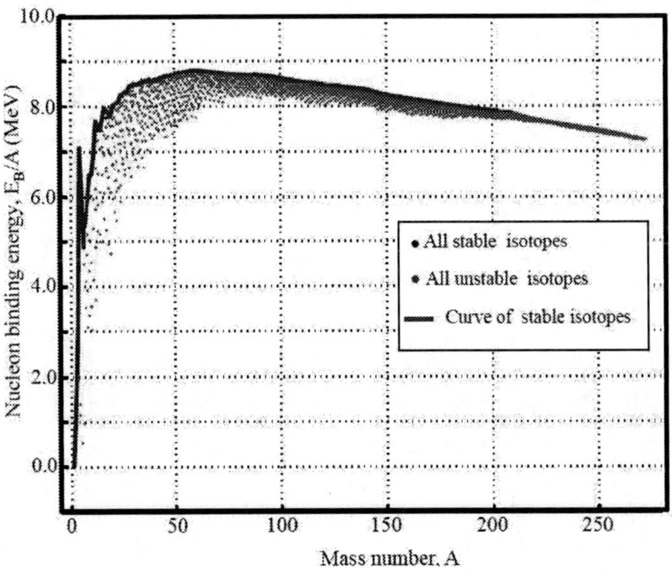

(b)

Figure 2.4. (a) Binding energy per nucleon ($\mathcal{U}_B/A$ in MeV versus A) of stable nuclides along with specific data points. (b) Binding energy per nucleon ($\mathcal{U}_B/A$ in MeV versus A) of stable nuclides (black) and unstable nuclides (red)—every element beyond an atomic number of 83 is unstable.

further neglect the electronic binding energy, $\mathcal{U}_b$, to finally obtain the expression for the nuclear binding energy:

$$\mathcal{U}_B(A, Z) = [Zm_\mathrm{p} + (A - Z)m_\mathrm{n} - (m_\mathrm{atom} - Zm_\mathrm{e})]\, c_0^2 \,. \qquad (2.3)$$

The binding energy is usually expressed and plotted as $\mathcal{U}_B/A$ (i.e., binding energy per nucleon), as shown in Tables 2.3 and 2.5 and Figure 2.4. We can see that the binding energy initially rises rapidly as more nucleons are added, and then the average binding energy becomes almost constant at about 8 MeV for mass 12 to mass 238. An iron (Fe) isotope (with $\mathcal{U}_B/A \approx 8.8$ MeV) has nuclear particles most tightly bound.

The dependence of $\mathcal{U}_B/A$ on A (and Z) is captured by the semi-empirical mass formula of Von Weizsacker (1935). This formula is based on a nucleus model in which all nucleons are uniformly distributed inside a nucleus and are bound together by the nuclear force while the Coulomb interaction causes repulsion among protons (the electric repulsive force). This also accounts for experimental data. The semi-empirical mass formula here is (Von Weizsacker, 1935)

$$\begin{aligned}\mathcal{U}_B(A, Z) &= a_v A - a_s A^{2/3} - a_c Z(Z-1)A^{-1/3} - a_\mathrm{sym}\frac{(A \quad 2Z)^2}{A} \\ &+ \delta a_p A^{-3/4} \end{aligned} \qquad (2.4)$$

where $a_v = 15.5$ MeV, $a_s = 16.8$ MeV, $a_c = 0.72$ MeV, $a_\mathrm{sym} = 23$ MeV,$a_p = 34$ MeV, and

$$\delta = \begin{cases} 1 & \text{for even } Z \text{ and even } N \\ 0 & \text{for odd } Z \text{ and even } N \text{ or vice versa (i.e., } A \text{ is odd)} \\ -1 & \text{for even } Z \text{ and even } N \end{cases} \qquad (2.5)$$

By taking the first derivative with respect to Z, we can calculate the optimal Z such that the mass is minimal (the smaller the mass the greatest the binding). We then obtain:

$$Z_\mathrm{opt} = \frac{A}{2}\left[\frac{1 + \frac{1}{4}A^{-1/3}\frac{a_c}{a_\mathrm{sym}}}{1 + \frac{1}{4}A^{2/3}\frac{a_c}{a_\mathrm{sym}}}\right] \approx \frac{A}{2}\left(1 - \frac{1}{4}A^{2/3}\frac{a_c}{a_\mathrm{sym}}\right) \,. \qquad (2.6)$$

The approximation and series expansion in (2.6) are made because $a_c \ll a_{\text{sym}}$. This formula allows us to predict the stable isotropic. It gives $Z_{\text{opt}} \approx A/2$ (i.e., $N/Z \approx 1$) at small A, but has a correction for larger A such that $Z_{\text{opt}} \approx 0.41A$ (i.e., $N/Z \approx 1.44$) for heavy nuclei.

So the predominant factor in nuclear stability is the neutron-to-proton ratio (N/Z) or equivalently the proton-to-mass ratio (Z/A). Figure 2.5 shows a plot of A versus Z. The stability lines, shown in red, contain stable isotopes. The N/Z ratio is close to 1 for atoms of elements with low atomic numbers (less than about 20 protons); that is, in light nuclei, the number of protons is almost equal to the number of neutrons, hence Z over A equals one half. The N/Z ratio steadily increases as the atomic number increases past element 20 (calcium) to about element 84 (polonium) at 1.5 neutrons for each proton. Every element with an atomic number beyond 83 is unstable; thorium and uranium are unstable (see Table 2.5).

Several other observations can be made regarding the relationship between the stability of a nucleus and its structure. Nuclides containing odd numbers of both protons and neutrons are rarely stable (see Figures 2.4 and 2.5). There are 250 stable nuclei. There are only four stable nuclei that have odd Z and odd N, namely ${}^2_1\text{H}$, ${}^6_3\text{Li}$, ${}^{10}_5\text{B}$, and ${}^{14}_7\text{N}$. Nuclides containing even numbers of both protons and neutrons (even Z and even N) are most stable, meaning less radioactive, as more than half (about 144) of the stable nuclei have even numbers of protons and neutrons, and are referred to as even-even nuclei. Approximately 20 percent (approximately 57) of the stable nuclei have even numbers of protons and odd numbers of neutrons (even Z and odd N), and approximately 50 the same number for stable nuclides containing odd numbers of protons and even numbers of neutrons (odd Z and even N).

Quiz: (1) Determine Z/A of Hydrogen, ${}^4_2\text{He}$, ${}^{60}_{27}\text{Co}$, and ${}^{235}_{27}\text{U}$. Explain the variation of this ratio.
(2) Calculate the average binding energy per nucleon of ${}^{93}_{41}\text{Nb}$. Predict the nuclear stability of ${}^{36}_{17}\text{Cl}$. (3) What is the binding energy of the oxygen isotope oxygen-18?

Nuclear half-life. Let $N(t)$ be the number of nuclei of a given unstable species at time t. The number of decaying atoms per unit

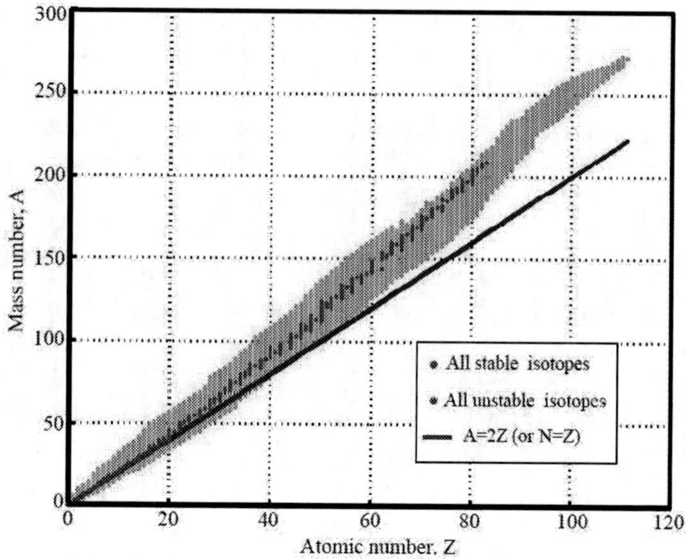

(a)

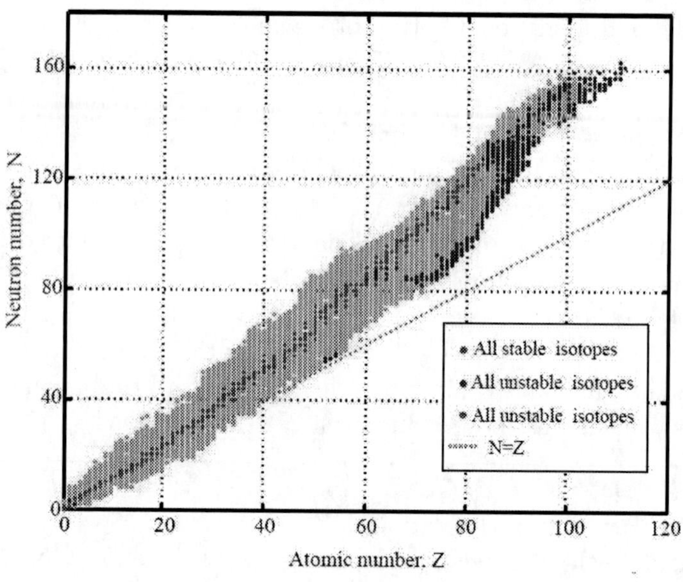

(b)

Figure 2.5. (a) The plot of A versus Z. The stable nuclides are on a jagged red line and the unstable nuclides are in green. The solid blue line indicates stable isotopes with a ratio $N/Z \approx 1$. (b) The plot of unstable isotopes based on their first decay. Blue dots represent unstable isotopes, which are first decayed as alpha. Green dots represent unstable isotopes, which are first decay as beta.

time is $dN(t)/dt$. It is found that this rate is constant in time and it is proportional to the number of nuclei themselves as follows:

$$\frac{dN(t)}{dt} = -\lambda N(t) \Longrightarrow N(t) = N_0 \exp(-\lambda t) \; , \tag{2.7}$$

where $N_0 = N(t = 0)$ is the number of nuclei at $t = 0$, and λ is the parameter characterizing the exponential. The implication here is that for all nuclear decay, the number of unstable nuclei of a given species diminishes exponentially with time, as all decay processes are subject to the basic law shown in (2.7). The negative sign signifies that $N(t)$ is decreasing with time and the decay rate is controlled by constant λ? Large values of λ mean rapid decay and small values of λ mean slow decay. Note that the decay rate is dependent only on the energy state of the nuclide, as it is independent of the history of the nucleus and essentially independent of external influences such as temperature, pressure, etc.

Radioactive decay is most commonly discussed in terms of the nuclear half-life, τ, which is defined as

$$N(\tau) = N_0/2 \Longleftrightarrow \exp(-\lambda\tau) = \frac{1}{2} \Longleftrightarrow \tau = \frac{\ln 2}{\lambda} \approx \frac{0.6931}{\lambda} \; . \tag{2.8}$$

This equation tells us how to convert λ to the half-life value. The half-life is the time necessary for the number of unstable nuclei of one species to diminish by one-half (see Figure 2.6). So after one half-life time, the number of X atoms left is $N_0/2$; after two half-lives, the number of X atoms left is $N_0/4$; and after three half-lives, the number of X atoms left is $N_0/8$, and soon. In more general terms, for n half-lives, there are only $N_0/(2^n)$ X atoms left. Half-lives can range from microseconds to billions of years (see Tables 2.5 and 2.6). For example, uranium-238 has a half-life of 4.47 billion years (an incredibly long time that is difficult to comprehend!), carbon-14 has a half-life of 5730 years, and radon gas has a half-life of 3.8 days. A sample of carbon containing one gram of carbon-14 today would contain 0.5 gram of carbon-14 after 5730 years, 0.25 gram 11,460 years from now, 0.125 gram 17,190 years from now, and so on. By determining the ratios of the different isotopes of carbon and other elements in biological samples and in rocks, scientists are able to accurately determine when these materials were formed. These decays take place at a steady rate, a rate (i.e., half-life) that has not changed over geological time, and is the same throughout the solar system. The Earth is not an isotope; it is included in Table 2.6, along with its age, as a reference point for the half-life numbers. The elemental composition of the Earth has remained essentially unchanged over its 4.5 billion year history.

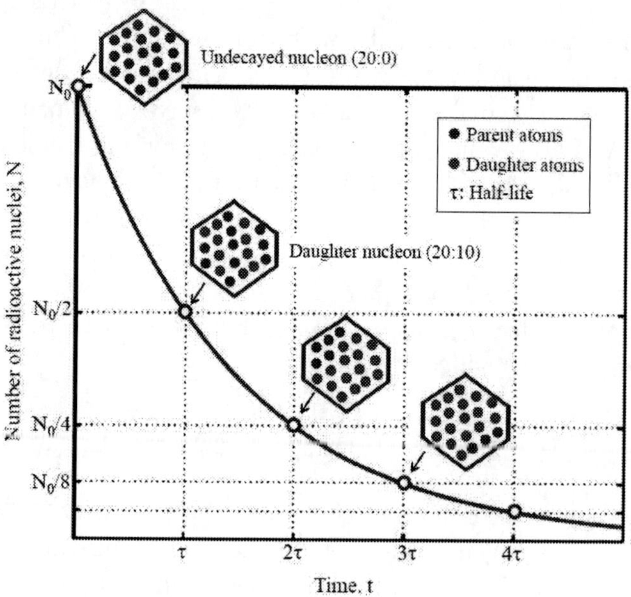

Figure 2.6. An illustration of the amount of radioactivity remaining after a given number of half-lives. For example, naturally occurring radon gas converts to lead with a half-life of $\tau = 3.8$ days. After two half-lives ($2\tau = 7.6$ days), 1/4 of the original radon and its radioactivity remain, and after 10 half-lives ($10\tau = 7.6$ days), only 0.1 percent of the original radioactivity remains and 99.9 percent of the radon has transformed into lead metal.

The activity, R, of a radionuclide is given by its disintegration rate:

$$R(t) = \frac{dN(t)}{dt} = -\lambda N(t) = \frac{\ln 2}{\tau} N \tag{2.9}$$

or

$$R(t) = R_0 \exp(-\lambda t) \ , \tag{2.10}$$

where

$$R_0 = \left(\frac{\ln(2)}{\tau}\right) N_0 \ . \tag{2.11}$$

The quantity R_0 is known as the decay rate or specific activity. It is often expressed as the number of disintegrations per second or more simply counts per second. The term *counts per second* is commonly used in the context of radiation detectors, as we discuss in Chapters 5 and 6. The other common units of the decay rate are Becquerel [Bq] and Curie [Ci] with 1 Bq = 1 disintegration/sec and 1 Ci = 3.7×10^{10} Bq = 3.7×10^{10} disintegration/sec. Each radioactive species has an intrinsic specific activity (ISA), which is the activity of a unit mass of pure material (the isotope). According to Adams and Weaver (1958), the relative activities of the elements potassium-40, uranium-238, and thorium-232 are 1, 1,300, and 3,600, respectively. One might think that potassium would be insignificant. However, potassium is much more abundant than the isotopes in the other two series, which makes its contribution to the overall radioactivity of a formation significant. Potassium is common in most clay and some evaporites, as described in the next section of this chapter.

During the last 60 or so years, techniques have been developed to determine the age of rocks on the basis of the decay rate of elements such as uranium, potassium, strontium, carbon, and several others, because the decay rate does not change over geological time and is the same throughout the solar system (see Ikelle, 2022a).

Quiz: Assume we have a box of 5×10^8 ^{24}Na atoms sitting on the table, how many will be left in the box after three hours and 30 minutes?

2.2. Compounds and Molecules

Matter is made up of one or more different types of chemical element. Under normal conditions (e.g., room temperature and pressure), no element exists as an independent atom in nature, except noble gases[1]

Table 2.6. Half-life of several isotopes—we can see here that the half-life varies considerably from isotope to isotope. ^{238}U, ^{235}U, ^{232}Th, and ^{40}K are long-lived radioactive elements, and ^{214}Po, ^{10}C, ^{24}Na, ^{222}Rn, ^{45}Ca, ^{60}Co, ^{26}Al are short-life radioactive elements. ^{26}Al with a half-life of 730,000 years, and are now gone after more than 6000 half-lives.

Isotope	Half-live time	location
Earth	4.5 billion years	
^{214}Po	160 μs	Crust
^{10}C	19 seconds	Laboratory
^{24}Na	15 minutes	Laboratory
^{18}F	110 minutes	Human brain
^{201}Tl	3 days	Human heart
^{222}Rn	3.8 days	Atmosphere
^{133}Xe	5 days	Lungs
^{131}I	8 days	Thyroid
^{125}I	59.4 days	Thyroid, prostate, and brain
^{45}Ca	163 days	Laboratory
^{60}Co	5.3 years	Laboratory and radiotherapy
^{14}C	5730 years	Atmosphere
^{235}U	700 million years	Crust
^{238}U	4.47 billion years	Crust
^{40}K	1.25 billion years	Crust/Mantle
^{232}Th	13.9 billion years	Crust/Mantle
^{26}Al	730,000 years	

[argon (^{18}Ar), neon (^{10}Ne), helium (^{2}He), xenon (^{54}Xe), krypton (^{36}Kr), radon (^{86}Rn), and oganesson (^{118}Og)], which have their outer shell occupied by eight electrons, with helium being the exception. This octet of electrons represents a particularly stable electronic arrangement. However groups of atoms are found to exist together as one species. Such groups of atoms are known as molecules and in some cases as compounds. The forces that hold these constituent

[1]The noble gases are colorless, odorless, tasteless, and nonflammable gases. Because noble gases are composed of only single atoms, they are monoatomic. At room temperature and pressure, they are unreactive gases. Xenon and Krypton in the atmosphere are used as signatures of nuclear and hydrogen explosions.

atoms together in molecules are called chemical bonds. Our objective in this section and in the next is to describe these basic combinations of atoms. We also explain why some atoms combine and others do not. Finally, we will discuss molecular diagrams and isomerism.

Our aim is to have you become more familiar with the various kinds of organic compounds and begin to see how scientists visualize molecules and correlate the diverse kinds of structures that you will have to deal with in your professional and day-to-day life.

2.2.1. Compounds, Molecules and Acids

A compound is a substance formed by two or more different types of chemical elements that are held together by energy in a stable association. Moreover, the elements are united chemically in a fixed proportion. Here are examples of compounds:

- Water $[H_2O] = 2\,\text{Hydrogen}\,(H) + \text{Oxygen}\,(O)$,
- Table salt $[NaCl] = \text{Sodium}\,(Na) + \text{Chlorine}\,(Cl)$,
- Carbon dioxide gas $[CO_2] = \text{Carbon}\,(C) + 2\,\text{Oxygen}\,(O)$,
- Calcium oxide $[CaO] = \text{Calcium}\,(Ca) + \text{Oxygen}\,(O)$,
- Baking soda $[NaHCO_3] = \text{Sodium}\,(Na) + \text{Hydrogen}\,(H) + \text{Carbon}\,(C) + 3\,\text{Oxygen}\,(O)$,
- Penicillin $[C_{16}H_{18}N_2O_4S] = 16\,\text{Carbon}\,(C) + 18\,\text{Hydrogen}\,(H) + 2\,\text{Nitrogen}\,(N) + 4\,\text{Oxygen}\,(O) + \text{Sulfur}\,(S)$.

Subscripts are used here to indicate multiple atoms or groups of atoms. These substances are compounds because each contains more than one kind of chemical element. The physical and chemical properties of a compound are generally different from the physical and chemical properties of its constituent elements. There are more than 50 million known compounds and more are being discovered daily.

The atoms in a compound are joined by chemical bonds; these bonds can occur when atoms with opposite charges attract (ionic bonds) or when atoms are bonded together by the sharing of electrons

Table 2.7. Average bond lengths (in Å$= 10^{-10}$ m) and bond energies for some common bonds. These bond lengths are also sometimes measured in picometers (1 pm $= 10^{-12}$); 100 pm = 1 Å. The dipole-moment unit is called a debye (D); 1D $= 3.33564 \times 10^{-30}$ C m. EN stands for electronegativity.

Bond	Length (Å)	Energy (kJ/mol)	Dipole moment (D)
C-N	1.47	305	0.22
C-C	1.54	347	0
C-O	1.43	358	0.86
C-H	1.09	413	0.4
C-Cl	1.78	339	1.56
C-Br	1.94	276	1.48
C-F	1.33	435	1.51
H-Cl	1.27	427	1.08
H-H	0.74	432	0
H-O	0.97	459	1.5
H-F	0.92	565	1.82
H-Br	1.41	363	0.82
O=O	1.21	498	0
C=C	1.34	614	0
C=O	1.23	745	0.7
C≡ C	1.21	839	0
N≡ N	1.10	945	0
C≡ O	1.13	1070	0.7

(covalent bonds). Consider, for example, sodium chloride (NaCl), which is the chemical name for common table salt. Sodium has 11 electrons: two in the inner energy level, eight in the next level, and one in the outer (valence) level—that is, Na^+. The chlorine atom has 17 electrons: two in the inner energy level, eight in the next level, and seven in the outer level. When placed together, metallic sodium and gaseous chlorine react swiftly and explosively, as the sodium atoms donate electrons to chlorine, forming an ionically bonded compound, NaCl, which is electrically neutral and a stable octet (see Figure 2.7a). On the other hand, water consists of an oxygen ion (O^-) bound to two hydrogen ions (H^+) by single covalent bonds (Figure 2.7b). Each of the two covalent bonds is made of two electrons: one electron from the hydrogen atom and one electron from the oxygen atom. The resulting compound is stable and attains the outer-shell noble gas configuration as a result of the sharing of electrons.

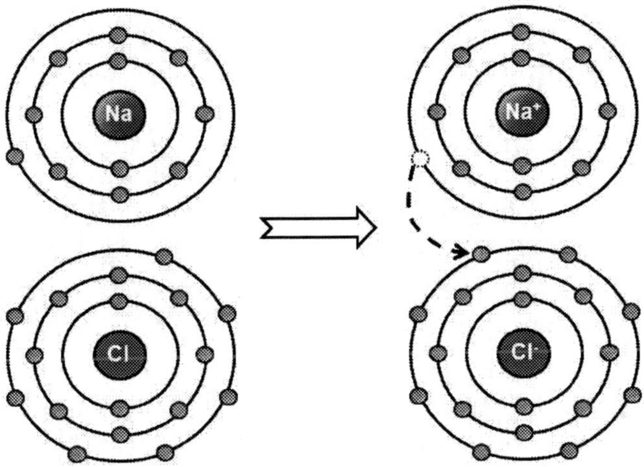

(a)

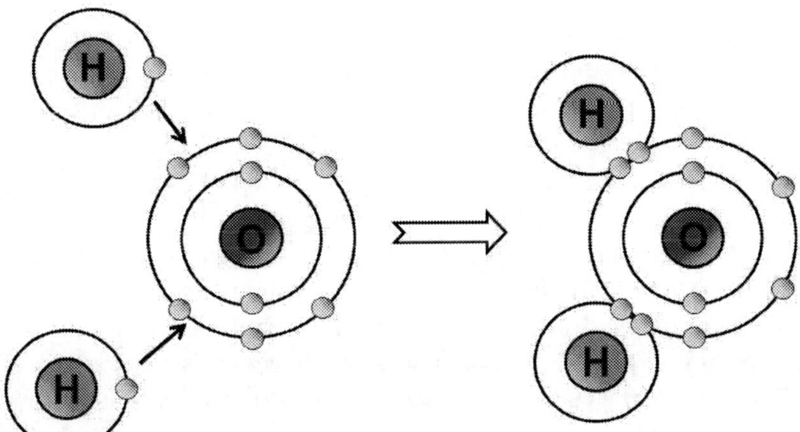

(b)

Figure 2.7. (a) The formation of ionic bonds by sodium chloride–when a sodium atom donates an electron to a chlorine atom, the sodium atom becomes a positively charged sodium ion, and the chlorine atom becomes a negatively charged chloride ion. Sodium chloride forms a highly regular lattice of alternating sodium ions and chloride ions. (b) Hydrogen and oxygen atoms share two electrons by covalent bond to form water.

Let us emphasize that in the formation of a compound only the outer-shell electrons (valence electrons) take part in chemical combinations. The inner-shell electrons are well-protected and are generally not involved in these processes of gaining or losing or sharing electrons. Also, the bonds forming compounds represent potential energy (see Box 2.1) because breaking them requires energy. Thus, the strength of chemical bonds is measured by the amount of energy required to break them (see Table 2.7)—note that oxygen has six valence electrons, so it only needs two more, which is the octet rule. Hydrogen is an exception to the octet rule; its nearest noble gas, helium, has only two electrons. Hydrogen nuclei form molecules with two nearby electrons, which is the duet rule.

Most compounds can be classified into two broad categories: organic and inorganic compounds. Naturally occurring organic compounds are made primarily of carbon and hydrogen (see Figure 2.8). They can also contain oxygen (O), nitrogen (N), sulfur (S), phosphorus (P) and halogens [i.e., iodine (I), chlorine (Cl), bromine (Br) and fluorine (F)]. Examples of organic compounds include table sugar known as sucrose ($C_{12}H_{22}O_{11}$), benzene (C_6H_6), methane (CH_4), and ethanol (C_2H_6O). Organic substances include coal (84 percent carbon + 6 percent hydrogen + 7 percent oxygen + ...), hair (45 percent carbon, 28 percent oxygen, 15 percent nitrogen, 7 percent hydrogen and 5 percent sulphur), and butter (80 percent fat and 20 percent water). Organic compounds consist only of carbon and hydrogen are called hydrocarbons (e.g., natural gas, acetylene, propane, and butane). Many hydrocarbons are found in plants, animals, and their fossils. We use hydrocarbons from fossil fuels as one of the primary sources of energy today because the covalent bonds between carbon and hydrogen are very energy-rich. Organic compounds are also key components of plastics, soaps, perfumes, sweeteners, fabrics, pharmaceuticals, fertilizers, agrochemicals, and many other substances that we use every day. So the building block of organic matter is the tetravalent carbon atom. With a few exceptions, carbon compounds can be formulated with four covalent bonds to each carbon, regardless of whether the combination is with carbon or some other element, as we can see in Figure 2.8.

Inorganic compounds are compounds without carbon bonded to

hydrogen in their compositions. Examples of inorganic compounds include table salt (NaCl), carbon dioxide (CO_2), argon (Ar), liquid sulfuric acid (H_2SO_4), solid phosphorus pentoxide (P_2O_5), water (H_2O), and glass (SiO_2). As we will discuss in Chapter 3, the Earth's crust contains most inorganic compounds. Solar radiation, without magnetosphere and atmosphere attenuation, can supply the energy needed to convert inorganic compounds into organics. In other words, while most organic compounds are made within living things, organic compounds found on Mars or within a nebula are not necessarily signs of extraterrestrial life because solar radiation can convert inorganic compounds into organics.

Table 2.8. Examples of one-atom, two-atom, three-atom, and five-atom molecules. Springs here indicate restoring forces; that is, when two atoms are pushed toward one another, the spring connecting them is compressed. Then the spring pushes back, trying to restore the atoms to their original separation. If two atoms are pulled apart, the spring again resists the force, this time pulling back toward one another.

Number of atoms	Molecule Structure	Molecule name
One-atom molecule		Argon
Two-atom molecule		Oxygen
		Carbon monoxide
Three-atom molecule		Carbon dioxide
Five-atom molecule	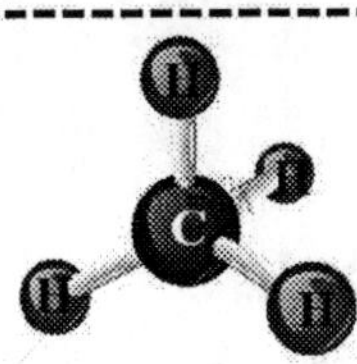	Methane

Pure elements are not compounds, so they are neither organic nor inorganic compounds. They include argon (Ar), silver (Ag), sulfur (S), and carbon (C).

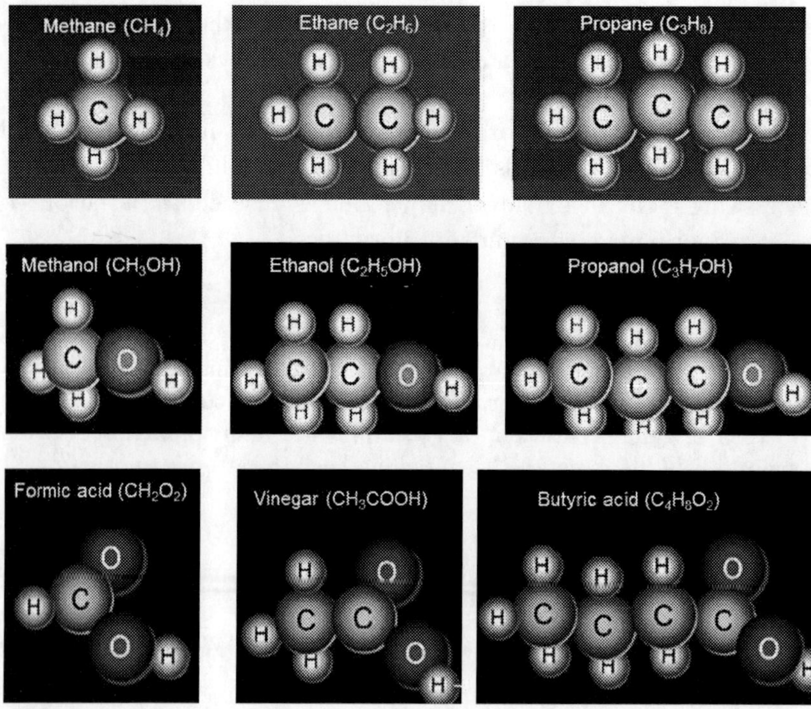

Figure 2.8. Examples of organic compounds: they always have carbon, but may contain other elements like hydrogen, oxygen, and nitrogen. The first row has hydrocarbons, the second row has alcohols, and the third row is organic acids. (adapted from www.chemistryland.com).

A molecule is a group of two or more atoms held together by chemical bonds in a stable association. Here are some examples:

- Water $[H_2O] = 2\,\text{Hydrogen}\,(H_2) + \text{Oxygen}\,(O)$,
- Table salt $[NaCl] = \text{Sodium}\,(Na) + \text{Chlorine}\,(Cl)$,
- Glucose $[C_6H_{12}O_6] = 6\,\text{Carbon}(C) + 12\,\text{Hydrogen}\,(H) + 6\,\text{Oxygen}\,(O_6)$,

- Calcium oxide $[CaO] = \text{calcium}\,(Ca) + \text{Oxygen}\,(O)$,
- Nitrogen gas $[N_2] = 2\,\text{Nitrogen}\,(N)$,
- Hydrogen gas $[H_2] = 2\,\text{Hydrogen}\,(H)$,
- Oxygen gas $[O_2] = 2\,\text{Oxygen}\,(O)$,
- Chlorine $[Cl_2] = 2\,\text{Chlorine}\,(Cl)$,
- Ozone $[O_3] = 3\,\text{Oxygen}\,(O)$.

Water, table salt, calcium oxide, and glucose (a type of sugar) are not only compounds, they are also molecules. However, nitrogen gas, hydrogen gas, oxygen gas, and ozone are molecules but they are not compounds since each only contains one kind of element (e.g., nitrogen gas contains only nitrogen and ozone contains only oxygen). In more general terms, X_nY_m, where X and Y are chemical elements and n and m are two integers, is a compound if $X \neq Y$, whereas X_nX_m, Y_pY_q, and X_rY_s are molecules. In other words, all molecules are not compounds, but all compounds are molecules; and compounds form a subset of the set of molecules. Figure 2.9 illustrates the construct of H_2 as an example of a molecule that consists of the atoms of a single chemical element. Each hydrogen atom has an unpaired electron and an unfilled outer energy level; for these reasons, the hydrogen atom is unstable (a hydrogen atom actually contains two electrons and a proton). When two hydrogen atoms are close to each other, however, each atom's electron can orbit both nuclei—that is, each hydrogen atom attains the configuration of the noble gas, helium (a duplet of electrons). The result is a diatomic (two-atom) molecule of hydrogen gas. Oxygen and nitrogen gases and chlorine molecules are also covalently bonded as hydrogen gas, as illustrated in Figure 2.9. Molecules that are composed of only two atoms, of either the same or different chemical elements, are characterized as diatomic (see Table 2.8). If a diatomic molecule consists of two atoms of the same element, such as hydrogen (H_2) or oxygen (O_2), then it is said to be homonuclear. Otherwise, if a diatomic molecule consists of two different atoms, such as carbon monoxide (CO) or nitric oxide (NO), the molecule is said to be heteronuclear. Molecules containing more than two atoms are called polyatomic molecules—e.g., carbon dioxide (CO_2) and water (H_2O).

Ionic compounds. The ions described in the previous sections are called monatomic ions—that is, they are ions formed from only one atom. We also found many polyatomic ions that act as discrete units and are mostly electrically charged molecules (i.e., a group of bonded atoms with an overall charge). Some of them are listed in Table 2.9. For example, consider the compound CO_3^-. A carbon atom forms a bond with more electron-negative oxygen. This occurs because the electrons in the carbon-oxygen bond are drawn ("lost") toward oxygen (oxygen and carbon are both electron-negative). Conversely, if the same carbon is bound to hydrogen ("gained"), the carbon atom is reduced in the H_3O^+ compound.

There is some regularity in the names of these polyatomic ions. Some nonmetals form a series of polyatomic ions with oxygen (all having the same charge): ClO^-, hypochlorite; ClO_2^-, chlorite; ClO_3^-, chlorate; ClO_4^-, perchlorate. The general rule for such series is: X_m^{n-}, stem + -ate (e.g., SO_4^{2-}; sulfate); X_{m-1}^{n-}, stem + -ite (e.g., SO_3^{2-}; sulfite); X_{m-2}^{n-}, hypo- + stem + -ite (e.g., SO_2^{2-}; hyposulfite); X_{m+1}^{n-}, per- + stem + -ate (e.g., SO_5^{2-}); and X^{n-}, stem + -ide (the monatomic ion) (e.g., S^{2-}; sulfide). In some cases, the *-ate* form has three oxygen, and in some cases four.

Acids, bases, and buffers. In the science of water, the term *acid* is used to distinguish the presence of the hydrogen ion, H^+, in solution. Any substance that dissociates in water to increase the concentration of H^+ ions is called an acid (Brönsted, 1923 and Lowry, 1923). One of the common descriptive features of acidic substances is that they have a sour taste (e.g., vinegar or lemon juice). The quantitative description of acidic substances consists of measuring the hydrogen-ion concentration of a solution by using the pH (power of hydrogen) scale, which is defined as follows:

$$pH = -\log_{10}(H^+) \tag{2.12}$$

and H^+ is measured in moles/liter; one mole of particles $= N_{Av}$ particles, where N_{Av} ($N_{Av} \approx 6.02 \times 10^{23}$) is the Avogadro number and the particles can be atoms, molecules, ions, or electrons. The pH scale ranges from 0 to 14 (see Figure 2.10). Note that the pH scale is logarithmic, which means that a difference of 1 on the scale represents a tenfold change in hydrogen ion concentration. For pure water, H^+ is 10^{-7} mole/liter and the pH is 7. Acidic solutions have pH values of below 7. The stronger an acid is, the more H^+ ions it produces and the lower its pH. For example, hydrochloric acid (HCl), which is abundant in your stomach, has a pH of 1, and orange juice is mildly acidic (pH $\approx$ 3.5).

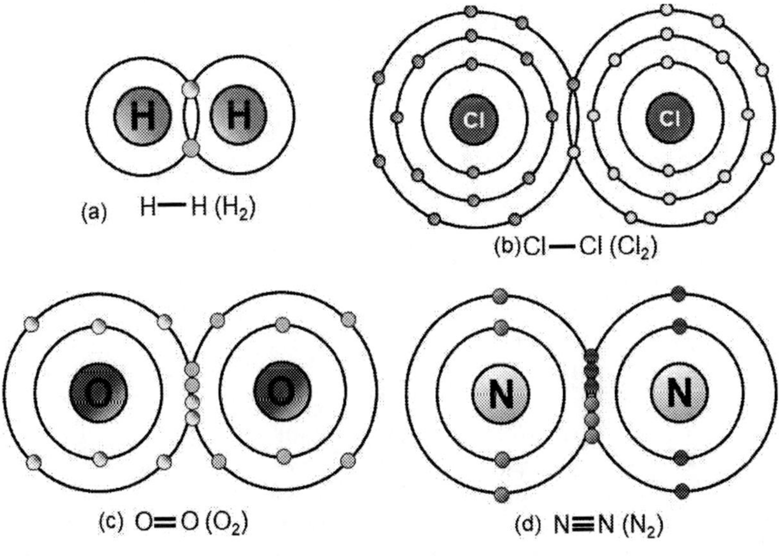

Figure 2.9. An illustration of single, double, and triple bonds—in (a) and (b), the two-electron bonds, which are illustrated by the hydrogen-hydrogen and chlorine-chlorine bonds, are single bonds. For example, each of the two atoms of hydrogen shares its electron in a single covalent bond. In (c), two electrons from each of the oxygen atoms are mutually shared, thereby producing two two-electron bonds, an arrangement that is called a double bond, in (d), three electrons from each of the nitrogen atoms are mutually shared, thereby producing three two-electron bonds, an arrangement that is called a triple bond.

Remember that hydrogen can form a cation (H^+) or an anion (H^-), as shown in Table 2.4. Compounds with hydrogen as an anion are rare. In acids, hydrogen forms hydrogen cations. Again, an acid releases hydrogen ions (H^+) when in solution. In a strong acid, such as hydrochloric acid (HCl), all hydrogen ions (H^+), and chloride ions (Cl^-) dissociate (separate) when placed in water and these ions are no longer held together by ionic bonding. In a weak acid, such as carbonic acid (H_2CO_3), only some of the ions dissociate into hydrogen ions (H^+) and bicarbonate ions ($HCO_3{}^-$), while others are still held together by ionic bonding.

Table 2.9. Ionic compounds are compounds composed of ions, charged particles that form when an atom (or group of atoms) gains or loses electrons. Again, covalent or molecular compounds form when elements share electrons in a covalent bond to form molecules. Molecular compounds are electrically neutral. There is one positive polyatomic ion. It is NH_4^+ and is called the ammonium ion.

polyatomic ion	Name	polyatomic ion	Name
NH_4^+	ammonium	$C_2H_3O_2^-$	acetate
H_3O^+	hydronium	$C_2O_4^{2-}$	oxalate
OH^-	hydroxide	BrO_3^-	bromate
CO_3^-	carbonate	PO_4^{3-}	phosphate
HCO_3^-	bicarbonate	SiO_4^{4-}	silicate
NO_3^-	nitrate	HS^-	hydrogen sulfide
NO_2^-	nitrite	ClO_3^-	chlorate
SO_4^{2-}	sulfate	IO_3^-	iodate
HSO_4^-	bisulfate	O_2^{2-}	peroxide
SO_3^{2-}	sulfites	PO_4^{3-}	phosphite

The term *base* is used to distinguish the presence of the hydrogen ion, OH^-, in solution (Lowry, 1923; Brönsted, 1923; Myers, 2003; and Masterton et al., 2011). In other words, a base is a substance that takes or accepts hydrogen ions (H^+) already present in the solution. Basic (or alkaline) solutions have pH values above 7. Very strong bases, such as sodium hydroxide (NaOH), have pH values of 12 or more. The blood in your veins is slightly alkaline (pH = 7.4) and baking soda is alkaline (pH = 9.0). Sodium hydroxide (NaOH) is a strong base because when placed in water, it dissociates completely into sodium ions (Na^+) and hydroxyl ions (OH^-), all of which are now released and dissolved in water.

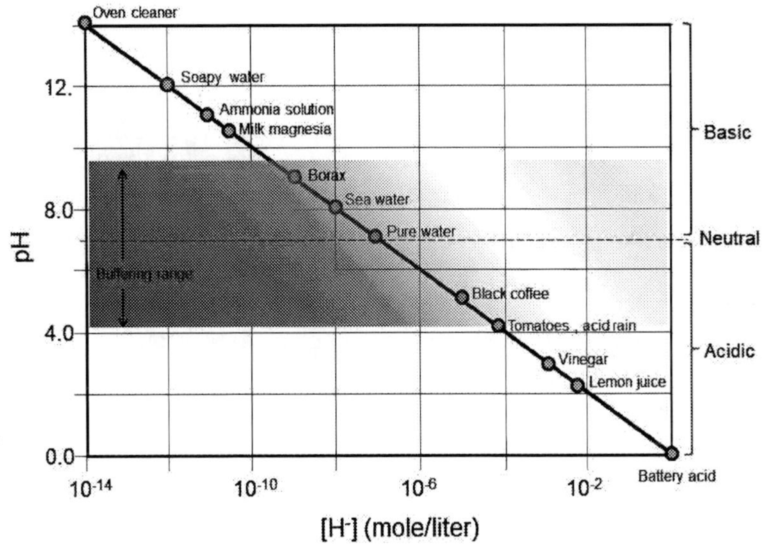

Figure 2.10. The pHs of common substances. Solutions with a pH less than 7 are acidic, while those with a pH greater than 7 are basic. Lemon juice is 100 times more acidic than tomato juice, and oven cleaner is 10 times more basic than soapy water.

Three other important notions related to acids and bases are conjugate acids, conjugate bases, and buffers. When an acid donates a proton (H^+), it forms a conjugate base. For example, when the strong acid HCl donates a proton, it produces Cl^-, which is its conjugate base. Conversely, when a base gains an H^+ ion, it produces a conjugate acid. The conjugate acid of OH^- is H_2O, the conjugate acid of hydrogen $HCO_3{}^-$ is H_2CO_3, and its conjugate base is $CO_3{}^{2-}$.

A mixture of a weak acid and its conjugate base and a mixture of a weak base and its conjugate acid are known as buffer solutions, or buffers. Buffer solutions resist a change in pH when small amounts of a strong acid or a strong base are added. A solution of acetic acid and sodium acetate ($CH_3COOH + CH_3COONa$) is an example of a buffer that consists of a weak acid and its salt. An example of a buffer that consists of a weak base and its salt is a solution of ammonia and ammonium chloride ($NH_3 + NH_4Cl$). Pyruvate ($C_3H_3O_3$) is a pivotal molecule in the functioning of humans and is a conjugate base of pyruvic acid. The chemical formula for pyruvic acid is $C_3H_4O_3$ and its pH is 3.1. Its deprotonated form is $C_3H_3O_3$ with pH = 7.2.

Quiz: (1) Why are the properties of CO_2 and SiO_2 different? (2) Again, an acid is a substance that gives up a proton. A base is a substance that accepts a proton. Using these definitions, rank the following molecules based on their acid strengths: H_2SO_4, H_2O, HCl, H_3O^+, and $NH_4{}^+$.
Similarly, rank the following molecules based on their base strengths: $HSO_4{}^+$, OH^-, Cl^-, H_2O, and NH_3 .
(3) Using the periodic table rank the bonds in each set in order of decreasing bond length and decreasing bond strength: (a) S-F, S-Br, S-Cl. (b) C=O, C-O.

2.2.2. Molecular Diagrammatica

The size and shape of molecules are as much an important part of molecular structure as is the order in which the component atoms are bonded. Contrary to the impression you may get so far from the formulas describing compounds and molecules, complex molecules are not limited to their formulas. They also have well-defined spatial arrangements that are determined by the lengths and directional character of their chemical bonds. A simple modification of the spatial arrangement of a molecule can change its chemical and/or physical properties. So it is important to familiarize yourself with the possible arrangements of bonds in space, including their visualization.

Double and triple chemical bonds. As described earlier, the atoms in a molecule are joined by chemical bonds. These bonds can result when atoms with opposite charges attract (ionic bonds), when two atoms share one or more pairs of electrons (covalent bonds), or when atoms interact in other ways. The two-electron bond, which is illustrated by the carbon-hydrogen bonds (C−H) in methane or ethane, the carbon-carbon (C−C) bond in ethane, and the hydrogen-hydrogen (H−H) bond in H_2, in Figure 2.9a, is called a single bond.

By convention, a single straight line connecting the atomic symbols is used to represent a single (two-electron) bond.

Double bonds are chemical bonds that allow two atoms to share two pairs of electrons. Ethene (ethylene), C_2H_4, in which two electrons from each of the carbon atoms are mutually shared, thereby producing two two-electron bonds (C=C), two shared electron pairs (O=O) in O_2 (Figure 2.9c), and two double bonds between the carbon and oxygen atoms (C=O) in the carbon dioxide molecule, illustrate double-bond arrangements. By convention, two straight lines are used to represent a double (four-electron) bond. The strongest covalent bonds are triple bonds, such as those that link the two nitrogen atoms of nitrogen gas molecules (N≡N) and C≡H in ethyne (acetylene), C_2H_2. By convention, three straight lines are used to represent a triple (six-electron) bond. These bonds are stronger than single and double bonds. More chemical energy is required to break a triple bond than a single or double bond (e.g., C−C single bond: 347 kJ/mol; C=C double bond: 614 kJ/mol C≡C triple bond: 839 kJ/mol).

There are a number of recurring features in organic compounds that are commonly known as functional groups. The classical functional groups are hydroxyl group (R−OH), amino group ($R-NH_2$), carboxyl group (R−COOH), amide group ($R-CONH_2$), methyl group ($R-CH_3$), and alkyl group ($R-CH_3$). For example, $R-CH_3$ means that compounds of this group include a branch consisting of $-CH_3$. The main reason for this classification is that compounds of a functional group have similar chemical behavior.

The length of a chemical bond is the average distance between the nuclei of two bonded atoms, regardless of where the bonding electrons happen to be. As is apparent in Table 2.7, bond lengths vary considerably depending on the identity of both atoms and the type of bonding (single, double, or triple). Triple and double bonds clearly are shorter than single bonds, but they are stronger. Note that atomic radii are generally very large compared to bond lengths, hence, the bonds connecting the atoms are not clearly visible when molecules are visualized with atomic radii and bond lengths at the same scale.

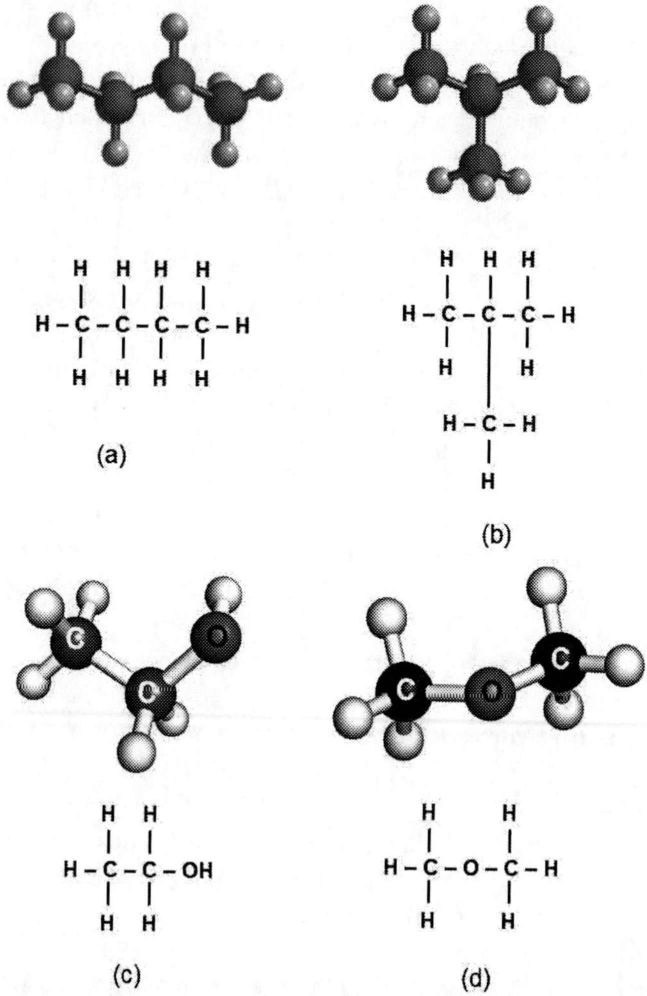

Figure 2.11. Here are two examples of isomers (compounds that have the same molecular formula but different structures)—that is, examples where there is more than one possible carbon skeleton. (a) Butane can be written as $C_4H_{10} = CH_3CH_2CH_2CH_3$ or as (b) ibutance as $C_4H_{10} = CH_3CHCH_3CH_3$. (c) Ethanol can be written as $C_2H_6O = CH_3CH_2OH$ or (d) Dimethyl ether as $C_2H_6O = CH_3OCH_3$.

Isomers: Some molecular formulas have multiple spatial representations in the way the atoms are arranged—that is, some combinations of atoms can be assembled in more than one way. These representations can have different chemical and physical properties, such as melting points, boiling points, and densities (mass/volume). For example, butane and 2-methylpropane (isobutane) have the same molecular formula, C_4H_{10}, but have different bonding arrangements between the atoms, as illustrated in Figures 2.11a and 2.11b. The melting point, boiling point, and density of butane are −138 degrees Celsius, −1 degrees Celsius, and 0.579 g/cm^3, respectively, whereas these same properties for isobutane are −159° C, −12 degrees Celsius, and 0.549 g/cm^3, respectively. Similarly, methoxymethane (dimethyl ether) and ethanol have the same formula, C_2H_6O, but different atom arrangements, as illustrated in Figures 2.11c and 2.11d. Yet, ethanol has a boiling point of 78.4 degrees Celsius, while dimethyl ether has a boiling point of −23.7 degrees Celsius. Compounds having the same number and kind of atoms are called *isomers*. Isomers do not have the same chemical and physical properties. As we will see in the next subsection, isomers are important parts of the execution of the functions of living things.

In more general terms, as the number of carbons goes up in some organic compounds, the number of isomers grows. For example, C_nH_{2n+2}, known as alkanes, has no isomers for $n \leq 3$, two isomers for $n = 4$ (butane and isobutane), three for $n = 5$, five for $n = 6$, and more than 62 trillions for $n = 40$. Most of the hydrocarbons in petroleum belong to the family of alkanes. Similarly, C_nH_{2n}, known as alkenes (alkane = alkene + H_2) and also as cycloalkenes, has no isomers for $n \leq 2$, two isomers for $n = 3$, five for $n = 4$, thirteen for $n = 5$, and twenty five for $n = 6$. Alkenes are the family of molecules containing carbon and hydrogen and at least one carbon-carbon double bond. These molecules can be in continuous chains, as we have seen so far, or in rings, as we will see later.

To summarize, isomers can differ from one to another by their rotations around a single bond, the relative orientations of their atoms in space, the location of a certain group of atoms, and the order in which the individual atoms are connected.

Polar and nonpolar bond: electronegativity. A molecule can be characterized as nonpolar or polar by determining the difference in the electronegativities of its chemical elements. The electronegativity (EN) of an element is a measure of the ability of this element to attract shared electrons in a covalent bond with another element. Electronegativity is measured on the Pauling scale, which ranges from 0 to 4 (Pauling, 1932; see also Appendix A). F, O, N, and Cl are the most electronegative elements and Cs and Fr with electronegative at 0.7 are the least electronegative elements. Noble gases have no electronegativity. Consider a generic diatomic molecule, XY, containing elements X and Y. When the absolute value of the difference in electronegativity of the two bonded atoms is smaller than 0.5 [i.e., $|EN(X) - EN(Y)| < 0.5$], XY is considered nonpolar. In these cases, both atoms exert almost equal attractions for the bonding electrons. Hence, diatomic molecules containing elements of identical electronegativity, such as H_2, O_2, Cl_2, and N_2, are nonpolar [i.e., $|EN(X) - EN(Y)| = 0 < 0.5$].

When the absolute value of the difference in electronegativity of the two bonded atoms is equal or greater than 0.5 and smaller than 1.7 [i.e., $0.5 \leq |EN(X) - EN(Y)| < 1.7$], XY is covalently polar; that is, we have a bond with an uneven distribution of electron density. The oxygen atom (EN=3.5) is more electronegative than the hydrogen atom (EN=2.1) and it therefore attracts electrons more strongly; the difference is 1.4 and O−H bond is strongly polarized. We have a bond with an uneven distribution of electron density because there is one pair of unshared electrons (i.e., one lone pair of electrons). C−H, with a difference of 0.4, is nonpolar; C−N bond, with a difference of -0.5, is relatively nonpolar; and C−O, with a difference of -1.0, is polar.

When the absolute value of the difference in electronegativity of the two bonded atoms is equal or greater than 1.7 [i.e., $|EN(X) - EN(Y)| \geq 1.7$], the bond between X and Y is ionic. Sodium has EN = 0.9, while chlorine has EN = 3. The difference is 2.1, which means that sodium chloride has an ionic bond.

Let us emphasize the concept of *lone pair*, as this concept is used throughout the remaining of this section, especially in the valence-shell electron-pair repulsion theory (VSEPR theory), which we will describe later. A lone pair refers to a pair of valence electrons that are not shared with another atom in a covalent bond and is sometimes called an unshared pair or non-bonding pair (McMurry, 2000; Fox and Whitesell, 1997; and Kumar et al., 2014). Lone pairs are located in the outermost electron shell of atoms. In Figure 2.7b there are two electron pairs not used in chemical bonding. Thus, there are two lone pairs.

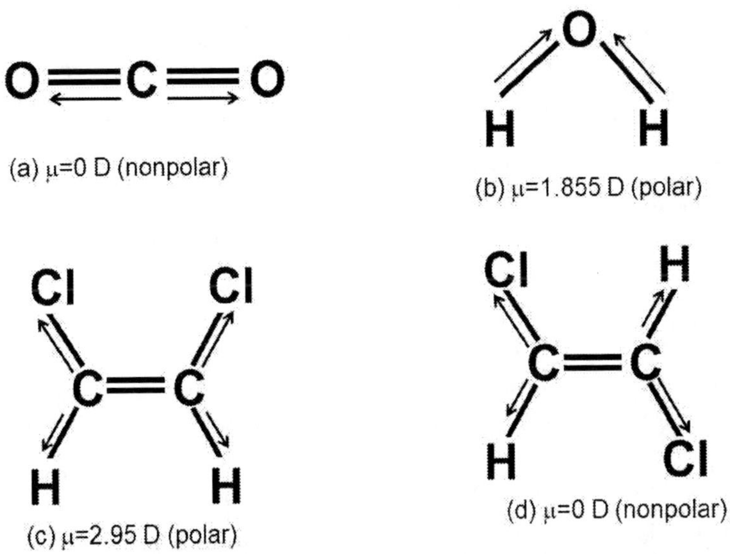

Figure 2.12. The molecules H_2O, CO_2, CH_4, and two isomers of $C_2H_2Cl_2$ at their resting states, along with their molecular dipole moments. H_2O and one of the isomers of $C_2H_2Cl_2$ have dipole moments at their resting state, whereas CO_2, CH_4, and the other isomers of $C_2H_2Cl_2$ do not have a dipole moment at a resting state. For example, CO_2 has no dipole moment because the two bond vectors cancel each other.

Polar and nonpolar bonds: molecular polarization. Another way to define molecular polarization is to use the notion of dipole moment which we will describe in more detail in later chapters. The dipole moment can be defined as follows:

$$\text{dipole moment} = \sum_{\text{bonds}} [\text{electrical charge} \times \text{bond distance vector}] .$$

The bond distance vector is the distance vector between the nuclei of two bonded atoms along the straight line joining the nuclei. The difference in electron densities between the two atoms indicates the direction of the bond distance vector (see Figure 2.12). The vector points from the lower electron-density atom to the higher electron-density atom. In other words, the dipole moment is a vector with magnitude and direction. Each bond has its dipole moment.

> **Quiz:** Is CCl_4 polar or nonpolar? Justify your answer. Which bond is more polarized, the H−F or H−Br bond?

For polyatomic molecules, the dipole moment is the vector sum of the dipole moments of various bonds. So by changing even the direction of a single bond in a polyatomic molecule, we change its dipole moment and we end up modeling one of the potential isomers of the given polyatomic molecule. Nonpolar molecules have zero dipole moment and polar molecules have nonzero dipole moments—the more polar the bond the higher the dipole moment. The dipole moments for molecules are generally determined by lab experiments. As we can see in Table 2.10, the dipole moment not only tells us if a compound is polar or nonpolar, it also allows us to determine the geometry of a molecule. Notice that the dipole moment of water (1.85 D; 1D $= 3.33564 \times 10^{-30}$ C m, where C is coulomb and m is meter) is greater than the dipole moment of the single bond (1.5 D; see Table 2.10) because the dipoles are oriented so that they reinforce the dipole of the bond, they are all in the same relative direction. Hence, the H_2O molecule is a polar molecule (see Figure 2.12). In contrast, the two bond dipoles in CO_2, although equal in magnitude, are opposite in direction. Therefore, the dipole moment of CO_2 is zero, even though the individual bonds are polar, making CO_2 a nonpolar molecule. Notice also that one isomer of $C_2H_2Cl_2$ is nonpolar and the other one is polar. The molecule $C_2H_2Cl_2$ in Figure 2.12d, is a symmetrical molecule in the vertical projection with respect to the carbon-atom bond. Thus, the symmetry of the molecule causes the bond dipole moments to cancel.

Table 2.10. Dipole moments, lone pairs, and geometries of selected molecules. (a) molecules in which the central atom has no lone pair. (b) molecules in which the central atom has one lone pair. (c) molecules in which the central atom has two lone pairs

Molecule	Example	Dipole Moment (D)	# lone pair	Geometry	Molecular diagram
AB_2	CO_2	0	0	Linear	O=C=O
	CS_2	0	0	Linear	S=C=S
	BeF_2	0	0	Linear	F—Be—F
	HCN	0	0	Linear	H—C≡N
AB_3	BF_3	0	0	Trigonal-planar	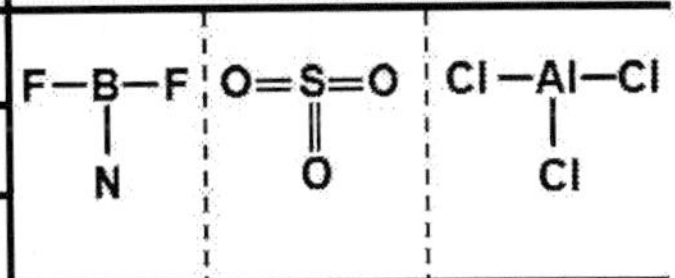
	SO_3	0	0	Trigonal-planar	
	AlCl3	0	0	Trigonal-planar	
AB_4	CH_4	0	0	Tetrahedral	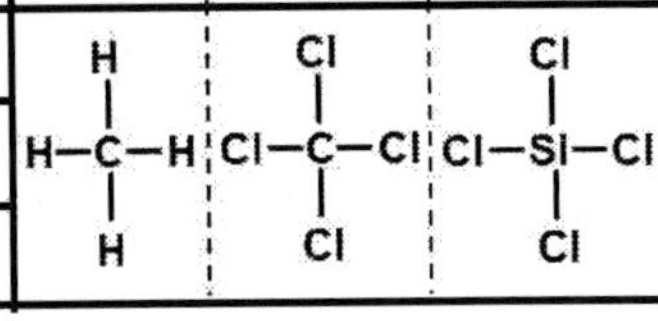
	CCl_4	0	0	Tetrahedral	
	$SiCl_4$	0	0	Tetrahedral	

(a)

Molecule	Example	Dipole Moment (D)	# lone pair	Geometry	Molecular diagram
AB_2	NO_2	0.316	1	Bend	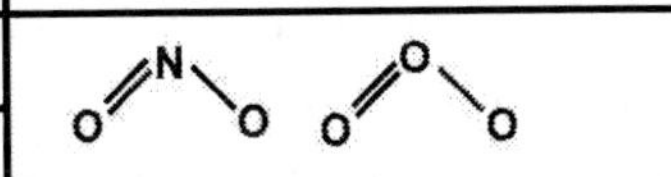
	O_3	0.530	1	Bend	
AB_3	NH_3	1.470	1	Trigonal pyramidal	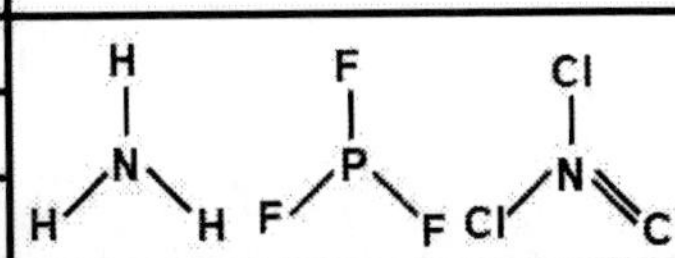
	PF_3	1.030	1	Trigonal pyramidal	
	NCl_3	0.390	1	Trigonal pyramidal	
AB_4	SF_4	0.632	1	Trigonal bipyramidal	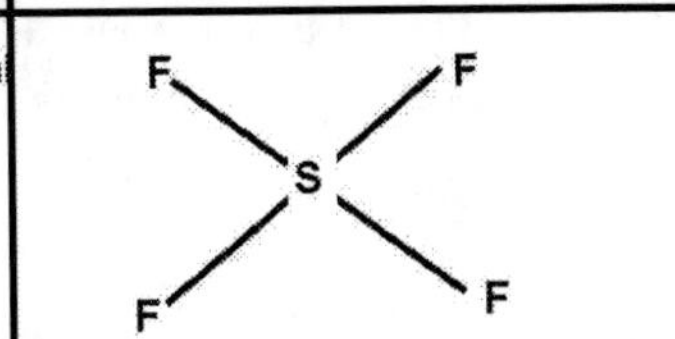

(b)

Molecule	Example	Dipole Moment (D)	# lone pair	Geometry	Molecular diagram
AB_2	H_2O	1.855	2	Bend	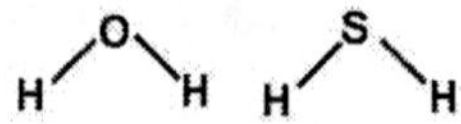
	H_2S	0.970	2	Bend	
AB_3	ClF_3	0.780	2	Trigonal pyramidal	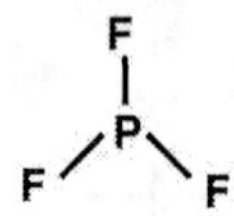
AB_4	XeF_4	0	2	Trigonal-planar	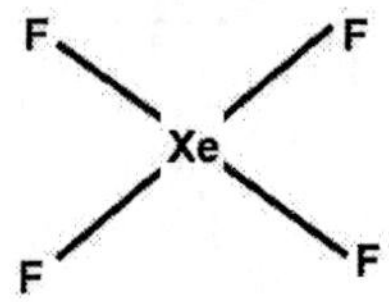

(c)

Box 2.2. Solubility of Organic Compounds

Two broad categories of matter are mixtures and pure substances. A pure substance has a constant composition. Table sugar (sucrose), which consists of 42.1 percent carbon, 6.5 percent hydrogen, and 51.4 percent oxygen by mass is an example of pure substances because any sample of sucrose has the same physical properties, such as melting point, color, and sweetness. A mixture is composed of two or more types of matter in varying amounts. It can be separated by physical changes, such as evaporation (evaporation describes the transformation from liquid to gas; see Ikelle, 2022a and 2022b). A mixture with a composition that varies from point to point is called a heterogeneous mixture. A household salad dressing made of oil and vinegar is an example of a heterogeneous mixture. When you bring oil and vinegar together, they do not mix. They eventually separate in layers—the mixture is not the same from one point to another point. A drink consisting of water and sugar is a homogeneous mixture (i.e., a solution) because each drop of this drink tastes the same. The relative amount of a given solution component is known as its *concentration*. The component of the solution with the highest concentration is called the *solvent* and may be viewed as the medium in which the other components are *dissolved*. A solution in which water is the solvent is called an aqueous solution.

Solutions are crucial to the processes that sustain life and to many other processes in our environment. Actually, most organic chemistry takes place in the solution phase. If the solvent is polar or ionic, like water, then a polar organic compound dissolves in this solution. Similarly a nonpolar solvent, such as hexane (C_6H_{14}), dissolves nonpolar compounds. A polar solvent will not dissolve a nonpolar compound. For example, biphenyl ($C_{12}H_{10}$) does not dissolve at all in water because it is a very non-polar molecule, with only carbon-carbon and carbon-hydrogen bonds. It is not able to form significant attractive interactions with very polar solvent molecules like water.

The small alcohols—methanol (CH_4O), ethanol (C_2H_6O), and propanol (C_3H_8O)—dissolve easily in water because water is able to form hydrogen bonds with the hydroxyl group in these molecules, and the combined energy of the formation of these water-alcohol hydrogen bonds is more than enough to make up for the energy that is lost when the alcohol-alcohol (and water-water) hydrogen bonds are broken up.

The longer-chain alcohols—hexanol ($C_6H_{14}O$), heptanol ($C_7H_{16}O$), and octanol ($C_8H_{18}O$)—are increasingly non-soluble in water. The same favorable water-alcohol hydrogen bonds are still possible with these larger alcohols. The difference is that larger alcohols have larger nonpolar regions. At about four or five carbons, the influence of the nonpolar part of the molecule begins to overcome that of the part, and water solubility is lost. For the same reason, macromolecules known as lipids (of which fats are an important) have areas that lack polar covalent bonds. In these cases, lipids or long-chain alcohols form a layer on top of the water.

Polar and nonpolar bonds: VSEPR. Sidgwick and Powell (1940) proposed a simple way of determining the molecular geometry of molecules of the form AB_n, where A and B are generic atoms and n is an arbitrary positive integer. This method is also based on the idea of repulsive interactions between electron pairs in the valence shell of the atoms and is now known as VSEPR. Gillespie and Nyholm (1957) refined the VSEPR method by dividing molecules into two categories: (1) molecules in which the central atom has no lone pair and (2) molecules in which the central atom has one or more lone pairs. Table 2.10 shows the molecular geometries of AB_n molecules for these two categories with $n = 2$, 3, and 4. We can see that there are five basic shapes: linear, trigonal planar, tetrahedral, trigonal pyramidal, and trigonal bipyramidal. These shapes depend on n and the number of lone pairs associated with the AB_n molecule. For example, in NH_3 (i.e., $n = 3$), there are five valence electrons in nitrogen and one each in hydrogen, so there are eight electrons total. The bonds for the molecule use up six of them, and the remaining two become a lone pair on the central atom, N. The molecular geometry is a trigonal pyramidal.

For molecules with no lone pairs, we can see that the arrangement of AB_2, AB_3, and AB_4 are linear, trigonal planar, and tetrahedral, respectively. For example, in BH_3, there are three valence electrons in boron, and one each in hydrogen. There are no lone pairs, so the molecular geometry of BH_3 is trigonal planar.

Quiz: Work out the shapes of PO_4, XeO_3, ClO_2, $XeOF_4$.

Consider more general molecules of the form A_mB_n, where A and B are generic atoms and m and n are arbitrary positive integers. In other words, molecules now have m central atoms instead of one central atom. The total number of valence electrons here is $\chi_T = (m \times k_A) + (n \times k_B)$. The VSEPR method can be extended to these molecules by requiring the number of valence electrons used in the bonding of A atoms between themselves to be equal or less than $\chi_D = (m \times k_A) - (n \times k_B)$, where k_A represents the valence electrons in A and k_B represents the valence electrons in B. The remaining electrons are then used in the bonding of A atoms with B atoms. Consider for example molecule C_2H_2—that is, $m = 2$, $n = 2$, $k_A = 4$, $k_B = 4$, $\chi_T = 10$, and $\chi_D = 6$. The bonding between the two carbon atoms can only use six of their valence electrons in a triple bond and the remaining electrons to form bonds with hydrogen atoms, as shown in Table 2.11. Similarly, in the molecule C_2H_4 (i.e., $m = 2$, $n = 4$, $k_A = 4$, $k_B = 4$, $\chi_T = 12$, and $\chi_D = 4$), the bonding between the two carbon atoms can only use four of their valence electrons in a double bond and the remaining electrons to form bonds with hydrogen atoms.

Table 2.11. Geometries of selected molecules with two central atoms

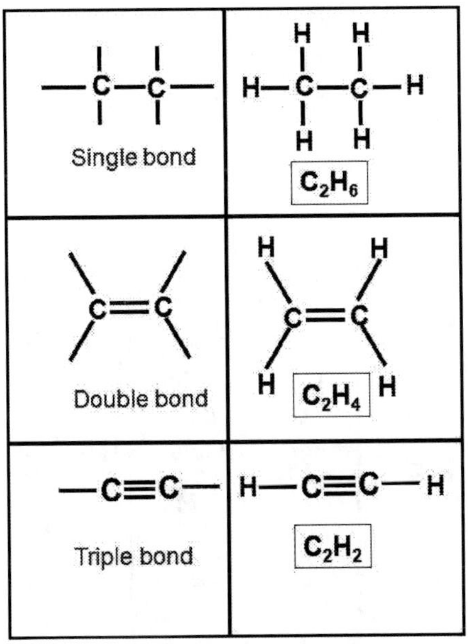

Consider now C_nH_{2n} molecules with $n \geq 3$ with n central carbons. As we discussed earlier, C_nH_{2n} has multiple isomers for $n \geq 3$. For $n = 3$ (i.e., C_3H_6), we have two isomers. One of them is $Ch_3CH{=}Ch_2$; that is, it is an open-chain molecular geometry (i.e., carbons are attached to one another to form chains), as we have described so far. In the other one, the central carbons form a ring shape and hydrogen are then bonded to these central atoms, as shown in Table 2.12. The ring shape varies n. For instance, cyclopentane, C_5H_{10}, is represented as a regular pentagon in which each apex represents a carbon atom. Two hydrogen atoms are bonded to each carbon atom, as described in Table 2.13. At each carbon, we have the standard trigonal planar, described earlier. Likewise, cyclopropane,

C_3H_6; cyclobutane, C_4H_8; and cyclohexane, C_6H_{12}, are drawn as regular polygons. The nomenclature for these compounds is similar to the open-chain compounds that we have used before, except that the prefix of cyclo- is added to the name of the corresponding open-chain hydrocarbon. Sometimes ring shapes are dispensed with the symbols for carbon and hydrogen atoms and leave only the lines in a structural formula. For instance, cyclopentane, C_5H_{10} often is represented as a regular pentagon in which it is understood that each apex represents a carbon atom with the requisite number of hydrogens. Likewise, cyclopropane, C_3H_6, cyclobutane, C_4H_8, and cyclohexane, C_6HI_2, are represented by polygons.

Table 2.12. Geometries of selected molecules with three and four central atoms

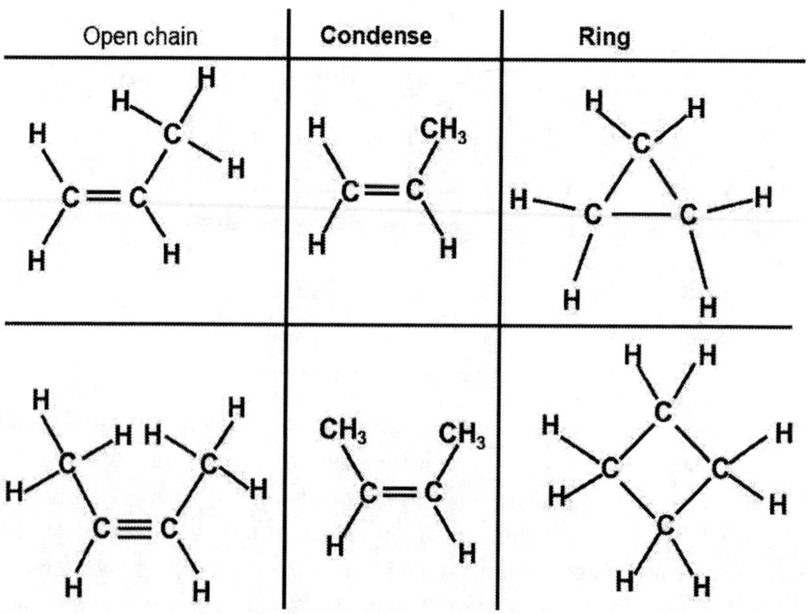

Benzene is produced during oil refining and has many industrial uses. A benzene molecule, C_6H_6, is often drawn as a ring of six carbon atoms, with alternating double bonds and single bonds, as depicted in Figure 2.13. Each carbon is bonded to one hydrogen only

because of the three alternating double bonds. In this configuration, each carbon is bonded to three others and one double bone (Kekulé, 1957 and 1958). Hence the molecular geometry at each carbon is trigonal planar. As shown in Figure 2.13, an alternative symbol used to represent the ring of benzene is a circle inside the hexagon.

Table 2.13. Geometries of selected molecules with five and six central atoms.

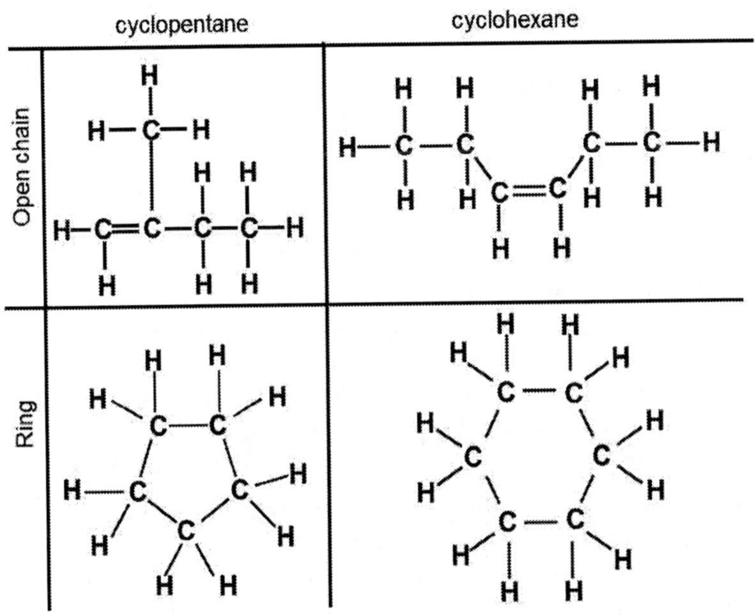

Ring shape is not limited to hydrocarbons. For example, glucose ($C_6H_{12}O_6$) is one of the most common molecules associated with the digestion of some living things. This molecule has open-chain representations as well as form rings. Figure 2.14 shows examples of such representations. The ring here consists of five carbon atoms and oxygen atoms (Haworth, 1937; and Moreno, 2012). It is in ring form that certain sugars actually occur in nature.

Box 2.3. Avogadro Number and Mole

In most practical cases, we are dealing with grams of sample and liters of samples instead of atomic mass A, which has unit amu. Sample A grams of element X is related to atomic mass A of X as follows:

$$A \text{ grams} = 6.02214179 \times 10^{23} \times (A \text{ amu}) = N_{Av}\,(A \text{ amu}) \ ,$$

$$1 \text{ gram} = 6.02214179 \times 10^{23} \times (1 \text{ amu}) = N_{Av}\,(A \text{ amu}) \ ,$$

or

$$1 \text{ mole of particles} = 6.02214179 \times 10^{23} \text{ particles} = N_{Av} \text{ particles} \ ,$$

where $N_{Av} = 6.02214179 \times 10^{23}$ is the Avogadro number. The Avogadro number relates atomic scale and human scale. A mole (abbreviated mol) is another reference to the Avogadro number. A mole of particles is equal to N_{Av} particles, just like a pair of particles is equal to two particles and a dozen particles is equal to twelve particles. For example, the molecular mass of water is 18.015 amu and the mass of one mole of water is 18.015 grams. Similarly, one mole of carbon-12 has 12g. In other words, the masses of one mole of different elements are different, since the masses of the individual atoms are different. It provides a specific measure of the number of atoms or molecules in a bulk sample of matter. The number of carbon-12 in a 10-kg piece of wood is

$$\begin{aligned} N &= \frac{N_{Av}\,(\text{atoms/mole}) \times \text{mass of the sample (grams)}}{A\,(\text{grams/mole})} \\ &= \frac{6.022 \times 10^{23}\,(\text{atoms/mole}) \times 10 \times 10^{3}\,(\text{grams})}{12\,(\text{grams/mole})} \\ &= 5.02 \times 10^{26}\,(\text{atoms}) \end{aligned}$$

The most common unit of concentration is molarity. Molarity is also the number of millimoles of solute present in exactly 1 milliliter of solution: molarity = moles of solute liters of solution.

2.2.3. Intermolecular Forces: Gases, Liquids and Solids

We are familiar with the gas, liquid, and solid phases of matter. One example is vapor water (gas) in the atmosphere, liquid water in the oceans, and solid water in glaciers. We here want to develop an understanding of the differences between these phases from atomic and molecular point of view.

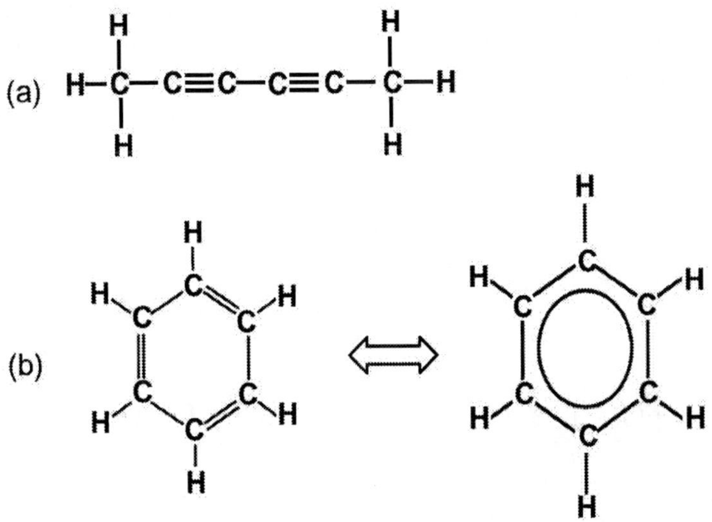

Figure 2.13. (a) Open-chain molecular geometry of benzene (C_6H_6). (b) Ring molecular geometry of benzene.

Again, a molecule with two or more atoms is held together by strong chemical bonds. These bonds can be covalent or ionic. They represent the *interatomic forces* (or *intramolecular forces*—that is, forces occurring within molecules) exerted within individual molecules to keep the atoms together (see Figure 2.16). The slightly positive charges in this figure are denoted by δ^+ and the slightly negative charges are denoted by δ^-. Also, remember that the forces are attractive when a negative charge interacts with a nearby positive charge and repulsive when the neighboring charges are the same, either both positive or both negative (see also Figure 2.3). An atom can participate in more than one chemical bond. The bonds arrange themselves around the atom to find their positions and the angles of least energy—that is, the principle of least energy. There is some flexibility in the angle of the bond, which leads to bending vibrations (i.e., vibrations that involve changes in the angle between two bonds). The notion of bending vibrations is important in regard to climate change as discussed in Ikelle (2022b and 2022c).

So, where do the charges in molecules come from? The charges exist in molecules that are held together by polar covalent bonds; that is, the charges in the molecules are associated with an uneven distribution of electrons between the bonded atoms of the molecules. It is actually this uneven electron sharing that creates a dipole. When two molecules with *polar covalent bonds* are near each other, they can form favorable interactions if the partial charges align appropriately, as shown in Figure 2.16, forming a dipole-dipole interaction.

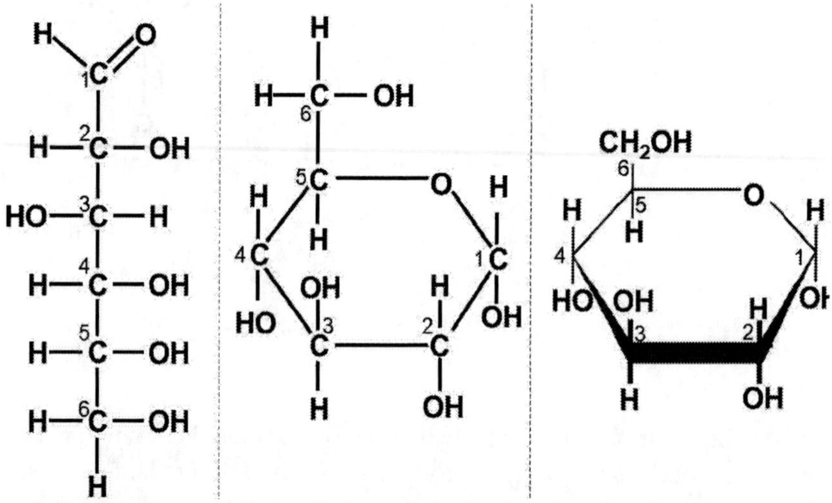

Figure 2.14. (a) and (b) Linear and ring shapes. The carbons are numbered 1 to 6, to facilitate the relation between the two shapes. (c) Abbreviated ring structure. Each corner represents carbon. The ring's thicker edge indicates that you are looking at the ring edge-on; the components attached to the ring lie above or below the plane of the ring.

Molecules in matter also interact with one another through electrical forces, known as *intermolecular forces* (that is, forces occurring between molecules), as depicted in Figure 2.16. These forces hold molecules to form, under appropriate temperature-pressure conditions, a solid (e.g., rocks, charcoal, table salt, sugar, and wood), liquid (e.g., water, gasoline, and vinegar), or gas (e.g., air and oxygen). These intermolecular forces allow individual molecules to move, but the molecules still confine gases, liquids, and solids to a specific volume. They determine many of the physical properties of a substance. For example, boiling points are a measure of how strong the attractive intermolecular forces are between molecules, liquids flow because the intermolecular forces between molecules are weak enough to allow the molecules to move around relative to one another, the intermolecular forces in a solid are so strong that they allow very little movement, and the molecules are so far apart that the intermolecular forces are effectively nonexistent and the molecules are completely free to move and flow independently. The strengths of intermolecular forces vary widely. For example, the transitions from solid to liquid, and to gaseous phases of a substance occur at high pressures and low temperatures because these conditions favor the increase of intermolecular forces.

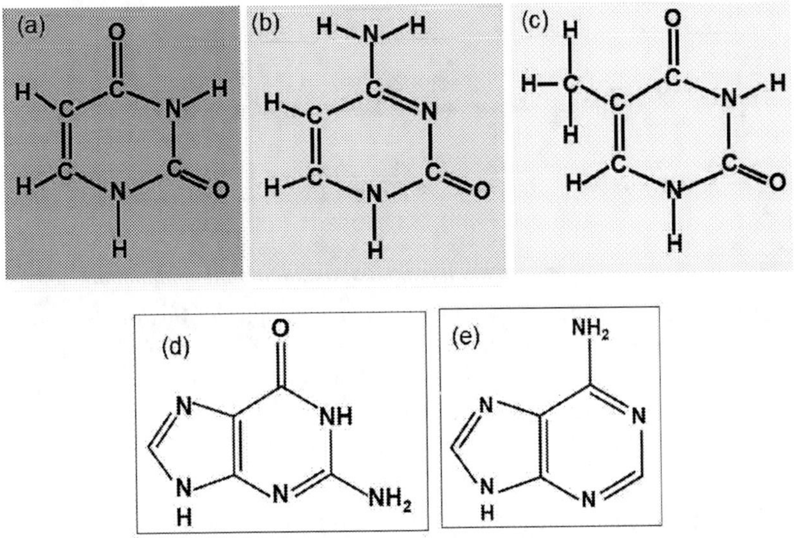

Figure 2.15. Ring shapes of five nitrogenous bases: (a) uracil, (b) cytosine, (c) thymine, (d) Guanine, and (e) adenine.

Note that usually the intermolecular forces between molecules are generally weak compared to the intramolecular forces that bond atoms together within a molecule.

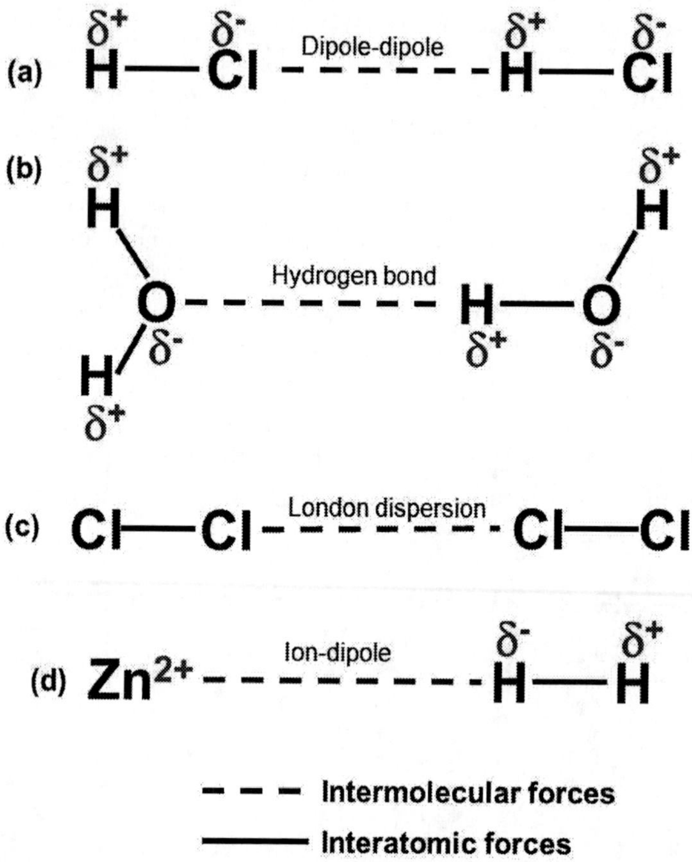

Figure 2.16. (a), (b), (c), and (d) show examples of intramolecular and intermolecular forces. Solid lines are intramolecular forces and dashed lines are intermolecular forces. The types of intermolecular forces are located just above the dashed lines. The slightly positive charges are denoted by δ^+ and the slightly negative charges are denoted by δ^-.

There are three major classes of intermolecular forces: ion-dipole forces, dipole-dipole forces, and London dispersion forces (see Table 2.14). They are all different manifestations of "opposite charges attract." The table actually includes four forces. The addition force is a particular case of dipole-dipole forces known as hydrogen bonding forces. It is the strongest of the dipole-dipole forces.

An ion-dipole force is a force between an ion and a dipole (polar) molecule, as depicted in Figure 2.16; a dipole molecule has one end slightly positive and the other slightly negative. The ion and the polar molecule align so that the positive and negative charges are next to one another, allowing for maximum attraction. The strength of the ion-dipole force is proportionate to ion charge. Cations would be attracted to the negative side of a polar molecule, and anions would be attracted to the positive side. Remember that an ion is an atom that has gained or lost one or more electrons and therefore has a negative or positive charge. A cation is an atom that has lost a valence electron and therefore has more positive protons than negative electrons, so it is positively charged. An anion is an atom that has gained a valence electron and is negatively charged.

Table 2.14. The types of intermolecular forces

Intermolecular forces	Occur between	Strength (kJ/mol)
London dispersion	Temporary or induced dipoles	0 – 3.5
Ion-dipole	Ion and polar covalent	12 – 75
Dipole-dipole	Polar covalent and polar covalent	3 – 13
Hydrogen bonding	Hydrogen atom and F, O, or N only	14 – 40

When the dipole of the molecule is not permanent, but rather a temporarily charged dipole, the ion-dipole force is sometimes called an ion-induced dipole. The temporary dipole is induced by the presence of the ion.

Dipole-dipole forces are intermolecular attractive forces that occur between two polar molecules (i.e., molecules that contain dipoles), such as water (H_2O), hydrogen chloride (HCl), and iodine monochloride (ICl). They occur when the slightly positive end of one polar molecule is attracted to the slightly negative end of another polar molecule (see Figure 2.16). These forces are always attractive and no electrons are shared between the two molecules. They are stronger than London dispersion forces.

Hydrogen bonding forces are a particular type of dipole-dipole forces that occur between a positive hydrogen on one molecule and a molecule containing the highly electronegative elements, either nitrogen (N), fluorine (F), or oxygen (O); that is, the attraction forces containing one of the very polar bonds of O−H, N−H, and F−H, as depicted in Figure 2.16. Hydrogen bonds are very important in biology because hydrogen bonds keep the DNA bases paired together, helping DNA maintain its unique structure.

London dispersion forces are the weakest intermolecular forces of all; i.e.,

$$\text{London dispersion} < \text{Dipole} - \text{dipole} < \text{Hydrogen bonding} < \text{Ion} - \text{ion} \,,$$

where the ion-ion force is an intramolecular force with a strength greater than 300 kJ/mol of interactions. They represent the attraction between two non-polar molecules due to induced (or instantaneous) dipoles in a molecule. That is, dipole molecules exist only momentarily in this case because they are caused by the correlated movements of the electrons of interacting molecules.

All molecules are attracted to other molecules via London dispersion forces because the electrons of any atom or molecule are in constant motion and at any moment in time, an atom or molecule can develop a temporary, instantaneous dipole. The presence of this dipole can, in turn, distort the electrons of a neighboring atom or molecule, producing an induced dipole. These two rapidly fluctuating, temporary dipoles thus result in a dispersion force. In general London dispersion forces are considered to be the weakest intermolecular force. However, they become very important for larger molecules. A big atom has many electrons moving around a large volume, so at any instant in time it is more likely to have an uneven distribution of electrons; thus, it is more likely to have a dipole. For hydrocarbons and other nonpolar molecules that lack strong dipoles, London dispersion forces are really the only attractive forces between molecules. Note that these temporary charges can also interact with permanent dipoles. These interactions are called dipole-induced dipole interactions. The partial charge of the polar molecule interacts with the electrons in the nonpolar molecule and induces them to move so they are not evenly distributed anymore, creating an induced dipole that can interact favorably with the polar molecule's permanent dipole.

Quiz: Identify the intermolecular forces in Figure 2.17.

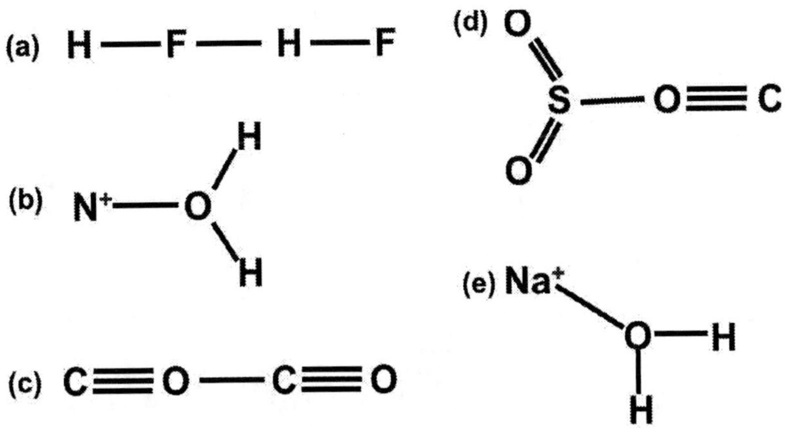

Figure 2.17. (a), (b), (c), (d), and (e) show examples of intramolecular and intermolecular forces.

Cohesive and adhesive forces. In summary, interactions between ions, dipoles, and induced dipoles account for many properties of molecules, including the condensation of gases to the liquid or solid states, stronger interactions that allow the solid and liquid states to persist to higher temperatures. Non-polar molecules show similar behavior due to London dispersion forces. The various intermolecular forces between identical molecules of a substance are examples of cohesive forces. The intermolecular forces of attraction between two different molecules are called adhesive forces. We discuss cohesive and adhesive forces in great detail in Ikelle (2022b).

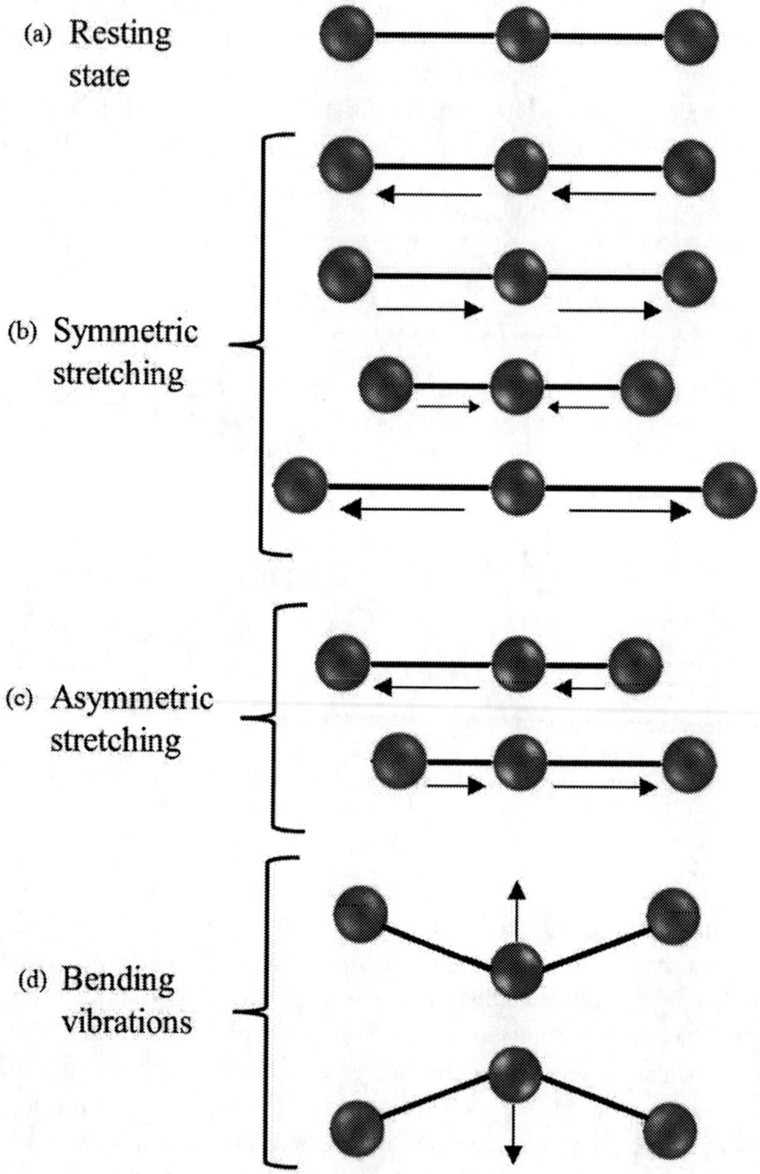

Figure 2.18. (a) A molecule at a resting state. (b) A symmetric stretch that allows molecules to move through space. (c) An asymmetric stretch that leads to an increase or decrease in bond length. (d) A bending vibration is a vibrational mode in which the angle between atoms changes. Only molecules with three or more atoms can experience bending vibrations.

> **Quiz:** What is the number of atoms in the following molecules: argon, hydrogen gas, nitrogen gas, oxygen gas, carbon monoxide, and carbon dioxide, and steam?

Molecular vibration and heat. Again, all matter is made of atoms and molecules that are constantly moving. A molecule can vibrate and move, depending on the number of atoms from which it is built. The bonds of atoms continually vibrate and move. They can vibrate in stretching motions or bending motions. The two-atom molecule (i.e., the diatomic molecule) can move in three ways. The entire molecule can move back and forth in a straight line—that is, each atom can move toward or away from the other atom, as illustrated in Figure 2.18. Stretching can be either symmetric or asymmetric. The two-atom molecule can remain in a fixed location, whereas its atoms rotate or vibrate around its center. Basically, a two-atom molecule can move linearly (translation), rotate, and vibrate. The three-atom molecule has even more possibilities. Complex molecules can move linearly (translation), rotate, and vibrate—all simultaneously, including bending vibrations—and vibrational modes where the angle between atoms changes. Water molecules and carbon dioxide, for example, can rotate and vibrate in a number of ways. Note that a polyatomic molecule is a single entity composed of at least three covalently-bonded atoms (for example, H_2O and CO_2). Note also that linear motions include symmetric and asymmetric stretching.

In summary, the macroscopic properties of matter (solids, liquids, and gases) are directly related to their molecular arrangements and to their intermolecular forces. Solids are ordered molecular arranged and have strong intermolecular forces; fluids are partially-ordered molecular arranged and have medium intermolecular forces; and gases are disordered molecular arranged and have weak intermolecular forces.

2.3. Living Things: Organic Compounds

Living things, also called organisms, include animals, plants, bacteria, archaea, and fungi (singular: fungus). Non-living things include water, air, and rocks. Living things distinguish themselves from non-living things by the fact that they can reproduce, grow, consume/eat, get rid of waste, respond to the environment, move without outside help (non-living things move only under external forces), and die. For example, animals move to find food and keep away from predators and plants move to face the light.

At one time it was thought that the makeup of living things was totally different from that of non-living things. A clue to this difference was how they responded to heat. For example, think of a heating coal sample. Coal is organic matter that contains mostly carbon. When it is heating in domestic stoves and furnaces, it turns into a smoky flame and it does not return to its original condition. However, heat up lead, for example, and it will melt, but upon cooling it returns to its original condition. So there seems to be a difference.

(a)

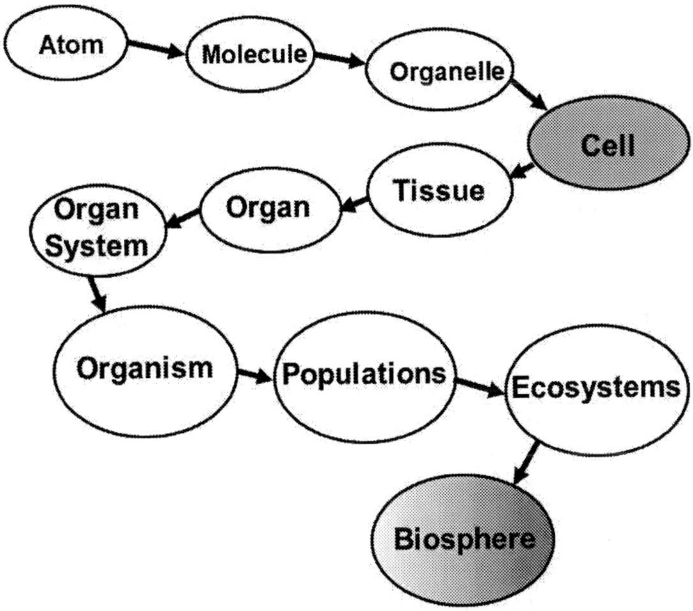

(b)

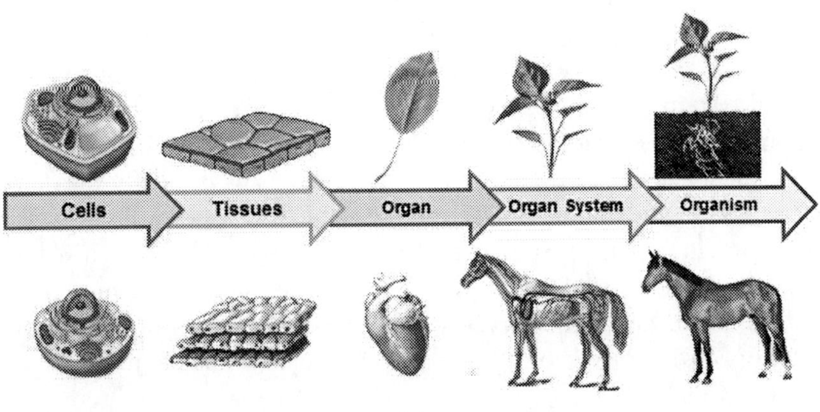

(c)

Figure 2.19. (a) The three groups of living things are: microbes (bacteria and fungi), plants, and animals. (b) The living world can be organized into different levels. Cell is the basic unit of structure and function of all living things. Ecosystem is all interacting communities of organisms and biosphere is the global ecosystem, including all the Earth's regions that can support life (land, air, water). (c) Plant and animal organizational scheme. They have cells, tissues, organs, and organ systems.

It was also believed that living things had a "life force" in them, which made the chemicals in living things unique. However, in 1828, a chemist named Friedrich Wöhler accidentally created urea, a compound that mammals produce to get rid of excess nitrogen. Urea is secreted in their urine. Wöhler created it using inorganic (non-living) salts. In other words, he created chemicals found in the body from chemicals found in the ground or the atmosphere (non-living sources). Thus, nowadays organic compounds are not defined as only those compounds from organisms, but compounds based on carbon.

2.3.1. Structures of Living Things

Living things are highly organized and structured, increasing in complexity from atoms to molecules and then in sequence to organelles, cells, tissues, organs, organ systems, organisms, and populations (see Figure 2.19). All living and non-living things, as all matter, are made of atoms. A molecule is a group of atoms that are found to exist together as one species, as described earlier. All living and non-living things also contain molecules. Cells are the smallest independent living things—that is, cells are only found in living things and all lives start with a cell. Some organisms are made of a single cell. Other organisms, especially larger organisms, contain large numbers of cells. In these cases, cells similar in structure and function come together to form tissues. For example, bone cells in our bodies form bone tissue and blood cells are part of blood tissue, a liquid tissue transporting food and oxygen throughout the body. Tissues form organs. Your heart, for example, is an organ made up of muscle tissue, blood tissue, and nerve tissue—the heart starts beating about 22 days after conception. The brain, stomach, kidneys, and skin are other organ examples. Plants have organs such as leaves, stems, and roots. Organs form organ systems, and organ systems combine to form an organism.

All the organisms of a species living within a specific area are collectively called a population. For example, all of the palm trees in a given rural area represent the population of palm trees in this area. Different populations may live in the same specific area. For example, the area may also include a population of guava trees and a population of insects. Notice that the requirement that a living thing is made of one or more cells makes viruses non-living things. Basically, viruses are not made of cells. New viruses are made by invading and hijacking the reproductive mechanism of a living cell; only then can viruses obtain the materials they need to reproduce.

We here focus on two aspects of living things: macromolecules, which are the molecules inside living cells, and the structures of the two types of cells found in living things.

2.3.2. Microbes

Because the understanding of microbes (or microorganisms) is critical to detecting, diagnosing, treating, and even preventing infectious diseases, and in addition to the fact that students' exposure to microbes is quite limited in pre-college education, let us elaborate a little bit on the characteristics of microbes. They are very tiny organisms that are everywhere in our environment; microbes are the smallest living entities in the world. Moreover, most of them are colorless, transparent, and difficult to see with the naked eyes. They are observed under a microscope. Without microbes, the recycling of essential nutrients on Earth would halt. Microbes are grouped into archaea, bacteria (also known as eubacteria), and eukarya. They are grouped into archaea, bacteria, and eukarya. Major groups of microbes are shown in Table 2.15. Each of these groups contains several sub-groups as we will discover as we go through the book.

Let us talk a little bit more about bacteria. The three basic shapes of bacteria (Figure 2.20) are (1) round or spherical bacteria (cocci), (2) rectangular or rod-shaped bacteria (bacilli), and (3) curved and spiral-shaped bacteria (spirilla). Some of the bacteria cause harm to humans and others provide certain benefits to humans. They flourish inside your yogurt, in your kitchen, in Antarctic ice, in hot springs and volcanic (hydrothermal) vents, kilometers deep in the ground.

The five-volume set of books entitled *Bergey's Manual of Systematic Bacteriology* contains descriptions of more than 5,000 validly named species of bacteria. Some studies suggest that this number represents only less than 1 percent of the number of bacterial species in nature.

Table 2.15. The three major broad groups of microbes and their subgroups: each subgroup contains multiple spices. For example, cyanobacteria have 78 species and proteobacteria have 1,644 species. Bacteria and archaea are both prokaryotes but differ enough to be placed in separate domains

Archaea	Bacteria	Eukarya
Euryarchaeotes	Proteobacteria	Baker's yeast (Saccharomyces cerevisiae)
Crenachaeotes	Chlamydias	Fission yeast (Schizosaccharomyces pombe)
Nanoachaeotes	Spirochaetes	Pathogenic yeast
Korachaeotes	Cyanobacteria	Eukaryotes

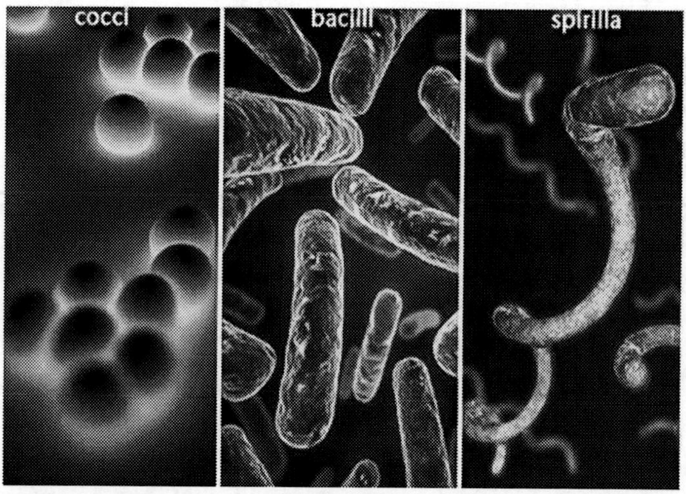

Figure 2.20. The three basic shapes of bacteria are (i) round or spherical bacteria (cocci), (ii) rectangular or rod-shaped bacteria (bacilli), and (iii) curved and spiral-shaped bacteria (spirilla).

Archaea are considered the most primitive form of living microbes on Earth. They have been found in all types of habitats, including in humans, deep-sea vents, saline waters, hot springs, below petroleum deposits, acidic, alkaline, hot, cold, salty, and/or very high pressure environments. They are found in rocks 7 kilometers below the Earth's surface, at temperatures as high as 130 degrees Celsius (266 degrees Fahrenheit) and as low as −17 degrees Celsius (1 degree Fahrenheit). Most other living things will be killed under these conditions. Yet, some archaea adapt to extreme environments. Microbial eukaryotes include algae, fungi, and helminths. We will discuss the differences between these microbe groups later after we have introduced macromolecules and cells of living things.

The classification of microbes into bacteria, archaea, and fungi is just the first, largest, and most inclusive group under which microbes are classified. They are known as domains. They are further divided into subgroups mainly based common properties (similarities), and how closely related they are to a common ancestor (evolutionary relationship). Subgroups include five kingdoms that are further split into *phylum*, *class*, *order*, *family*, *genus*, and *species*, which is the smallest group. For example, a genus contains one or more species. Intermediate minor rankings are not shown. Five major kingdoms have been described and include prokaryota (e.g. archaea and bacteria), protoctista (e.g. protozoa and algae), fungi, plantae, and animalia. The science of classifying organisms is called taxonomy and the groups making up the classification hierarchy are called taxa. The names assigned to microbes are in Latin. Classification of microorganisms has been largely aided by studies of fossils and recently by DNA sequencing.

2.3.3. Macromolecules: Large Organic Molecules

A molecule is made of two or more atoms, even thousands of atoms, that are held together by strong chemical bonds. Molecules containing thousands of atoms are known as macromolecules. Molecules of sugar, fats, oils, insulin, and egg white are examples of macromolecules. They are mostly organic—that is, their chemical compositions contain the carbon atom, which has the ability to form four strong covalent bonds and can also react with other atoms to form long chains of molecules associated with macromolecules. In addition, they may contain hydrogen (H), oxygen (O), nitrogen (N), sulfur (S), phosphorous (P), and other minor elements. Moreover, macromolecules almost always synthesize living things, from bacteria to elephants.

Box 2.4. Pathogenicity

Some bacteria enter our bodies through skin (e.g., a cut, abrasion, and puncture) and mucous membranes (e.g., gastrointestinal tract, respiratory tract, or urogenital tract) and cause diseases such as plague, tuberculosis, anthrax, and malaria. These bacteria are called pathogens. They represent a small percentage of the microbial world. Moreover, archaea are non-pathogens. In fact, our life would not be possible without some prokaryotic microbial species. For example, they are used in the fermentation of food, the growth of plants, the development of novel antibiotics, and have positive effects on our ecosystem. The bacteria that inhabit our skin and gastrointestinal tract can protect us from infectious diseases, help us digest our food, produce some of our vitamins and other nutrients, regulate our moods, and control weight by affecting our food choices. Moreover the protective mechanisms in our body allow us to cough, urinate, and sneeze some of the effects of microbes. An infectious disease occurs when there is a parasitic relationship between a host and microbes (i.e., microbes start living off of the host by multiplying).

Macromolecules are organized into hundreds of smaller molecules, known as monomers, which are linked together into long chains (Figure 2.21). Hence, macromolecules can be organized based on the smaller molecules that form their chains. These groups are carbohydrates, lipids (also simply called fats), proteins, and nucleic acids (see Table 2.16). Carbohydrates and lipids are made of only carbon, hydrogen, and oxygen (CHO). Proteins are made of carbon, hydrogen, oxygen, and nitrogen (CHON). Nucleic acids such as DNA and RNA contain carbon, hydrogen, oxygen, nitrogen, and phosphorus (CHONP). We here just briefly describe these macromolecules. We will further describe them in later chapters as we improve our understanding of living things

Table 2.16. Elements that make the various kinds of macromolecules. The immense diversity of life is based on a few elements. Also included here are carbohydrate-, protein-, and lipid-rich food

	Elements	Examples
Carbohydrates	H C O	Sugars 
Lipids	H C O	Fats 
Proteins	H C O N S	Enzymes 
Nucleic acids	H C O N S P	DNA and RNA

Carbohydrates (with glucose as the key subunit) are important in the functioning of humans, animals, and plants. They provide energy to the body, particularly through glucose, a simple sugar. The general chain of carbohydrates is $(CH2O)_n$, where n is the number of carbons. The ratio of carbon to hydrogen to oxygen in carbohydrate molecules is 1:2:1. The component carbon (*carbo*) and the component water (*hydrate*) give the name to this group of organic molecules. Carbohydrates store energy in their many carbon-hydrogen (C–H) bonds. Simple sugars like glyceraldehyde and its isomer, dihydroxyacetone, contain three carbon atoms (i.e., $n = 3$); ribose and its isomer, ribulose, are five-carbon sugars (i.e., $n = 5$); and fructose (found in fruit), and its isomers, galactose (part of lactose, or milk sugar), and glucose are six-carbon sugars that correspond to $n = 6$ (i.e., $(CH2O)_6$ or $C_6H_{12}O_6$). Figure 2.22 shows three of these carbohydrates. Note that these carbohydrates have isomers. For example, fructose, galactose, and glucose are isomers. As discussed earlier, the composition, as well as the arrangements of atoms affects the chemical properties of these molecules. Your taste buds can tell the difference, as fructose tastes much sweeter than glucose, despite the fact that both sugars have the same chemical composition (i.e., the same molecular formula).

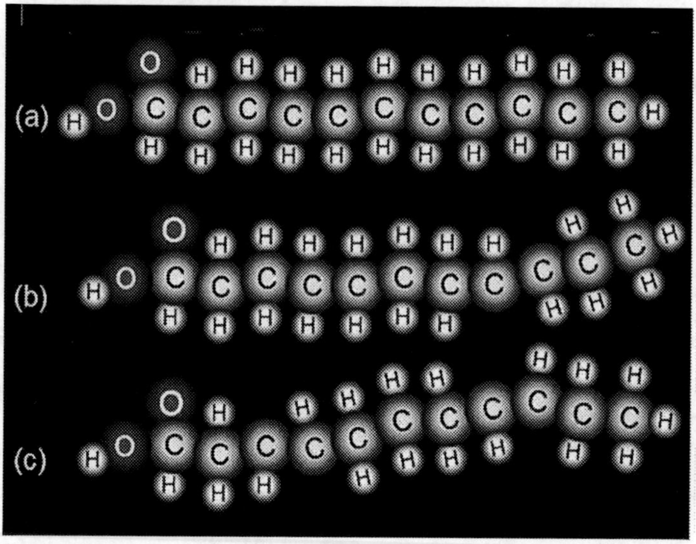

Figure 2.21. Fatty acids—(a) a fully saturated fatty acid with no carbon-carbon double bonds, with melting points at 70° C, (c) and (b) unsaturated fatty acids with melting points at 13° C and −17° C, respectively.

Carbohydrates are generally grouped into three subtypes: monosaccharides, disaccharides, and polysaccharides. *Monosaccharides*, or simple sugars, include glucose, galactose, and fructose. These simple sugars combine to make *disaccharides* [double sugars such as sucrose (table sugar), which is composed of the monomers glucose and fructose; lactose, which is naturally found in milk; maltose (malt sugar)] and *polysaccharides* (long chains like starch, cellulose, chitin, and glycogen)—disaccharides often play a role in the transport of sugars, as we will discuss in later chapters.

The relative sweetness of sugars to sucrose is as follows: 1.73 for fructose, 0.74 for glucose, 0.22 for galactose, 1.00 for sucrose, 0.32 for maltose, and 0.16 for lactose.

Lipids (with glycerol and fatty acids as the key subunits) are the building blocks of many hormones and are an important constituent of the plasma membrane. They have a very high proportion of nonpolar carbon-hydrogen (CH) bonds, as we can see in Figure 2.23, which makes them insoluble in water because carbon-carbon or

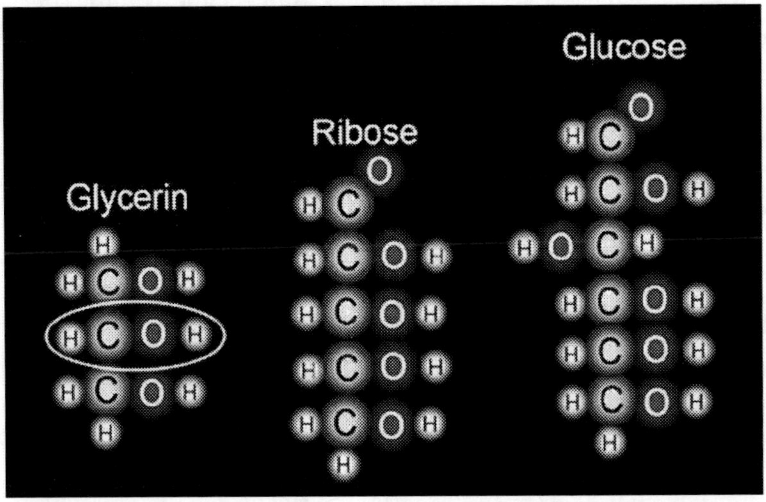

Figure 2.22. The molecular structure of glycerin ($C_3H_8O_3$) and two sugars [ribose ($C_5H_{10}O_5$) and glucose ($C_6H_{12}O_6$)].

carbon-hydrogen bonds nonpolar—the contribution of the polar head (COOH) is negligible, in the overall polarity of lipids, compared to that of the long, very nonpolar tails. By comparing Figures 2.21 and 2.22, we have C−H (nonpolar) in lipids instead of C−OH bonds in carbohydrates. When placed in water, many lipid molecules will spontaneously cluster together and expose polar groups. This spontaneous assembly of lipids is of paramount importance to cells, as it underlies the structure of cellular membranes. Another function in cells performed by lipids is that cells store energy for long-term use in the form of lipids called fats.

The most familiar lipids are fatty acids, to which an acidic carboxyl group is attached, hence the name *fatty acid.* The number of carbons in fatty acids may range from four to 36; the most common are those containing 12-18 carbons. Fatty acids may be saturated or unsaturated. In a fatty acid chain, if there are only single bonds between neighboring carbons in the hydrocarbon chain (Figure 2.21a), the fatty acid is saturated with hydrogen; in other words, the number of hydrogen atoms attached to the carbon skeleton is maximized. Animal fats with stearic acid and palmitic acid contained in meat, and fat with butyric acid contained in butter, are examples of saturated fats. Other examples of saturated fatty acids include molecules of coconut, palm, lard, and yellow grease. When the hydrocarbon chain contains a double bond, the fatty acid is an unsaturated fatty acid. If there is one double bond in the molecule, as in Figure 2.21b, then it is known as a monounsaturated fat (e.g., olive oil), and if there is more than one double bond, as in Figure 2.21c, then it is known as a polyunsaturated fat (e.g., canola oil). The bends in Figure 2.21b and Figure 2.21c are caused by double bonds. They prevent fatty acids from packing tightly. Unsaturated fats help to improve blood cholesterol levels, whereas saturated fats contribute to plaque formation in the arteries, which increases the risk of a heart attack. In common language, solid fatty acids are called fats and liquid ones are called *oils*.

A triacylglycerol molecule (also known as a triglyceride) is a combination of glycerol ($C_3H_8O_3$) and three fatty acids (see Figure 2.23). During this covalent bond formation, three water molecules are released. In human medicine, a common test for heart disease risk factors is the measurement of triglyceride levels in the blood. Also, lipid are the most common type of lipids found in our body fat tissues and in our diet, and serve mostly as an energy store. Phosphatidates are glycerophospholipids with two fatty acid groups.

Proteins (with amino acids as the key subunits) are one of the most abundant macromolecules in living systems. They make up the bodies of organisms and protect the body from foreign pathogens. Each cell in a living system contains thousands of proteins, each with a unique function. Enzymes, which living cells produce, are catalysts in biochemical reactions (like digestion) and are usually complex or conjugated proteins, as discussed in Ikelle (2022a).

Proteins are all, however, amino acid polymers (i.e., macromolecules built by linking together a large number of small, similar chemical subunits, called amino acids, like railroad cars coupled to form a train) arranged in a linear sequence with a central carbon atom bonded to an amino group (NH_2), a carboxyl group ($COOH$), and to a hydrogen atom (Figure 2.24). The key elements of an amino acid are carbon (C), hydrogen (H), oxygen (O), and nitrogen (N), although other elements are found in the side chains of certain amino acids. The generic formula is $H_2NCHRCOOH$ in most cases where R is an organic substituent often called a side chain and is specific for each amino acid (Nelson and Cox, 2005).

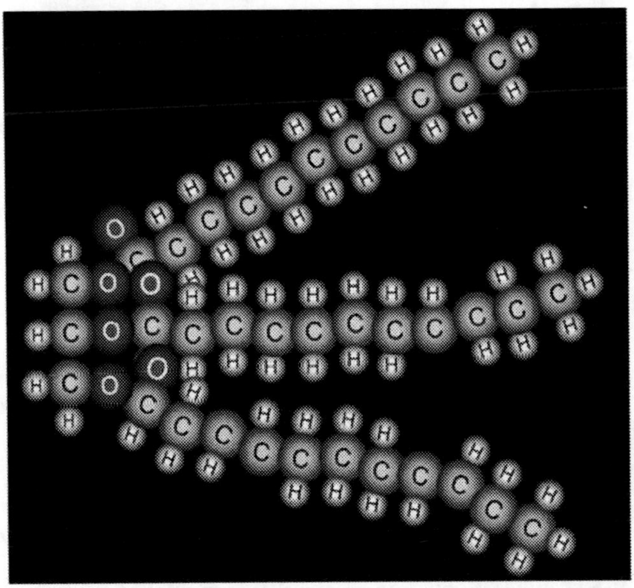

Figure 2.23. The structure of a triacylglycerol lipid–it formed by conjugation of a glycerol to three fatty acids.

In Figure 2.24b, we have $R{=}CC_2H_4NH_2O$ and this amino acid is known as glutamine; R=H and this amino acid is known as glycine; $R{=}CH_3$ and this amino acid is known as alanine; $R{=}CH_2HS$ and this amino acid is known as cysteine (see also Ikelle, 2022a). There are 20 common amino acids present in proteins. Nine of these are essential amino acids for humans because the human body cannot produce them and we obtain them from our diet. Insulin is a protein hormone that helps regulate blood glucose levels. Others may be conditionally essential for certain ages or medical conditions. Essential amino acids may also differ between species.

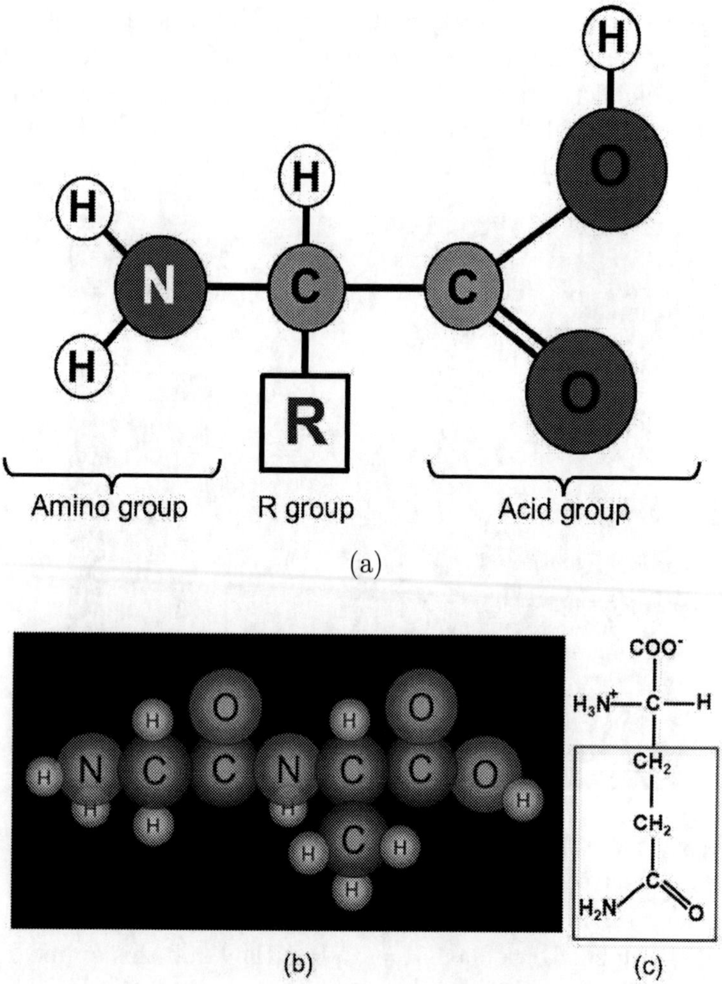

Figure 2.24. (a) the molecular structure of amino acids—amino acids have a central asymmetric carbon to which an amino group, a carboxyl group, a hydrogen atom, and a side chain (R group) are attached. (b) Glutamine—one of the 20 amino acids commonly found in proteins. (c) The R group of glutamine in the red box.

Because of their biological significance, amino acids are important in nutrition and are commonly used in nutritional supplements, fertilizers, feed, and food technology. Industrial uses include the production of drugs, biodegradable plastics, and chiral catalysts. Moreover, amino acids are central to some theories of the origin of life. These theories contend that amino acids were among the first molecules formed on the early Earth.

Nucleic acids (with nucleotides as the key subunits) are the most important macromolecules for the continuity of life. They carry the genetic information in a cell and are made of nucleotides, which contain: a phosphate group (i.e., a molecule R attached to $PO_4{}^{3-}$), a five-carbon sugar, and a nitrogenous base; i.e.,

$$\text{nucleic acids} \xrightarrow{\text{broken down into}} \text{nucleotides} \xrightarrow{\text{broken down into}} RPO_4{}^{3-}$$

$$+\text{nitrogen base} + \text{pentose sugar}$$

The phosphate acid (H_3PO_4) is included in this phosphate group. In this case, R is H_3. The nitrogenous bases (bases containing nitrogen atoms) found in nucleotides are adenine ($C_5H_5N_5$), guanine ($C_5H_5N_5O$), thymine ($C_5H_6N_2O_2$), cytosine ($C_4H_4N_2O_2$), and uracil ($C_4H_4N_2O_2$), which can be considered to be demethylated thymine. Adenine is often represented by the capital letter A, guanine by the capital letter G, thymine by the capital letter T, cytosine by the capital letter C, and uracil by the capital letter U. The three main types of nucleic acids are deoxyribonucleic acid (DNA), ribonucleic acid (RNA), and adenosine triphosphate (ATP).

The nucleotide ATP ($C_{10}H_{16}N_5O_{13}P_3$) is a direct and rapid energy source for most cellular activities, such as muscle contraction, nerve impulse propagation, condensate dissolution, and chemical synthesis. It consists of the nitrogen base adenine, the sugar ribose, and three phosphates, as depicted in Figure 2.25. The part that is really critical is the phosphorous part—the triphosphate. If you remove just one of these phosphate groups from the end, the energy is sufficient to liberate about 7,000 calories per mole, about the same as

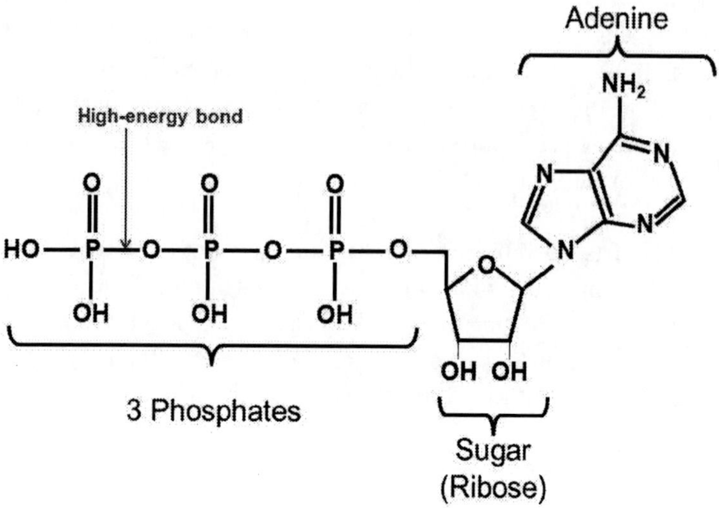

(a)

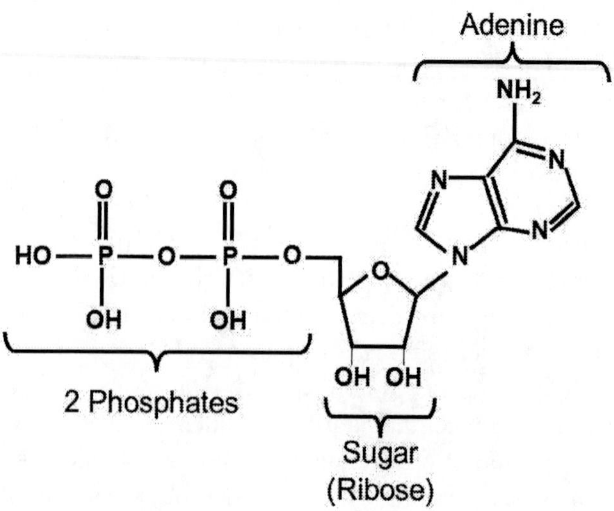

(b)

Figure 2.25. (a) ATP is another chemical compound we need to become familiar with. It is composed of a nitrogenous base (an adenine molecule) bonded to a ribose (sugar) molecule and three phosphate groups. (b) Adenosine diphosphate (ADP), along with ATP, is essential to the flow of energy in living cells. It contains one less phosphate group than does ATP.

the energy in a single peanut. When the ATP lopse, one of its phosphorous groups, it forms from an adenosine diphosphate ADP). Through the use of food, living things convert the ADP back to ATP, so that the energy is again available to do needed work, as discussed in Ikelle (2022a).

As illustrated in Figure 2.26, DNA and RNA are nucleotides. The nitrogenous bases adenine (A), guanine (G), cytosine (C) and thymine (T) occur in DNA, and A, G, C, and uracil (U) occur in RNA—the pentose sugar is ribose in the case of RNA and deoxynucleotide for DNA—both use phosphoric acid (H_4PO_4).

DNA is the genetic material in all living organisms, ranging from single-celled bacteria to multicellular mammals. DNA contains all the instructions for making every protein needed by a living thing. This generic information is stored as a code made up of adenine, guanine, cytosine, and thymine. Human DNA consists of about 3 billions of these bases, which are linked covalently into a polynucleotide chain, as shown in Figure 2.27. This chain forms a spiral called a double-helix with two strands of DNA running in opposite directions to each other. The structure of the double helix is somewhat like a ladder, with the base pairs forming the ladder's rungs and the sugar and phosphate molecules forming the vertical side pieces of the ladder. The order, or sequence, of these bases determines the information available for building and maintaining an organism, similar to the way in which letters of the alphabet appear in a certain order to form words and sentences.

RNA copies and transfers this genetic information so that proteins can be made. While the structure of DNA is a double-helix, RNA is typically single-stranded and comes in various forms. The single-stranded structure of RNA allows this molecule to fold back on itself and form various stable secondary structures as necessary. RNA also tells our bodies what to do and then follows through with those instructions. There are three types of RNA, messenger(mRNA), transfer(tRNA), and ribosomal(rRNA). Messenger RNA is used because DNA cannot "speak" directly to proteins. mRNA takes the information into the cytoplasm (an organelle of the cell that we will describe later), where the information is read, tRNA takes the nucleic acids and decides the proper sequence of amino acids for protein synthesis. rRNA is an enzyme that helps in protein synthesis, and this enzyme helps to produce peptide bonds in a polypeptide.

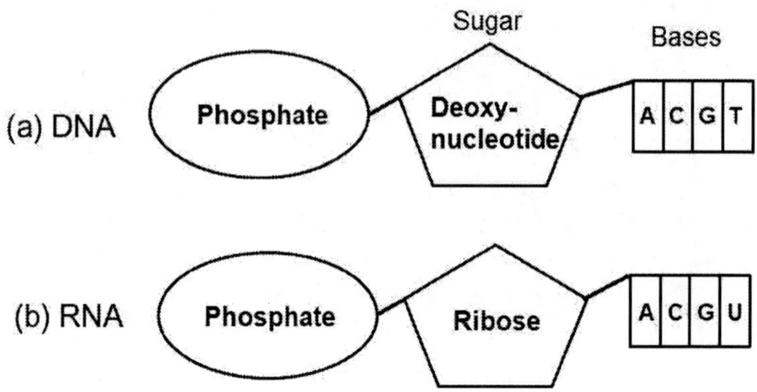

Figure 2.26. (a) The DNA nucleotide consists of phosphoric acid, deoxyribose, and A, C, G, or T bases. All organisms' DNAs has the same bases. (b) The RNA nucleotide consists of phosphoric acid, deoxyribose, and A, C, G, or U bases.

Biomarcomolecules. Not all macromolecules are pure nucleic acids, pure protein, pure carbohydrate, or pure lipid. Conjugated proteins are protein molecules combined with another kind of macromolecule. For example, proteins combine with lipids to form lipoproteins. Lipoproteins are found in the blood, where they act as carriers for less-soluble molecules such as cholesterol. Proteins combined with carbohydrates form glycoproteins. Lipids bound to carbohydrates become glycolipids. Glycoproteins and glycolipids are important components of cell membranes.

2.3.4. Cell Types and Cell Membranes

Again a cell (e.g., nerve cells, skin cells, muscle cells, blood cells, and brain cells) is the smallest and basic microscopic unit of living things capable of existing independently. Cells contain inorganic compounds (water and minerals) and organic compounds (proteins, carbohydrates, nucleic acids, and lipids). The motion of charged particles (e.g., electrons and ions) in human bodies drives communication between cells, on the one hand, and between cells and the environment, on the other, as we will discuss in the later chapters of this book series. They are central to the functioning of our five main senses (smell, touch, sight, taste, and hearing) and the communication of our brains and minds with the rest of our bodies. Without the motion of these charged particles in our bodies, you could not even read this chapter or stand up straight.

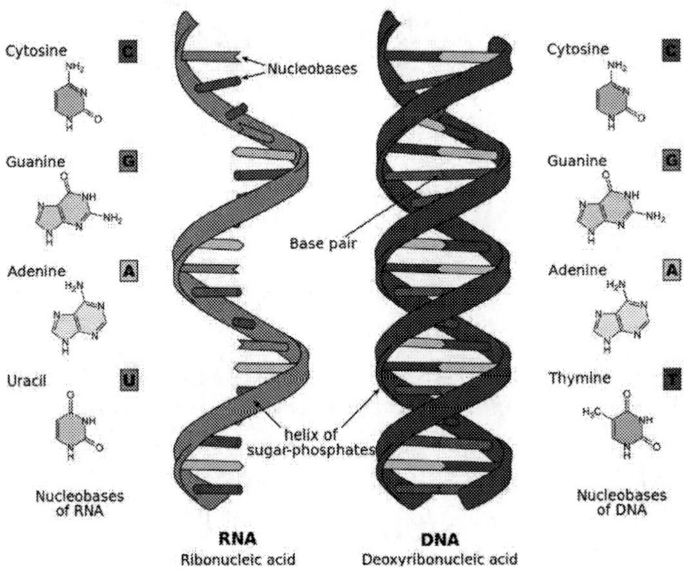

Figure 2.27. In modern cell illustrations, RNA (light blue, center) is made from a DNA template (purple, left) to create proteins (green, right). DNA and RNA share three nucleotides in their composition (cytosine, guanine, and adenine), and they differ in uracil (found only in RNA) and thymine (found only in DNA). The RNA is single strand, whereas DNA is double strand [source: Art by Open Learning Initiative CC-BY-NC-SA].

Quiz: Figure 2.28 shows sea anemones. Is this creature an animal or a plant? Justify your answer.

Figure 2.28. A sea anemone is an example of living things. Sea anemones are found throughout the world's oceans, most of the time in shallow tropical waters. Some species of anemone live at depths more than 10,000 meters below sea level.

Prokaryotic and eukaryotic cells. Cells fall into one of two broad categories: prokaryotic and eukaryotic (see Figure 2.29). Bacteria and archaea are single-celled prokaryotes. All other living organisms on Earth are eukaryotes; animal cells, plant cells, fungi, and protists are eukaryotes. Eukaryotes can be either single-celled (e.g., yeast) or multicellular (e.g., mold). Prokaryotes and single-celled eukaryotes are typically referred to as microorganisms or microbes.

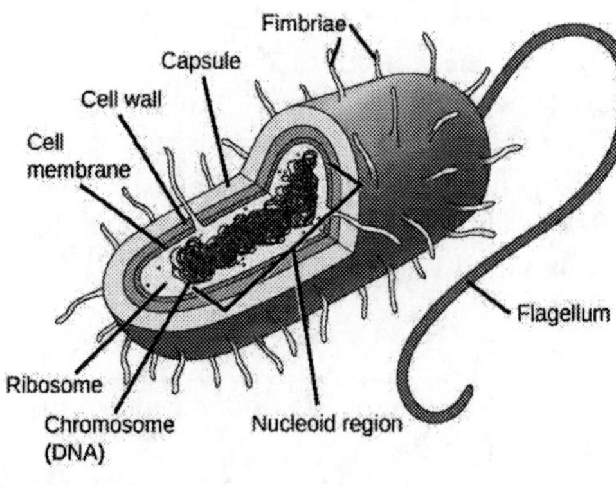

(a)

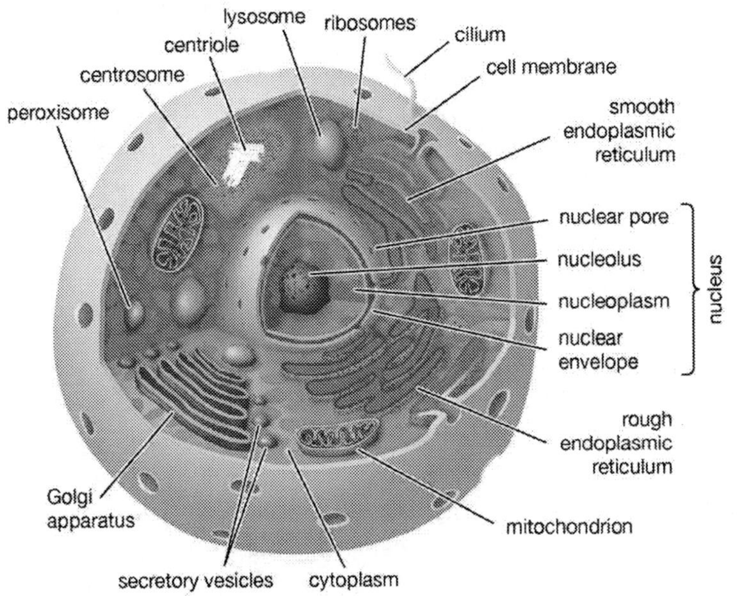

(b)

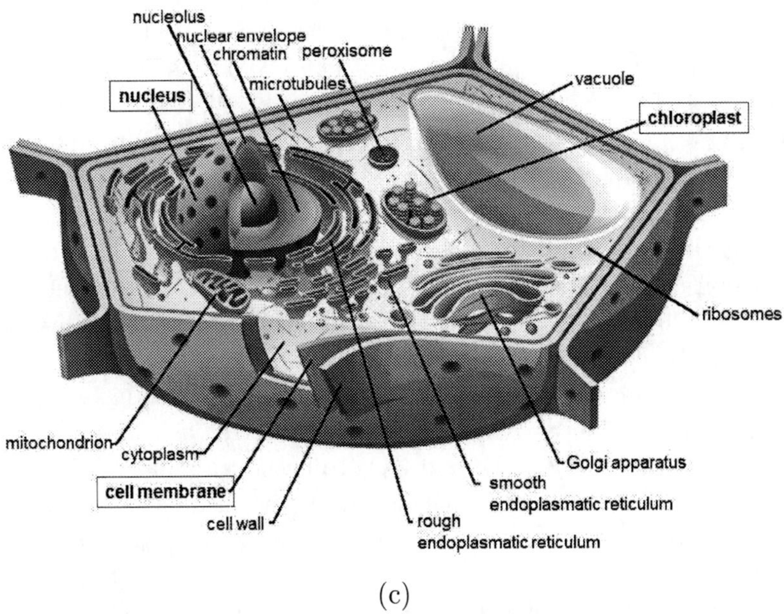

(c)

Figure 2.29. Artistic illustrations of (a) prokaryotic cells, (b) eukaryotic (animal) cells with their key organelles, and (c) eukaryotic (plant) cells with their key organelles.

Prokaryotic cells were first on Earth and for billions of years were the only form of life on Earth. The oldest fossils of bacteria and associated prokaryotic cells are estimated to be about 3.7 billion years old. Eukaryotic cells appeared on Earth about 1.4 billion years ago, long after prokaryotic cells. However, they are much more complex than prokaryotic cells.

All these cells have three main parts: the plasma membrane (outer covering of the cell) that acts as a wall between the cell's inside world and what's going on outside, the cytoplasm (an inner liquid filling) that supports all metabolic functions within a cell, and the nucleus that contains genetic information (DNA), as depicted in Figure 2.29. The plasma membrane (often called the cell membrane) is a thin flexible barrier that separates the inside of the cell from the environment outside the cell and regulates what can pass in and out of the cell. Macromolecules found in the membrane include cholesterol, proteins, glycolipids, and glycoproteins. The cytoplasm is where most functions of the cell are carried out. It contains the DNA (genes) that code for proteins necessary for the cell to function and ribosomes, which are used to build proteins. In eukaryotic cells, it looks a bit like mixed-fruit jelly, where the watery jelly is called the cytosol; and the different fruits in it are called organelles. The cytosol also contains many molecules and ions involved in cell functions. Different organelles also perform different cell functions and many are also separated from the cytosol by membranes. Mitochondria make the cell's energy—the more energy the cell needs, the more mitochondria it has. The largest organelle in eukaryotic cells is the nucleus. It is separated from the cytoplasm by a nuclear envelope (membrane). It houses the cell's DNA. The nucleus' nuclear envelope has many nuclear pores through which mRNA, and proteins can pass.

The cell membrane is a complex barrier separating every cell from its external environment. It is characterized as semi-permeable because it allows H_2O, O_2, and CO_2 to pass into and out of the cell and preventing the passage of some to maintain internal conditions. In other words, the cell membrane acts more like a gatekeeper that controls which molecules can enter and leave the cell. It consists of a phospholipid lipid bilayer (two layers of fat and particles can not

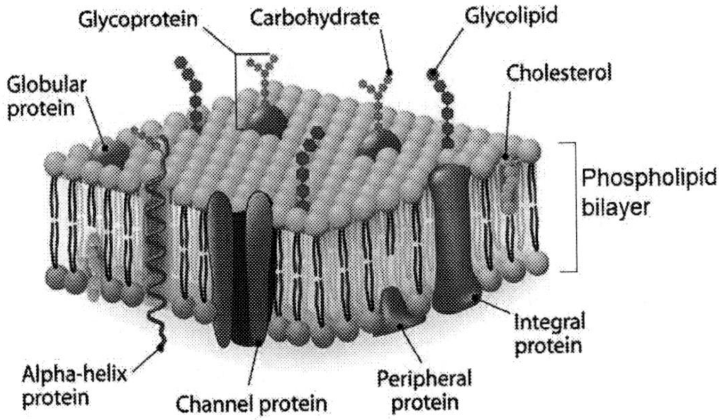

(a)

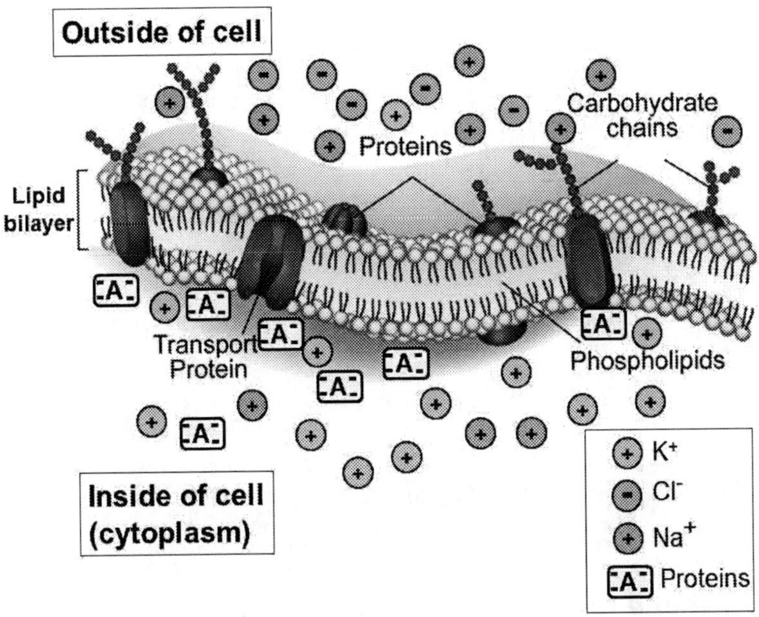

(b)

Figure 2.30. (a) A cell membrane is a phospholipid bilayer with embedded proteins that separate the cell from its environment. Membranes regulate the movement of materials into and out of the cell and facilitate electrical signaling between them. (Adapted from https://microbenotes.com/membrane-carbohydrate/). (b) A detailed structure of the cell membrane with an example of ion distribution. Membranes are semi-permeable because they allow certain molecules but not others to enter into the cell.

pass through this lipid bilayer) with embedded proteins, as illustrated in Figure 2.30. The lipid bilayer is composed of phospholipids and a phospholipid is a lipid molecule composed of two fatty acid chains, a glycerol backbone, and a phosphate group.

Note that prokaryotic cells do not have organelles, including a nucleus. In contrast, they have flagella for swimming and pili for sticking to things. In fact, flagella for swimming and pili are the most recognizable extracellular bacterial cell structures and are responsible for bacterial motility (i.e., movement). Figure 2.29 and Table 2.17 capture the differences and similarities between eukaryotic and procaryotic cells.

Table 2.17. Differences between eucaryotic and procaryotic cells

	Prokaryotic	Eukaryotic
Examples	Bacteria, yeast	All other organisms, humans
Size (diameter)	0.2–2 μm	10–100 μm
Cell membrane	Present	Present
Cytoplasm	Present	Present
Ribosomes	Present	Present
Nucleus	Absent (no nucleus)	Present
Membranous organelles	Absent	Present
DNA	DNA is bunched up in the center of the cell (Single circular chromosome)	DNA found in the nucleus (Multiple linear chromosomes)
Chromosomes	single chromosome	> one chromosome
Cell division	binary fission (simple division)	mitosis (fission meiosis or budding) and meiosis

Animal and plant cells. While animals and plants may look very different, their cells (i.e., building blocks) have a lot of similarities. However plant cells have three extra organelles, the vacuole, chloroplast, and cell wall, as described in Table 2.18 and Figure 2.29.

It has been estimated that humans contain about 40 trillion (40×10^{12}) cells (e.g., Bianconi et al., 2013). The human brain accounts for around 86 billion of these cells. There are hundreds of different types of cells in the human body, which vary in shape and size (average diameters are from 0.1 μm to 100 μm).

Table 2.18. Similarities and differences between plant and animal cells

Organelle	Plants	Animals	Function
Cell shape	Square or rectangular	Irregular or round	
Cell wall	Present	Absent	support (grow tall) and protection.
Chloroplast	Present	Absent	photosynthesis
Central vacuole	Present	Absent	its liquid content provides provides turgor pressure
Centrosome	Absent	Present	some role in cell division
Lysosomes	Absent	Present	some role in cell division
Nerve cells	Absent	Present	Send and receive signals to and from the brain and nervous system

Microbes: As described above, microbes (or microorganisms) are prokaryotes, single-celled organisms lacking a true nucleus and other membrane-enclosed organelles, and single-cell eukaryotics. Prokaryotic microbes were generally considered synonymous with bacteria, and archaea were referred to as archaebacteria until 1977, when sufficient differences between archaea and bacteria were discovered and archaea stopped being referred to as bacteria. Nowadays, microbes are grouped into archaea, bacteria (also known as eubacteria), and eukarya (see Figure 2.31). Archaea and bacteria differ in cell-wall characteristics and membrane lipid composition. More importantly, archaea contain hundreds of genes with no counterparts in bacteria or even eukarya (Woese and Fox, 1977; Woese et al., 1978; Woese, 1987). Thymine is absent in the tRNA (transferase RNA; thymine is absent in the tRNA) of archaea but present in the tRNA of bacteria. Unexpectedly, ribosomal proteins from archaea were found to be more similar to those of eukarya than to bacterial ribosomal proteins. So, archaea and eukarya seem more closely related than bacteria and eukarya (see Table 2.19).

Microbial eukaryotes can be unicellular or multicellular. Algae are plant-like organisms that can be either unicellular or multicellular. Fungi (singular: fungus) are also eukaryotes. Some multicellular fungi, such as mushrooms, resemble plants, but they are actually quite different. Fungi's cell walls are usually made of chitin rather than cellulose. Unicellular fungi—yeasts—are included in the study of microbiology. There are more than 1000 known species. Yeasts are found in many different environments, from the deep sea to the human navel. Some yeasts have beneficial uses, such as causing bread to rise and beverages to ferment; but yeasts can also cause food to spoil. Some even cause diseases, such as vaginal yeast infections and oral thrush. Helminths are multicellular parasitic worms. They are considered microbes because their eggs and larvae are often microscopic.

Table 2.19. Some Key differences between archaea and bacteria

	Archaea	Bacteria
Another name	Archaebacteria	Eubacteria
Shape	Cocci, Rods, spiral, plates, or coils	Cocci, Rods, spiral, plates, or coils
Cell wall	No peptidoglycan	Peptidoglycan
Environment	Unusual environment	Found everywhere
RNA	Three RNA	Single RNA
tRNA	Absent	Present
Reproduction	Asexual reproduction	Asexual reproduction
pathogens	Non-pathogens	Pathogens
Examples	Pyrolobus fumarii	Escherichia coli

Protists. As most groupings in life, there are always someones left out. In the case of living things, protists are left out because they are not animals, plants, or fungi. For sure, they are not bacteria or archaea because all protists are eukaryotes, meaning each of their cells has a nucleus. Yet, they are one of the key producers of atmospheric oxygen through the photosynthesis process (see also Ikelle, 2022a). Most protists are unicellular aquatic organisms, with the exception of some multicellular algal protists, though they can survive in most environments, including soils and human bodies. Many protists have flagella or cilia-hair-like appendages that propel them through water; some use pseudopodia, or false feet, to move. Protists are sometimes classified as animal-like, plant-like, and fungus-like protists based on their resemblance to one of the three classical groups.

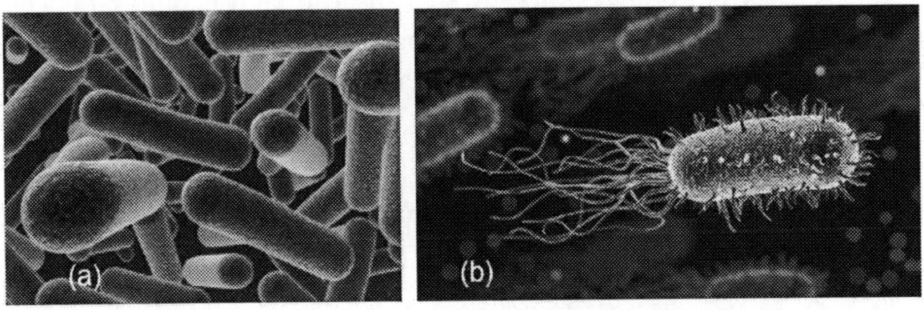

Figure 2.31. (a) An example of archaea and (b) bacteria. Both microbes are single-celled microorganisms with an undeveloped nucleus. Archaea can survive at very high temperatures whereas bacteria cannot live at 100 degrees Celsius temperature. Also archaeal growth cannot be inhibited by antibiotics whereas bacterial growth can be inhibited by the use of antibiotics in case of illness.

Quiz: (1) The large molecules of all living things fall into just four main classes. Name them. (2) What are isomers?

2.3.5. Cell Reproduction Mechanisms

Living things are constantly replenishing their cell supply by creating new cells to replace those that are old or damaged, as well as making cells to be used to create new organisms. Cell theory states that all cells arise from pre-existing cells through cell division—that is, cell division is the reproduction mechanism whereby living things grow, develop, and produce offspring. Even in multicellular organisms, every cell is ultimately derived from a single cell. In addition, cells contain hereditary information (DNA), which is passed from cell to cell (e.g., from parent cells to daughter cells) during cell division. Daughter cells are a result from the cell divisions and parent cells are the input of cell division processes.

Before we describe cell-division processes, let us introduce two important groups of cells that characterize cell reproduction: diploid and haploid cells. Cells that contain two complete sets of chromosomes (i.e., two copies of each chromosome) are known as diploid cells and cells that contain only one set of chromosomes are known as haploid cells. Nearly all the cells in the human body are diploid: one set of chromosomes is inherited from the mother and the other set is inherited from the father. The only non-diploid cells are egg and sperm cells (i.e., sex cells, which are also known as gametes). They are haploid cells.

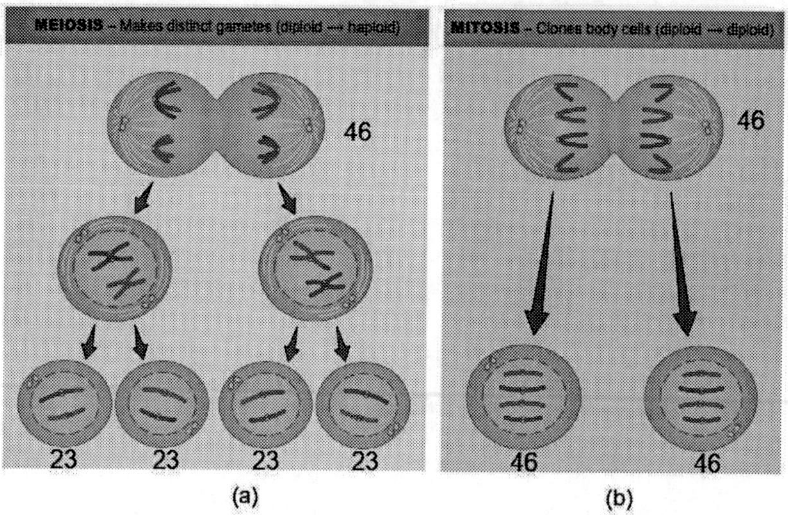

Figure 2.32. (a) Meiosis and (b) mitosis and meiosis are two processes by which eukaryotic cells divide (reproduce). Mitosis results in two genetically identical daughter cells, whereas meiosis results in four genetically different cells.

Mitosis and meiosis are the two basic cell division mechanisms. In mitosis, a single cell divides forming two equally-sized daughter cells, as illustrated in Figure 2.32a. Cells produced by mitosis are genetically identical (clones). The division process begins with cell elongation and an increase in cell volume. The cell then replicates its DNA in preparation for having two copies of its chromosome, one for each newly formed cell. After the nucleoids are segregated to each end of the elongated cell, the division is completed by dividing the elongated cell into two equally sized daughter cells. This reproduction mechanism is asexual; no union of cells or nuclei of cells is involved.

Mitosis is used by single-celled organisms to reproduce. It is also used for the organic growth (culture) of tissues, fibers, and membranes. Mitosis is critical for producing new cells and keeping the organism alive and healthy because all cells go through a process of formation, growth, division, and eventual death. Most cells in the human body only last a few days to a few weeks (an exception is brain cells, which typically last your whole life), so your body must constantly be making new cells through mitosis. This process is also behind the growth of children into adults, the self-healing of cuts and bruises, and even the regrowth of skin, limbs, and appendages in animals like geckos and lizards. Mitosis can take over 10 hours for mammalian cells in culture (Jett, 2015), budding yeast can take about 80 minutes to complete a cell cycle (Brewer et al., 1984), while bacteria can divide every 20 minutes.

Note that cancer is characterized by a disorderly proliferation of cells that can invade adjacent tissues and spread via the lymphatic system (network of vessels, through which lymph drains from the tissues to blood and lymph is a fluid containing white blood cells) or blood vessels to other parts of the body. In other words, cancer cells are produced by uncontrolled mitosis. They can produce three or more daughter cells from a single parent cell instead of normal two daughter cells and have a faster rate of reproduction than normal cells. Due to this irregular division of cancer cells, we often end up with too many daughter cells or not enough chromosomes.

Meiosis is a sexual reproduction in which the male and female sex cells (i.e., sperm in men and eggs in women) combine to create new offspring that are never genetically identical to the parent cell. It is this reproduction process that has increased and continues to increase genetic diversity in the world. This diversity is caused by the pairing up of chromosomes from both parents and the exchange DNA in a process known as crossing over (see Figure 2.32). It is this crossover process that enables children to be related among themselves and to their parents but still different from their two parents and siblings.

Only sex cells go through the meiosis process and it only occurs in eukaryotic animal cells, plant cells, and fungi. Meiosis has various timescales in different organisms, which can be affected by several factors, including the temperature and environment of the organism, and the amount of nuclear DNA. The process lasts six hours in yeast but can last more than 40 years in human females, one to two days in male fruit flies, and about 24 days in human males (Bennett, 1977).

Table 2.20. Similarities and differences between mitosis and meiosis processes

	Meiosis	Mitosis
Type of reproduction	Sexual	Asexual
Parent cells	Always diploid	Haploid or diploid
Occurs in	Humans, animals, plants, and fungi.	All living things.
Genetically	Different offspring	Identical (exact clone).
Crossing over	Yes, mixing of chromosomes can occur.	No, crossing over cannot occur.
Daughter cells produced	Four haploid cells	Two diploid cells
Chromosome number	Reduced by half.	Remains the same.
Disease	Uncontrolled mitosis occurs in cancer	Trigger miscarriage
Discovered by	Oscar Hertwig	Walther Flemming

A disease can cause the meiosis process to go wrong, leading to a wrong number of chromosomes, ending up in germ cells (sex cells), for example. A miscarriage may ensue. Note that germ cells are either a sperm or an egg; all other human cells are called somatic cells. Somatic cells reproduce by mitosis and germ cells produced by meiosis.

Table 2.20 captures the similarities and differences between the mitosis and meiosis processes. The end goal of these processes is the creation of daughter cells that can be used to either keep the organism alive (mitosis) or help create a new organism during sexual reproduction (meiosis). Mitosis as a reproductive process for single-cell organisms may have started on Earth around 3.7 billion years ago. Meiosis reproduction may have started on Earth around 1.4 billion years ago. From an evolutionary point of view, it is very likely that cells appeared first and the virus followed, as they require cells for their replication.

Microbe reproduction. Bacteria and archaea reproduce by the mitosis process which is asexual reproduction. The entire process can take as little as 20 minutes for an active culture of E. coli bacteria. After a certain number of generations, the mother cells age and die.

Viruses are extremely small, with a diameter ranging in size from 10 to 300 nm although some like the Ebola virus can have 1μm in diameter. They are observed using electron microscopes, which have resolutions on the order of 0.1 nm. Again, they are not considered living things because they do not have cell units. They are generally called acellular (not composed of cells) microbes, infectious particles, viroids, and prions are also used to refer to viruses. However, the vast majority of viruses possess either double-stranded DNA or single-stranded RNA, but they do not possess both while all living things possess both.

Viruses are not alive; they do not replicate (multiply) on their own. Viruses invade live host cells in order to replicate; that is why living things are not safe from viral infections. Each time the host's cells go through replication, the virus gets replicated as well. In humans, viruses can cause many diseases. They infect humans, animals, plants, fungi, protozoa, algae, and bacterial cells. Some viruses called oncogenic viruses or oncoviruses cause human cancers such as the common cold, lymphomas, carcinomas, some types of leukemia, and deadly Ebola.

Quiz: Why are viruses not transmitted through cooked food?

Plant reproduction. Plants dominate our landscape and influence the climate. Our nutrition and medicinal compounds, papers, and textiles depend on them. The list of plant benefits for human societies is almost unlimited. Earth is the only planet in our solar system with plants visible with the naked eye. They are eukaryotic multi-celled organisms. Their organs include leaves, stems, flowers, and roots. The tissues that form these organs are known as xylem and phloem. Xylem is made up of a structural protein called lignin and carries minerals and water from roots to other parts of plants. Phloem carries the products of photosynthesis that is glucose. There are close to 300,000 species of cataloged plants, of these, about 260,000 are plants that produce seeds (i.e., structures that can grow into new plants). Seeds include corn, beans, and nuts that you are regularly eat. Seeds are the primary means by which annuals such as field crops, flowering bedding plants, and vegetables reproduce. In other words, plant reproduction is about seed production.

Plants are produced by the mitosis and meiosis processes. Meiosis plant reproduction needs flowers to reproduce. A flower has two parts that help it make seeds, a male part and female part. The male part makes a powder called pollen. The female part makes tiny eggs. When pollen and an egg come together, a seed can form. In flowering plants, fruit forms around the seeds—that is, fruit is a structure that holds seeds. Transport of pollen to egg can be through wind blowing pollen from one flower to another. Animals such as hummingbirds, bees, and bats can carry pollen, too. Pollen clings to the bodies of some animals which then carry the pollen to another flower. This reproduction method is termed seed reproduction or sexual reproduction.

Many plants use asexual reproduction (mitosis). This method does not require plants to produce flowers, attract pollinators, or find a means of seed dispersal. Plants can grow new plants from a fragment of a stem, a leaf, or a root in a process called regeneration. It is common horticultural practice to propagate desirable varieties of garden plants by means of plant fragments, or cuttings. Naturally fallen branches of willows (Salix) and poplars (Populus) root under suitable conditions in nature and eventually develop into trees. So, asexual reproduction produces plants that are genetically identical to the parent plant because no mixing of male and female gametes (sex cells) takes place. Traditionally, these plants survive well under stable environmental conditions when compared with plants produced by sexual reproduction because they carry genes identical to those of their parents. Many perennials such as landscape plants, turfgrasses, and fruit crops reproduce quicker and more efficiently by asexual means. Tubers (potatoes), rhizomes (onions, ginger), and runners (strawberries) are examples of plants asexually reproducing. Both asexual and sexual reproductions are important to the agricultural industry, specifically the commercial production of horticultural, agronomic, and forestry crops.

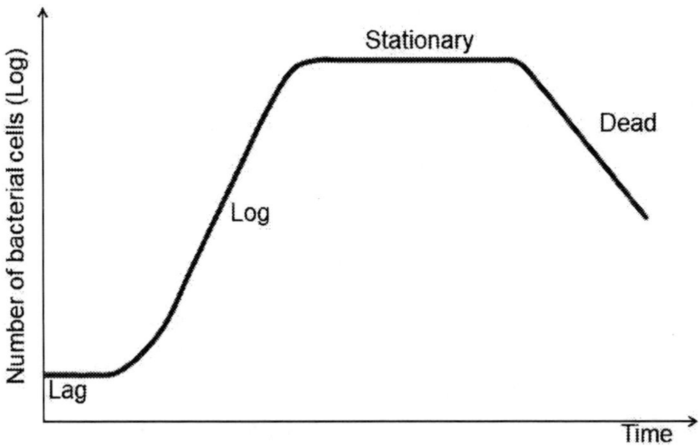

Figure 2.33. The bacterial growth curve. It represents the number of live cells in a bacterial population over a period of time. It has four phases: the lag, exponential, stationary, and death phases.

Animal reproduction. Almost all animal species are capable of reproducing sexually through the meiosis process. For many animal species, meiosis is the only mode of reproduction possible. During this reproduction, the male and female sex cells of a species combine in a process called fertilization. Fertilization may be internal, especially in land animals, or external, as is common in many aquatic species. After fertilization, a developmental sequence ensues as cells divide and differentiate. Many of the events in development are shared in groups of related animal species, and these events are one of the main ways scientists classify high-level groups of animals. During development, animal cells specialize and form tissues, determining their future morphology and physiology. In many animals, such as mammals, the young resemble the adult. Other animals, such as some insects and amphibians, undergo complete metamorphosis,

Box 2.5. Bacterial Growth

Microbes become a problem for humans if they can grow (i.e., increase in population). Their growth is a key factor in them becoming a positive or negative thing for humans. While the growth of multicellular organisms such as humans is the size of a single organism, microbial growth is measured by the increase in population. Bacterial growth in the laboratory suggests that bacteria have four phases: lag phase, exponential phase, stationary phase, and death (e.g., Buchanan et al., 1918). Here are these phases (see Figure 2.33):

- *Lag phase production* lasts almost two hours and no reproduction occurs. It is the period in which microbes adapt to their environment conditions.

- *Exponential phase* is a period of growth, a series of doublings of the population, where one cell becomes two cells, becomes four, becomes eight, etc. After n divisions, the number of cells is $N = N_0 2^n$, where N_0 is the initial cell number. Cells from the division do not necessarily survive. However, if the number of surviving cells is large, the bacterial population undergoes exponential growth that can be described as follows:

$$\frac{dN}{dt} = \mu N \Longleftrightarrow \frac{dN}{N} = \mu_p dt \Longrightarrow N = N_0 \exp[\mu_p(t - t_0)] \ ,$$

  where t is time, N_0 is the number of bacteria at a time t_0, μ_p is the growth rate (1/second). Environmental factors, such as temperature and oxygen levels, influence the value of μ_p. For N to be doubled, we have

$$\frac{N}{N_0} = 2 \Longrightarrow N = N_0 \exp[\mu_p(t - t_0)] = N_0 2^{(t-t_0)/\tau} \ ,$$

  where $\tau = t - t_0$ is the generation time of the cell population. Most bacteria of medical interest have generation times of 15 minutes to an hour or so.

- *Stationary phase* can be defined as a state of no net growth; i.e.,

$$\frac{dN}{dt} = 0 .$$

Cells still grow but growth is balanced by dying cells. The reasons why the growth curve reaches the stationary phase include a complete elimination of essential nutrient and the source of nutrients becomes dying cells, waste products build up to a point where cell growth is no longer possible, and lack of physical space. New cells are smaller in size than their parents due to starvation conditions (Bridges et al., 2001).

- *Death phase* can be described by the following equation:

$$\frac{dN}{dt} = \mu_d N ,$$

where μ_d is the specific death rate. The rate of cell death is usually slower than the rate of growth during the exponential phase but it rarely reaches 100 percent because cells do mutate.

in which individuals enter one or more larval stages. For these animals, the young and the adult have different diets and sometimes habitats. In other species, a process of incomplete metamorphosis occurs in which the young somewhat resemble the adults and go through a series of stages separated by molts (shedding of the skin) until they reach the final adult form.

A number of animal species, especially those without backbones, but even some fish, amphibians, and reptiles, are capable of asexual reproduction. Asexual reproduction, except for occasional identical twins, is absent in birds and mammals.

Mechanism of heredity and gene mutation. Chromosomes, which are the structure within the cells, as described earlier, contain segments of DNA that contain the code associated with each person. These segments are referred to as *genes*. As the chromosomes, genes exist in every cell of the body. From the point of view of genetics (or heredity), it is only chromosomes in the sex cells (germ cells) that are important. Each human body's cells has normally 46 chromosomes, made up of two similar (but not identical) sets of 23 chromosomes each. However each sex cell contains only one set of 23 chromosomes. In sexual reproduction, the union of an egg cell, produced in the ovaries of the mother, with a sperm cell, originating in the testes of the father produces cells with each cell carrying a set of 23 chromosomes, one representing the characteristics of the mother and the other set those of the father. The resulting fused cell has 46 chromosomes. As the embryo develops, the cells reproduce themselves. In general, the 46 chromosomes are duplicated without change. Note that microbes have only one chromosome, fruit flies have four pairs, and a plant called Adder's Tongue Fern has over 1,000 chromosome pairs. The number of chromosomes is not what makes an organism genetically unique, though.

Mutations occur when, instead of a chromosome duplicating itself in every respect, one or more of the genes change. Such changes are rare but unfortunately permanent and chromosomes are duplicated with genes in their altered forms. If a mutation occurs in a body cell, there may be some effect on the individual, but the change is not passed on to the next generation. But, if the mutation occurs in a sex cell of either parent, a new characteristic may appear in a later generation. Note that exposure to nuclear radiation and biological agents such as viruses can cause mutations. So far we have only observed that radiation increases the spontaneous mutation rate, but it does not produce any new mutations. So the hideous creatures supposedly produced by radiation in science-fiction literature and cinema have not been observed in humans, maybe because large cell mutations kill cells.

Summary about cells. To summarize, cells, such as neurons and muscle cells, are the basic building blocks of all living things. Humans are composed of trillions of cells. These cells provide the structure of our bodies. They take in nutrients from food, convert these nutrients into energy, and carry out specialized functions that we will be discussed in the coming chapters. Cells also contain the living thing's hereditary materials and can make copies of themselves. Cells are separated from the outside environment by the cell membrane. The cell membrane controls the movement of substances in and out of cells organelles, which are specialized structures that perform certain tasks within the cell. The inside of the cell is called the intracellular space. The outside of the cell is called the extracellular space. It is truly remarkable that learning about the multiplicity of living species comes down to learning about a cell unit.

2.3.6. Where the First Cells Came from?

If you cannot make something that you have, you bought it, it was given to you, or you stole it. Cells cannot be made, they can only be divided. So who gave them to us or where did we get them from? We cannot emphasize enough the importance of this age-old question about the origin of life, which nowadays amounts to the search for the origin of the first cells. This question is at the heart of who we are as humans. Do not expect any answer here, but we hope that the background developed in this chapter and subsequent chapters will help you or future generations of scientists and engineers to answer it. Although the origin of the first cells on Earth is unknown, we can make some hypotheses based on the background we are developing in this book. Based on the description made in Chapter 1, and that we will expand on in Ikelle (2022a), we know that all chemical elements introduced in this chapter were already present in the first years of the formation of our solar system. They came from the interstellar medium (i.e., the space between stars). Some of these elements were put into the interstellar medium by dying stars. As discussed in Ikelle (2022a), only some isotopes of these elements are man-made, mostly for medical purposes.

Table 2.21 shows the predominant chemical elements of life, solid Earth, atmosphere, sun, and our universe. All major living and non-living things are predominantly controlled by 11 chemical elements, which were present in our solar system for more than 4 billion years, well before the occurrence of living things on Earth. Also, all nine planets, including their atmospheres, 170 or so moons, asteroids, and comets are made largely from the same six elements—hydrogen (H), helium (He), oxygen (O), carbon (C), nitrogen (N), and sulfur (S). These are the most abundant elements in the sun and other stars. With the exception of helium, they are again the main elements of living things. Other essential elements of life include molybdenum, copper, and zinc. Life cannot be sustained without them around us. For example, molybdenum (Mo) is found in low concentrations in plants, animals, and within the human body. It was fundamental to the occurrence of life on Earth and it is contained in RNA. It is very abundant in the ocean today, but was less so in times past. The amount of molybdenum in ancient rocks is sometimes used to determine the variations in the amount of oxygen in the oceans and atmosphere over time.

Table 2.21. Chemical elements that predominantly compose the universe: the sun, Earth's atmosphere, Earth's crust, human bodies, and cells in percentages.

	H	He	N	O	Si	Al	Fe	Ca	Na	K	Mg	C
Universe	77	21										
Sun	75	23										
Atmosphere			78	21								
Crust				46	27.7	8.1	5	3.6	2.8	2.6	2.1	
Human bodies	9.5		3.6	65								18
Cell	60		5	25								12

Some molecules were also present in the early years of our solar system and Earth. They include nitrogen gas (N_2), carbon dioxide (CO_2), water (H_2O), hydrogen gas (H_2), ammonia (NH_3), methane (CH_4), hydrogen sulfide (H_2S), carbonous acid (H_2CO_2), formaldehyde (H_2CO), carbon monoxide (CO), ethylene (C_2H_4), formonitrile (HCN), and hydrazine (N_2H_4). In other words, organic and inorganic molecules were present in the early years of our solar system. Table 2.22 shows some additional molecules that may have formed in the interstellar medium or in planetary atmospheres. There is no evidence yet that macromolecules were present. However, nitrogen and some of its compounds were present. Because nitrogen is a key ingredient of amino acids that constitute nucleic acids (e.g., DNA and RNA), and proteins, and makes up 14 percent of dry cell weight, we cannot exclude the possibility of the presence of some macromolecules on the very young Earth, billions of years ago. On the other hand, some bacteria known as clostridium are able to ferment amino acids as well as carbohydrates (see Ikelle, 2022a). In any case, the jump from molecules to the origin of the first cells is an unproven big leap. More arguments and evidence will be discussed in the next chapters.

Table 2.22. Some of the molecules that (may) occur in the interstellar or circumstellar medium and/or in planetary atmospheres (Müller et al., 2005; Endres et al., 2016; and https://cdms.astro.uni-koeln.de/cdms/portal/)

# atoms	2	3	4	5	6	7	8	9
	H_2	C_3	c-C_3H	C_5	C_5H	C_6H	CH_3C_3N	CH_3C_4H
	AlF	C_2H	l-C_3H	C_4H	l-H_2C_4	CH_2CHCN	$HCOOCH_3$	CH_3CH_2CN
	AlCl	C_2O	C_3N	C_4Si	C_2H_4	CH_3C_2H	CH_3COOH	$(CH_3)_2O$
	C_2	C_2S	C_3O	l-C_3H_2	CH_3CN	HC_5N	C_7H	CH_3CH_2OH
	CH	CH_2	C_3S	c-C_3H_2	CH_3NC	$HCOCH_3$	H_2C_6	HC_7N
	CH^+	HCN	C_2H_2	CH_2CN	CH_3OH	NH_2CH_3	CH_2OHCHO	C_8H
	CN	HCO	CH_2D^+	CH_4	CH_3SH	c-C_2H_4O	CH_2CHCHO	
	CO	HCO^+	HCCN	HC_3N	HC_3NH^+	CH_2CHOH		
	CO^+	HCS^+	$HCNH^+$	HC_2NC	HC_2CHO			
	CP	HOC^+	HNCO	HCOOH	NH_2CHO			
	CSi	H_2O	HNCS	H_2CHN	C_5N			
	HCl	H_2S	$HOCO^+$	H_2C_2O	HC_4N			
	KCl	HNC	H_2CO	H_2NCN				
	NH	HNO	H_2CN	HNC_3				
	NO	MgCN	H_2CS	SiH_4				
	NS	MgNC	H_3O^+	H_2COH^+				
	NaCl	N_2H^+	NH_3					
	OH	N_2O	SiC_3					
	PN	NaCN	C_4					
	SiO	CO_2						
	SiS	NH_2						
	HF	SiCN						
	CS	H_3^+						

So, as we can see in Table 2.21 (sixth row), humans are carbon-based life forms. Actually all living things on Earth, including all animals (dogs, cats, horses, etc.), are carbon-based. Any kind of laser surgery on animals that passes through tissues shows carbon residues under a microscope. In other words, at a microscopic level, we always observe powdered grains of carbon. Thus, laser surgery residue is one of the proofs of the presence of carbon in animals, including humans. Note that the overall carbon-based biomass composition of Earth's biosphere is about 550 gigatons of carbon with about 450 gigatons for plants, 2 gigatons for animals, and 77 gigatons for microbes (Bar-On et al., 2018).

Every model of the origin of the first cell must answer the following questions:

- What are the materials of the first cells and where they came from?
- How and where did the assemblage of the first cell took place?
- What are the atmospheric conditions, solar-output conditions, Earth-heat conditions, volcanism and plate tectonic conditions, and Earth magnetic field conditions under which the assemblage of the first cell took place?
- Is this assembly a series of spontaneous chemical and/or nuclear reactions under the right conditions.
- Why this assemblage appears to occur only twice in 4.6 billion years time?
- What are special conditions in the periods of 3.7 billion years ago and 1.4 billion years when the first prokaryotic and eukaryotic cells, respectively appeared on Earth?
- What are other types of assemblages occurring on other planets in our solar system?

Quiz: Does silicon-based life exist in our universe as an alternative to carbon-based life?

2.3.7. Artificial Intelligence: Binary Classification of Matter into Organic and Inorganic

Human intelligence[2] can be defined as a general mental ability for reasoning, problem-solving, and learning. It integrates cognitive functions such as perception, attention, huge memory, language (e.g.,

[2]Modern humankind is also known as homo-sapiens-sapiens—wise men—because our mental capacities are so important to our everyday lives and our sense of self. This species dates back about 150,000 years ago in Africa. Then, based on archeological findings, they migrated to Asia, Australia, Europe, and the Americas.

Akkadian, Sanskrit, and Latin), planning, creation, and common sense. Moreover, it is the creation of supernatural forces. In contrast, an AI (artificial intelligence) system is a set of techniques that are aimed at approximating some aspects of human or animal cognition by inserting computer software into robots and other related systems. It tries to automate some intellectual tasks normally performed by humans.

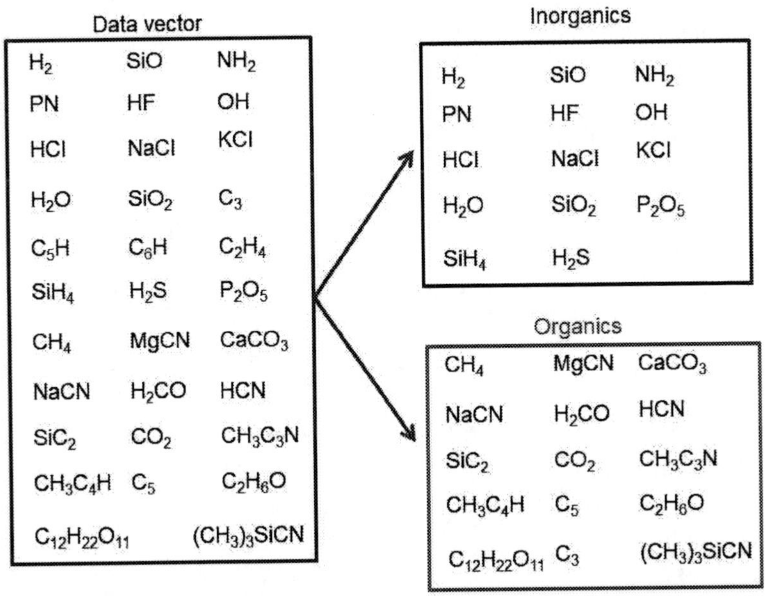

Figure 2.34. Matter classification using the presence or lack of carbons as the feature.

AI systems merge ideas from neuroscience and biology, statistics and mathematics, and physics and chemistry to make computers and robots that can learn. They are inherently multidisciplinary; they are clear indicators that we have entered an area of crossfertization of scientific disciplines. Even philosophers and psychologists, who are also concerned with intelligence, have now joined the AI community. Over 2000 years or so, philosophers and psychologists have been and are still trying to understand how seeing, learning, remembering, and reasoning are carried out by humans. They have developed multiple intelligence tests over the years to assist medical doctors, educators, and policy makers in determining people requiring extra medical, academic, and/or social assistance; and also in identifying people with exceptional abilities. However, unlike philosophy and psychology, AI systems, which became a part of human intelligence studies in the 1950s, aim to build mechanical/biological entities as well as understanding human intelligence.

One of the key features of human intelligence is the ability to distinguish several things based on our eyes, our touch, our body reactions, and/or our feelings. All these sensors and visions include features such as colors, hardness, temperature, pressure, gravity, past experiments, and memories. The challenge is to translate all the features into a form that computers can distinguish things just like humans.

The injection of science into our AI systems will enable them to capture some human experience and some degree of common sense. Advances in science also allow them to adapt to new circumstances. Another aspect of our AI systems is the terabytes of data collected in the fields that are difficult for humans to understand and interpret. Learning from these data allows AI systems to discover new rules and therefore, in a limited way, they can invent. In other words, AI systems can surprise us. Some of our AI systems also include reasoning and decision-making capabilities.

Table 2.23. Some differences between organic and inorganic compounds.

Properties	Organic	Inorganic
Carbons (C)	Presence of C	Lack of C
C-H bonds	Presence of C-H bonds	Lack of C-H
Nature of compound	Covalent	Electrovalent, ionic, Covalent
Melting and boiling	Low	High
Solubility in water	Insoluble	Soluble
Solubility in organic solvents	Soluble	Insoluble
Combustibility	combustible	Non-combustible
Phase	Solid, liquid, and gas	solid and liquid

Machine learning. There are various ways machines, including computers, can learn. We say that a machine is learning if we can give it the correct solutions to some problems and new problems, the machine can get the correct solutions without our help (generalize). This way of learning is known as supervised learning. Basically, a supervised learning algorithm takes a known set of input data (the learning set or training set) and known responses to the input data. It learns a model that can generate reasonable predictions of responses to new input data. The term *supervised learning* stems from the fact that during learning a model can obtain the correct value from the teacher, the supervisor. All supervised learning techniques are a form of classification or regression. Classification algorithms are trained to classify data into categories. They predict discrete responses—for example, whether a matter is organic and inorganic (i.e., binary classification), or whether a volcanic eruption will produce hot ash and gas that will engulf a small village, a major town, or worldwide (multiclass classification).

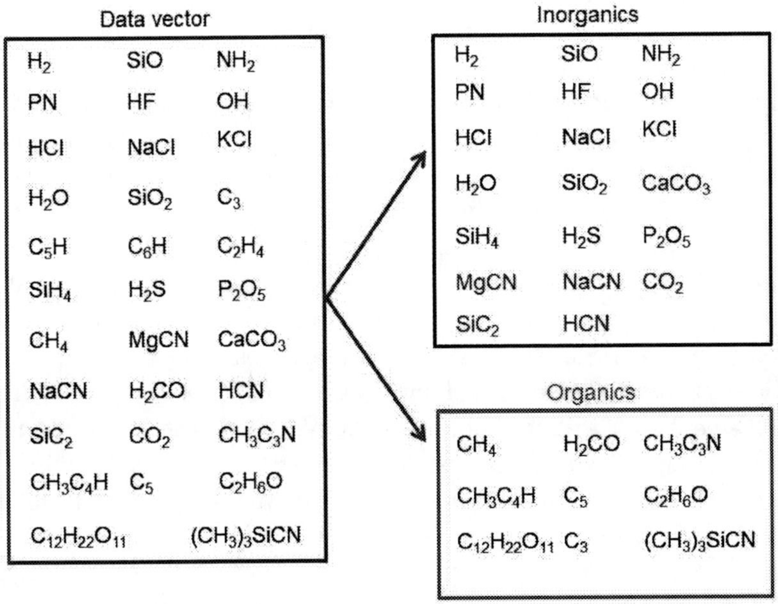

Figure 2.35. Two features are used to classify materials: the presence or absence of carbons and the presence or absence of carbon-hydrogen bonds.

Alternatively, we can tell the machine whether or not its answer was correct, but not how to find the correct answer. In this case, we have to provide the machine tools and data so that it can search for the right answer.

Matter classification. So classification consists of taking input vectors and deciding which of N classes they belong to, based on training from examples of each class. The most critical point about the classification problem is that it is discrete-each example belongs to precisely one class, and the set of classes covers the whole possible output space. As described above, most compounds can be classified into two broad categories: organic and inorganic compounds. At compound and molecule levels, this separation is not obvious. Remember that naturally occurring organic compounds are made primarily of carbon and hydrogen (see Figure 2.8). With the proliferation of compounds and molecules, we sometimes wish that the distinction between organics and inorganics could be made without human involvement. This binary classification can be done based on the presence or lack of carbons. The key parameter is known as *feature* which is here carbon. It is the feature that makes classification.

As we will see through this book, it is the features of learning algorithms that determine the effectiveness or non-effectiveness of a classification. For example, classification of matter into organic and inorganic may not be effective when compounds such as silicon carbide (SiC_2), calcium carbonate ($CaCO_3$), and sodium cyanide (NaCN) are included the input-data vector because these compounds are inorganic despite the fact that they contain carbons (see Figure 2.34). However, we can use one of more of the additional features listed in Table 2.23 for a more accurate classification while avoiding making a category of compounds that are neither organic or inorganic. For example, organic compounds are inflammable (combustible) and highly volatile while inorganic compounds are non-volatile as well as not inflammable (non-combustible). Also the fact that inorganic compounds lack carbon-to-hydrogen bonds can be added as a second *feature* (see Figure 2.35). The classification becomes multiclass.

Technology-wise, the machine for classifying organic and inorganic matter can include an infrared spectrometer (see Ikelle, 2022c).

Such a spectroscopy allows us to determine matter with carbon-hydrogen bonds.

To summarize, our use of AI in this book series excludes some functions and characteristics of the human brains and some of their interactions with the body. For example, humans are born and grow in ways that we cannot yet replicate by making computers grow on their own. Also we know that the brain is not an isolated organ. It is part of a much larger ecosystem that we cannot yet fully integrate into AI systems. We focus in developing AI systems that can learn how to intellectually solve specific problems, provide us new understandings, and make new discoveries about problems under consideration. The basic mathematics and statistical tools of such AI systems and their limitations are described in Ikelle (2009, 2017a), including clustering data, compressing data, visualizing data, filtering data, selecting relevant components from data, extracting dependencies between data components, modeling the data generating systems, constructing noise models for the observed data, integrating data from different sensors or sources. Do not lose sight of the fact that artificial intelligence systems are still about producing algorithms, as you have probably done in the past.

2.4. Classic and Quantum Mechanics

In this chapter, we have introduced macroscopic matter that we can see with our naked eyes such as the sun, water, rocks, human bodies, animals, and some microbes. We also introduced microscopic matter (i.e., particles) that we can only see through microscopes, such as electrons, neutrons, and protons. Experiments have shown these two types of matter do not behave in the same way. Quantum mechanics takes into account particle behavior and is therefore consistent with nature. Classical mechanics generally focuses on macroscopic matter behavior while ignoring any particle behavior that may be related to the macroscopic behavior. In other words, classical mechanics is an approximation of quantum mechanics or just a particular case of quantum mechanics. Because most scientific developments are still based on classical mechanics, many erroneous scientific predictions, including climate changes, magnetic reversal and Creighton events, are attributed to its use. The two mechanics were developed almost independently by Galileo di Vincenzo Bonaiuti de' Galilei (1564-1642), Sir Issac Newton (1643-1727), Pierre-Simon Laplace (1749-1827), André-Marie Ampère (1775-1836), Micheal Faraday (1791-1867), Lord Kelvin (1824-1907), James Clerk Maxwell (1831-1879), Max Karl Ernst Ludwig Planck (1858-1947), Albert Einstein (1879-1955), Max Born (1882-1970), Niels Henrik David Bohr (1885-1962), Erwin Schrodinger (1887-1961), Louis de Broglie (1892-1987), Werner Karl Heisenberg (1901-1976), Paul Adrien Maurice Dirac (1902-1984), John von Neumann (1903-1957), Richard Phillips Feynman (1918-1988), John Stewart Bell (1928-1990), and of course many others.

We humans, even animals, have naturally evolved with a certain intuition in the ways of thinking about our physical world. Our intuition is dominated by parameters such as forces, velocities, momentum, and energy. These parameters are in line with our biology, thus they constitute common human standard intuition. Exercise, lifting bodies, farming, fruit harvesting, mixing matter for food production and cooking, and launching projectiles are examples of physical activities that come naturally to us. They fit well with our intuitions about the physical world, and are essential to our existence. They are also the basis of classical mechanics. Quantum mechanics extends beyond these basic intuitions that we can see and observe around us to include parameters that we do not see or experience, such as electrons, uncertainty principle of position and velocity of particles, the wave-particle duality (i.e., particles can behave both as particles and waves), and the speed of any object is always smaller than the speed of light.

A particle has a position and trajectory at every instant of time, and therefore a velocity is the ratio of the distance between two points along that trajectory over the time interval based on classical mechanics; i.e.,

$$v = \frac{\text{distance between two points}}{dt} , \tag{2.13}$$

where dt is the change in time. One can also obtain the momentum, $p = mv$, where m is the mass of the object. There is inherent uncertainty in measuring variables of a particle. For example, it is impossible to determine both the momentum and the position of a particle at the same time. In other words, if one knows the precise momentum of the particle, it is impossible to know the precise position, and vice versa; i.e.,

$$\Delta x \Delta p \geq \frac{h}{4\pi} \quad \text{or} \quad \Delta x \Delta p \geq \frac{\hbar}{2}\,, \tag{2.14}$$

where Δx is uncertainty in position, Δp is uncertainty in momentum, and $h = 2\pi\hbar$ is Planck's constant. You can never simultaneously know exactly where a particle is, and how fast it is moving. This limitation is only critical when dealing with small particles such as electrons. But it does not matter for ordinary-sized objects such as a soccer ball, the Earth, and the Sun—that is, the uncertainty principle, which is known as the Heisenberg uncertainty principle, does not matter in classical mechanics. Suppose that we can accurately measure the position to 1 nanometer, the uncertainty of the momentum is $\Delta p \geq 5.27 \times 10^{-25}$ kg m/s. For a ball of 1 kg, the uncertainty of the velocity is $\Delta v_{\text{ball}} \geq 5.27 \times 10^{-25}$ m/s and that of electron with mass of 9.1×10^{-31} kg is $\Delta v_{\text{electron}} \geq 5.79 \times 10^{+5}$ m/s. The uncertainty of the ball's speed is negligible and that of the electron speed is not. As we discussed in Ikelle (2022a, 2022b, 2022c, and 2022d), the uncertainty principle allows particles to break the rules of classical mechanics by passing over the energy barriers. This phenomenon is known as *quantum tunneling.* One day, it may allow interstellar travel to be reduced from millions of years to months, or even to days.

Another key phenomenon in quantum physics that does not exist in classical physics is that quantum mechanics is based on statistical behavior (i.e., randomness). It predicts various outcomes for a given experiment while it still obeys the law of energy conservation. Classical mechanics is unable to predict the future because we always lack complete information about the system at present. Even if we had such knowledge (which is actually not possible according to Heisenberg's uncertainty principle), we still cannot tell what is the exact result of a future observation because we live in a random (non-deterministic) world. In the real world, there are a variety of outcomes. In spite of the fact that quantum mechanics does not predict exact results, it is the most accurate theory. The same experiment can give different results even if the initial conditions are exactly the same. Offspring from meiosis reproduction, for example, have very different anatomy, physiology, and souls despite sharing germ cells (sperm and egg) from the same origin. In such a world a classical theory can only predict one result, while we observe multiple outcomes.

To summarize, nature is made of quantum mechanics. So, like most modern science books, one of our goals in this book series is to develop your intuition of the quantum world.

Chapter 3

Inorganic Compounds, Minerals, Rocks, and Plate Tectonics

The Earth is the place where we live. It provides us with food through the farming of its soil. It provides us with energy resources (e.g., water, oil, gas, and coal) and minerals (e.g., gold, diamonds, uranium, and thorium). The sources of these resources and minerals in the subsurface are rocks.

Rock is a naturally occurring solid. It is a cohesive aggregate of grains of one or more minerals. A mineral is a naturally occurring inorganic (not formed by organisms), crystalline solid (not a liquid or gas) compound with a definite composition (or range of compositions) (e.g., Hurlbut, 1963). Ions of individual elements combine to form mineral compounds by chemical bonding. The minerals are the same in all parts of the world. The colors and textures (size, shape, and arrangement of grains) of the minerals in rocks tell us how rocks were formed and provide us with clues about the past. Biotite, clay, calcite, diamonds, feldspars, gold, halite, olivine, quartz, and silt are examples of minerals (see also Figures 3.1 and 3.2). A rock can also include non-minerals, such as organic matter within a coal bed. Our first key objective in this chapter is to help readers familiarize themselves with the set of inorganic compounds, minerals, and their aggregates, rocks, which constitute the Earth's crust. The solid Earth, with its complex rock arrangements, also kills, and even more frequently these days; earthquakes are one of the major causes of death.

Figure 3.1. An illustration of most minerals in the crust. About 92 percent of the crust is composed of silicate minerals and 8 percent of non-silicate minerals (carbonates, oxides, sulfides, etc.). Silicate minerals are the building blocks of most common rocks, including sandstone, basalt, and granite. Carbonates rocks (e.g., limestone and dolomite) and evaporates are notable exceptions. Quartz (silicon and oxygen), feldspar (silicon, oxygen, aluminum, sodium, and potassium), amphibole (silicon, oxygen, aluminum, iron, potassium, and magnesium), mica (silicon, oxygen, aluminum, iron, calcium, and magnesium), pyroxene, mica, and clay are all silicate minerals in order of hardness, with quartz being the hardest and clay being the softest. Alkali feldspar-K is also known as K-feldspar.

The 1960 magnitude-9.5 Valdivia earthquake (or the Great Chilean earthquake) shook the ground for approximately 10 minutes and killed more than 1,600 people. It is the biggest earthquake ever recorded. Many of the deaths were caused by the tsunami that ensued from the quake. The tsunami waves raced across the Pacific Ocean and devastated Hilo, Hawaii. Waves as high as 11 meters were recorded 10,000 kilometers from the epicenter and as far away as Japan and the Philippines—they also affected eastern New Zealand, southeast Australia, and the Aleutian Islands in the northern Pacific Ocean in the Alaska- Siberia region. The 1976 magnitude-7.8 Tangshan (China) earthquake shook the ground for about 23 seconds and killed around 750,000 people. So far, it is the largest quake of the twentieth and twenty-first centuries in terms of the death toll. The 2010 magnitude-7 Haiti earthquake shook the ground for about 90 seconds. It was followed two months later by a magnitude-6 aftershock, which completely destroyed the already fragile infrastructures by the main shock and years of economic neglect. More than 160,000 people died. The 2004 magnitude-9.1 Indonesia earthquake was the third-largest earthquake in the world in the twentieth and twenty-first centuries so far. The quake killed more than 225,000 people, the majority of them on the nearby Sumatra, Indonesia, mainland; many of them were washed out to sea by the powerful tsunami waves ensuing from the quake. These waves crisscrossed the Indian Ocean and ravaged the coastal regions of 12 Asian countries, including Thailand, Sri Lanka, and India, reaching the coasts of East Africa.

The March 11, 2011, Tohoku, Japan, magnitude-9.1 earthquake and the ensuing tsunami near the east coast of Honshu caused nearly 20,000 deaths and more than 300 billion dollars in damage, ranking as one of thc worst natural disasters ever recorded. The tsunami waves were responsible for most of the damage and most of the lives lost. These waves also hit the Fukushima Nuclear Power Plant and overwhelmed its nuclear reactors. The waves were 15 m high when they hit the nuclear reactors. The 6-m seawall was insufficient. A nuclear disaster ensued in addition to the damages and lost lives associated with the earthquake and tsunami. Our second key objective in this chapter is to help readers familiarize themselves with the dynamics of the solid Earth, which leads to these global tectonic events. Because volcanoes and earthquakes are both linked to movements of the Earth's tectonic plates, our discussion here will also include volcanism.

Key sentences: *Large earthquakes have effects far larger than 100 kilometers around the epicenter; very large earthquakes, like the 2010 Maule earthquake, are global tectonic events and can be monitored thousands of kilometers away from the eventual epicenter. On the* *moon (Earth's moon), more than 2,000 moderate moonquakes were recorded in six years by the Apollo program, which is more than 160 moonquakes in six months of a magnitude greater than 3. On Earth, we record on average more than 13,000 moderate earthquakes of a magnitude greater than 3 every year. InSight (NASA mission on Mars) has recorded about four or so moderate quakes (i.e., M_W between 3 and 4) so far in more than two years.*

Figure 3.2. An illustration of clay minerals. They are fine-grained soft minerals and common in soil and sedimentary rocks (source: en.wikipedia).

3.1. Minerals

3.1.1. Silicates and Non-Silicates

Again, rocks are the Earth's materials made of minerals. Most rocks have more than one kind of mineral. For example, granite rocks are made of several minerals, including potassium feldspar (alkali feldspar), plagioclase feldspar, quartz, hornblende, and diorite. Some rocks are made of one mineral and they are characterized as monomineralic. For example, glacial ice is water-ice monomineralic, limestone is calcite monomineralic, and rock salt is halite monomineralic.

Also, minerals possess crystal structures and cannot be broken into simpler components because they result from natural geological processes. They are able to maintain a set shape nearly indefinitely, they have a definable chemical composition, and they are generally inorganic. As defined in Chapter 2, liquids and non-crystalline materials such as glass are not minerals. Manmade minerals are called synthetic minerals (e.g., industrial diamonds).

Many different kinds of minerals exist, as illustrated in Figure 3.1 and Table 3.1. There are about 50 minerals in the crust and 98.5 percent of the crust mass of these minerals are composed oxygen (O), silicon (Si), aluminum (Al), iron (Fe), Calcium (Ca), sodium (Na), potassium (K), and magnesium (Mg), as described in Chapter 2. They can be divided into two broad classes: silicate and non-silicate minerals, based on the Si/O-ratio. The Si/O-ratio of silicate minerals varies from 0.25 (i.e., SiO_4) to 0.5 (i.e., SiO_2). About 92 percent of the crust is composed of silicate minerals and 8 percent of non-silicate minerals. All silicate minerals contain silicon and oxygen, the two most abundant elements in the Earth's crust, in varying proportions, ranging from SiO_2 to SiO_4. Quartz, feldspar, amphibole, pyroxene, mica, and clay are all silicate minerals. Non-silicate minerals include oxides (hematite), sulfides (pyrite), sulfates (gypsum), halides (halite), and carbonates (calcite and dolomite). Native minerals are single-element minerals, such as gold (Au), copper (Cu), diamond (C), graphite (C), and sulfur (S).

Electron microscopes and X-ray diffraction machines (Bish and Post, 1989; Moore and Reynolds, 1997; Erni et al., 2009; and Rudenberg and Rudenberg, 2010) are the main tools used to identify minerals. For example, an electron microscope allows us to see atoms by passing a beam of electrons through material with a detector at the opposite end. Atoms scatter electrons, which pass between them. A shadow on the detector indicates a row of atoms. We can therefore determine the shape of the crystal lattice (lattice atoms are held in place by atomic bonds) that controls the shape and many properties of minerals.

As we cannot all have X-ray diffraction machines or electron microscopes in our hands all the time, we sometimes identify minerals by visual chemical properties and physical properties, including color, hardness (the measure of a mineral's ability to resist scratching), weight (or density), and cleavage (the tendency of a mineral to break along a plane of weakness in the crystal lattice). For example, hematite (Fe_2O_3) can have various colors, but its streak (the color of the pulverized powder of a mineral) is always red-brown (rust) and quartz is not red. Quartz is three times harder than gypsum, twice as hard as calcite, and just slightly harder than alkali feldspar. Corundum is the hardness mineral in Table 3.1.

The minerals can also be identified based on their reactions to acid and to the presence of a magnetic field, their salty taste, and their striations. Calcite reacts with hydrogen chloride (HCl) and gives off CO_2, quartz does not react with acid, magnetite (Fe_3O_4) is strongly magnetic (i.e., magnetite will strongly attract a magnet), halite (NaCl) has a salty taste, and pyrite (FeS_2) has striations.

As shown in Table 3.1, non-silicate minerals are classified as oxides (minerals forming with O^{2-}), sulfides (minerals forming with S^-), sulfates (minerals forming with ${SO_4}^{2-}$), halides (minerals forming with Cl^- or F^-), and carbonates (minerals forming with ${CO_3}^{2-}$).

Note that the zeolite minerals in Table 3.1 play an important role in the construction of modern materials with voids, as described in Ikelle (2022b and 2022c). The chemical compositions of zeolites include a variable cation M which can be Na^+, K^+, Li^+, and so on. This cation has a valence n with x being 2.0 or more, and y being the mole of water in the voids. For example, the formula of sodium zeolites (also known as hydrated aluminum silicate) may be represented by $NaAlSi_2O_6 \cdot H_2O$. The sodium-zeolite mineral contains a relatively large central open area, which will allow large ions or other molecules to move through or reside within its framework (e.g., Coker and Davis, 1999).

As described in Chapter 2, we can also use artificial intelligence machines to facilitate the identification of various minerals and rocks. The mineral composition of Earth, Mars, Venus, etc., are all unique because they formed in different regions of the solar system at significantly different temperatures; Mercury's surface is most dominated by metal oxides (e.g., AL_2O_3), that of Venus, by iron-nickel alloys and silicate minerals, that of Mars, by Triolite (FeS) and hydrated minerals containing water, and that of Jovian planets, by water, ammonia, and methane. Likewise, different asteroids have different chemical compositions, which means they must have formed in different regions of the asteroid belt.

Table 3.1. Some of Earth's minerals—the vast majority of the minerals that make up the rocks of the Earth's crust are silicate minerals. These minerals include a great variety of clay minerals (e.g., kaolinite, illite, and smectite). Quartz is very common. Most beach sand is composed of SiO_2. Halide, sulfide, oxide, carbonate, and sulfate are non-silicates. Quartz is very common. Most beach sand is composed of SiO_2. The oldest mineral, zircon, is about 4.4 billion years old.

Minerals	Composition	Class	ρ (g/cm^3)
Quartz	SiO_2	Silicate	2.5 – 2.7
Plagioclase feldspar	$CaAl_2Si_2O_8$ or $NaAlSi_3O_8$	Silicates	2.6 – 2.8
K-feldspar	$KAlSI_3O_8$	Silicates	2.6
Olivine	(Fe or Mg)$_2SiO_4$	Silicates	3.2 – 4.4
Amphibole	$Ca_2Fe_5Si_8O_{22}(OH)_2$	Silicates	2.9 – 3.6
	or $Ca_2Mg_5Si_8O_{22}(OH)_2$	Silicates	
Pyroxene	(Ca , Mg, or Fe)SiO_3	Silicates	2.8 – 3.7
Kaolinite	$Al_2(Si_2O_5)(OH)_2$	Silicate	2.2 – 2.7
Muscovite	$KAl_3(Si_3O_{10})(OH)_2$	Silicate	2.9
Biotite	(Mg or Fe)$[(AlSi_3O_{10})(OH)_2]$	Silicate	2.9 – 3.4
Zircon	$ZrSiO_4$	Silicate	4.5–4.6
Zeolites	$M_{x/n}Si_{1-x}Al_xO_2.yH_2O$	Silicate	2.2 – 2.44
Halite	NaCl	Halide	2.2
Sylvite	KCl	Halide	1.9 – 2.0
Galena	PbS	Sulfide	7.4 – 7.6
Pyrite	FeS_2	Sulfide	4.9 – 5.2
Corundum	Al_2O_3	Oxide	4.0 – 4.1
Hematite	Fe_2O_3	Oxide	4.9 – 5.3
Water-ice	H_2O	Oxide	0.88 – 0.92
Calcite	$CaCO_3$	Carbonate	2.6 – 2.7
Dolomite	$MgCO_3$, $Mg(CO_3)_2$	Carbonate	2.7
Anhydrite	$CaSO_4$	Sulfate	2.98
Gypsum	$CaSO_4.2(H_2O)$	Sulfate	2.2 – 2.6
Barite	$BaSO_4$	Sulfate	4.5

Notice also that the metal elements (e.g., aluminum, iron, calcium, potassium, sodium, and magnesium) used to make things (e.g., airplanes, bus, car, and bike) are all found in minerals within rocks in the Earth's crust. Metal elements make up a relatively small percentage of the Earth's crust compared to non-metals. Oxygen and silicon alone represent more than 74 percent of the Earth's crust (by weight). Most of the metal elements are combined with other elements in compounds, usually with oxygen, silicon or other non-metals. In addition to these compounds, there are minerals. Minerals containing metals of value are called mineral ores [e.g., Magnesite ($MgCO_3$), Bauxite (Al_2O_3), Hematite (FeO_3), sphalerite (ZnS), and calaverite ($AuTe_2$)]. Mining is the process of finding underground rocks that contain minerals, bringing them to the surface, and then extracting the metal elements from the mineral ores and/or the rocks.

Table 3.2. Key oxides used to describe chemical compositions of rocks

Oxides	Composition	Density (g/cm^3)
Silica	SiO_2	2.65
Titanium dioxide	TiO_2	4.23
Aluminum oxide	Al_2O_3	3.99
Hematite	Fe_2O_3	5.27
Iron oxide	FeO	5.75
Manganese oxide	MnO	5.43
Magnesium oxide	MgO	3.60
Calcium oxide	CaO	3.34
Sodium oxide	Na_2O	2.27
Potassium oxide	K_2O	2.13
Water	H_2O	1.00
Carbon dioxide	CO_2	1.98
Phosphorus pentoxide	P_2O_5	2.39
Uraninite	UO_2	11.0
Thorianite	ThO_2	10.0

3.1.2. Oxides and Rare Earth's Minerals

There are particular minerals called oxides. These oxides allow us to identify similarities in the chemical compositions of minerals and therefore of rocks. For example, orthoclase (feldspar)

($K_2O\,.\,Al_2O_3\,.\,6SiO_2$), biotite ($K_2O\,.\,Al_2O_3\,.\,6(MgFe)O$), and muscovite ($K_2O\,.\,Al_2O_3\,.\,6SiO_2\,.\,2HO_2$) include potassium oxide (K_2O) in their chemical compositions. Thus, potassium oxide can be used to describe potassium concentrations in these minerals and consequently rocks containing potassium, a radioactive element (see Chapter 2) used in generating nuclear energy and nuclear bombs. In other words, rocks can be characterized by their mineral compositions which can themselves be characterized by oxides. Therefore, the chemical compositions of rocks can directly be expressed in terms of oxides. Table 3.2 shows some of these key oxides. Oxide minerals are common in all rock types. They are the source of the rust-red color of many rocks.

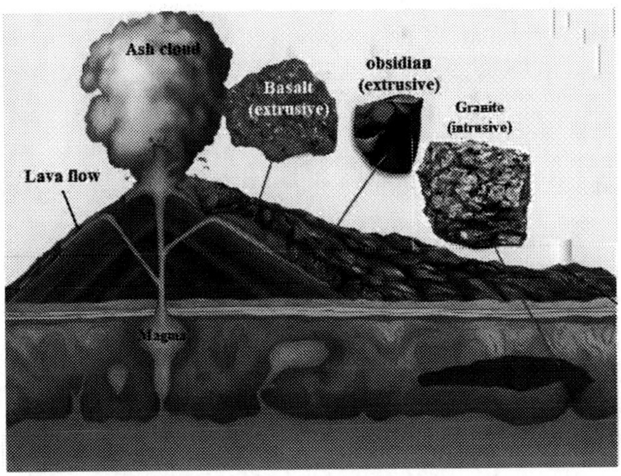

Figure 3.3. The magma below the surface of the Earth solidifies as intrusive igneous rocks (also plutonic rocks). Granite is an example of an intrusive igneous rock. The lava cools on the surface of the Earth as extrusive igneous rocks (also known as volcanic rocks). Rhyolite is an example of volcanic rock. In general, intrusive rocks are coarse-grained because they cool slowly and extrusive rocks are fine-grained because they cool rapidly.

3.2. Rock Types

3.2.1. Igneous Rocks

There are three main types (or classes) of rocks: igneous, sedimentary, and metamorphic. Igneous rocks are related to the upward motion of magma and volcanic eruptions (see Figure 3.3). They are mostly formed from the cooling and crystallization of magma below the Earth's surface (crust, at depths of several hundreds of meters to several tens of kilometers). These igneous rocks are known as intrusive igneous rocks or plutonic igneous rocks.

Igneous rocks are also formed at the Earth's surface. When magma reaches the surface of the Earth or sea bottom, it forms glassy and microcrystalline volcanic lavas, which cool and crystallize at the Earth's surface. These igneous rocks are known as extrusive igneous rocks or volcanic igneous rocks. Remember that a molten rock below the surface of the Earth is called magma and a molten rock on the surface of the Earth is known as lava.

There are extrusive igneous rocks resulting in ash (pyroclastic debris) from explosive volcanic eruptions. While some lava is produced during volcanic eruptions, the majority of the eruption is in the form of volcanic ash. The ash ends up falling back to the Earth's surface as small fragments, which cool and crystallize as igneous rocks. This type of extrusive igneous rock is called pyroclastic.

Intrusive igneous rocks are coarse-grained (because of the slow cooling of magma in the subsurface) and they include diorite, gabbro, granite, pegmatite, and peridotite. Extrusive igneous rocks are fine-grained (because of the rapid cooling of lava at the Earth's surface) and they include andesite, basalt, dacite, obsidian, pumice, rhyolite, scoria, and tuff. The predominant minerals in these rocks are feldspars and quartz. Igneous rocks are so tightly solidified, leaving little free space, that they almost contain no pore space to hold fluid (they have almost zero porosity) and therefore they very rarely contain hydrocarbons. However, because of their solidity, igneous rocks are heavily used in civil engineering for the construction of housing, roads, tunnels, canals, etc.

The above classification of igneous rocks is mainly based on texture (grain size and shape), magma location, and volcanic eruptions.

These rocks can also be classified based on their mineralogical compositions and therefore on their chemical composition. Actually, the variety of igneous rocks indicated above is due to the fact that magmas do not always have the same chemical composition. There are four main chemical groups of magmas (Le Maitre, 1989):

- Acidic (felsic) magmas are magmas rich in silicon dioxide (SiO_2), sodium oxide (Na_2O) (greater than 70 percent), and potassium oxide (K_2O). Rocks produced from such magmas may have up to 78 percent by weight of SiO_2 and are known as acidic or felsic rocks. Their mineral composition includes quartz, feldspars, muscovite, and biotite. Granite (intrusive) and rhyolite (extrusive) are examples of acidic rocks; quartz is an essential component of granite.

- Intermediate-acidic magmas are magmas rich in silicon dioxide (SiO_2), sodium oxide (Na_2O), and potassium oxide (K_2O), as well as calcium oxide (CaO) and aluminum oxide (Al_2O_3). Rocks produced from such magmas have SiO_2 values in the range of 65 to 70 percent by weight and are known as intermediate-acidic. Their mineral composition includes amphibole, clinopyroxene, and plagioclase feldspar. Granodiorite (intrusive) and dacite (extrusive) are examples of intermediate-acidic rocks.

- Intermediate magmas are magmas rich in silicon dioxide (SiO_2), sodium oxide (Na_2O), and potassium oxide (K_2O), as well as calcium oxide (CaO) and aluminum oxide (Al_2O_3). Rocks produced from such magmas have SiO_2 values in the range of 55 to 65 percent by weight and are known as intermediate. Their mineral composition includes amphibole, clinopyroxene, and plagioclase feldspar. Diorite (intrusive) and andesite (extrusive) are examples of intermediate rocks.

- Basic (mafic) magmas are magmas rich in calcium oxide (CaO), magnesium oxide (MgO), and iron oxide (FeO). Rocks of this type have SiO_2 values in the range of 45 to 55 percent by weight and are known as mafic or basic rocks. Their mineral composition includes clinopyroxene and plagioclase feldspar. Gabbro (intrusive) and basalt (extrusive) are examples of basic rocks.

- Ultrabasic (ultramafic) magmas are magmas poor in sodium oxide (SiO_2), but with large amounts of magnesium oxide (MgO), and iron oxide (FeO). Ultrabasic rocks may have SiO_2 values as low as 38 percent by weight and are also known as ultramafic. Their mineral composition includes olivine and orthopyroxene. Peridotite (intrusive) and pyroxenite (intrusive) are examples of basic rocks.

Thus, the mineralogy of an igneous rock is controlled primarily by the composition of the magma from which it is formed. The key to chemical classification in igneous rocks is the amount of Silica (SiO_2) in the magma. If magmas do not have much silica, their minerals are dominated by magnesium (Mg) and iron (Fe), hence, the term mafic, or even ultramafic for the really silica-poor varieties. Silica-rich magmas have feldspar in them and are called felsic as a result. The words *acidic* and *basic* are used for felsic and mafic, respectively, but the terms have nothing to do with pH (a measure of the acidity or alkalinity of water-soluble substances). One can often tell how much silica is in a rock just by its color. The more silica, the lighter the color.

Notice that minerals forming igneous rocks are also sometimes divided into essential minerals and accessory minerals. An essential mineral is a mineral whose presence in an igneous rock is essential to defining the root name of that rock. For example, olivine and pyroxenes are essential to defining the root name *ultramafic* and hence are essential minerals. The essential minerals of mafic rocks are plagioclase feldspar, clinopyroxene, olivine, and amphibole. Those of intermediate rocks are amphibole, clinopyroxene, and plagioclase feldspar and acidic rocks are K-feldspar (orthoclase), quartz, biotite, and amphibole. Accessory minerals are any mineral in an igneous rock not essential to the name of the rock. Accessory minerals of igneous rocks include apatite, chlorite, epidote, ilmenite, magnetite, muscovite, serpentine, titanite, and zircon.

Note also that NASA's Pathfinder spacecraft, which landed on Mars on July 4, 1997, provided us with the first Martian rock that turned out to contain many of the same minerals that were common on Earth, such as quartz, feldspar, and amphibole. The rock composition was similar to andesite, a rock formed by volcanic activity. Andesite on Mars was an unexpected find as volcanic eruptions are rare on the red planet, especially in the last hundred million years.

Table 3.3. Average chemical compositions of igneous rock types

Oxide	? (%)	? (%)	? (%)	? (%)
SiO_2	71.3	57.94	49.2	42.26
TiO_2	0.31	0.87	1.84	0.63
Al_2O_3	14.32	17.02	15.74	4.23
Fe_2O_3	1.21	3.27	3.79	3.61
FeO	1.64	4.04	7.13	6.58
MnO	0.05	0.14	0.2	0.41
MgO	0.71	3.33	6.73	31.24
CaO	1.81	6.79	9.47	5.05
Na_2O	3.68	3.48	2.91	0.49
K_2O	4.07	1.62	1.1	0.34
H_2O	0.77	1.17	0.95	3.91
CO_2	0.05	0.05	0.11	0.30
P_2O_5	0.12	0.21	0.35	0.10

Quiz: Determine whether the compositions of rock samples associated with the columns in Table 3.3 correspond to felsic, intermediate, mafic, or ultramafic.

3.2.2. Metamorphic Rocks

Metamorphic rocks are formed by the modification of pre-existing igneous, sedimentary, and even metamorphic rocks by high temperatures (i.e., intense heat) and/or high pressures. Most metamorphic rocks form at a depth of 40 to 100 kilometers, where pressures are 10,000 to 30,000 times greater than at the surface of the Earth and temperatures are $1,000$ degrees Celsius and higher. They are the results of a solid-state process in the subsurface. Materials under increasing temperature tend toward being liquid (melting) and under increasing pressure, they tend toward being solid; materials under decreasing temperature tend toward being solid, and under decreasing pressure, they tend toward being liquid. This dynamic keeps metamorphic rocks solid in the crust; the pressures on the Earth's surface are not high enough to cause rocks to melt and they increase fast enough to keep the metamorphic rocks solid deep in the crust, despite the high temperatures.

Metamorphism means to change form. It leads to change in the mineralogy, texture, and sometimes in the chemical composition of rocks through the intense heat and high pressure. Heat provides the energy to drive chemical reactions that result in the recrystallization of existing materials and/or the formation of new materials. The increasing pressure causes minerals to squeeze together and the developments' foliation. Some minerals may keep their composition but change their atom packing due to increased pressure. Metamorphic rocks are grouped into two main categories: foliated metamorphic rocks and nonfoliated metamorphic rocks. With foliated metamorphic rocks, there is a repetition of planar surfaces, or layers, as shown Figure 3.4b. The foliated metamorphic rock shown in this figure is gneiss. It is made up of granular mineral grains. It is commonly formed from shale, which is a clastic sedimentary rock, or from granite, which is an intrusive igneous rock.

Layers can be paper-thin (schist) or meters thick (gneiss). Examples of foliated metamorphic rocks include gneiss (a coarse-grained granular rock in which foliation results from alternating layers of light and dark minerals), which is the result of the metamorphism of shale, schist (a coarse-grained metamorphic rock that exhibits an ease of splitting due to the parallel orientation of coarse visible micas), which is the result of the metamorphism of shale at lower temperature and pressure than those resulting in the formation of gneiss, and slate, which is the result of the metamorphism of shale at lower temperature and pressure than those resulting in the formation of schist (see Table 3.4). Granite and rhyolite will be little affected by metamorphism until very high-grade conditions are reached because most of their minerals (like quartz) are stable over a wide range of temperatures.

At very high temperatures of metamorphism, there will be some recrystallization, and gneissic textures will likely be developed. Basalt will be metamorphosed into chlorite schist. Basalt is metamorphosed into an amphibole-rich rock (amphibolite) at higher temperatures. Quartz sandstone will not be significantly affected at low temperatures, but at higher grades the individual quartz grains will develop quartz overgrowths to produce the very hard rock known as quartzite. At various temperatures, limestone will undergo recrystallization to marble—a rock made up of interlocking crystals of calcite or dolomite, but with little or no foliation.

Figure 3.4. Examples of metamorphic rocks. (a) shows marble, which is a non-foliated metamorphic rock that is produced by the metamorphism of limestone or dolomites. (b) shows gneiss, which is a foliated metamorphic rock. It is commonly formed from shale or granite. (c) shows schist, which is a foliated metamorphic rock. It is the result of the metamorphism of shale at lower temperatures and pressures than those resulting from the formation of gneiss. (d) shows slate, which is also a foliated metamorphic rock. It is the result of the metamorphism of shale at lower temperatures and pressures than those resulting in the formation of schist.

Non-foliated metamorphic rocks are those without layering. They occur when heat is high but pressure is low without rocks melting. This dynamic prevents dominant minerals from showing visible alignments. Examples of non-foliated metamorphic rocks include marble, which is the result of the metamorphism of limestone, and quartzite, which is the result of the metamorphism of quartz sandstone. Virtually all igneous rocks and most metamorphic rocks are made of silicate minerals.

Table 3.4. Typical products of metamorphism for different rocks under different temperature conditions. Slates and phyllites are characterized by chlorite, muscovite, and biotite. Schist often has garnet and staurolite. The temperatures are in degrees Celsius.

Original rock	200 – 350°	300 – 550°	500 – 650°	above 650°
Shale	slate	phyllite	schist	gneiss
Granite	no change	no change	no change	gneiss
Basalt	chlorite schist	chlorite schist	amphibolite	amphibolite
Sandstone	no change	little change	quartzite	quartzite
Limestone, dolomite	marble	marble	marble	marble

One of the commonly observed textural changes in all metamorphic rocks is the elimination of porosity (holes). Marble is used in statues and the walls and floors of buildings (inside and outside). They are usually highly polished. The exterior of the famous Taj Mahal in India is constructed mainly of metamorphic rock marble. Slate is also used in buildings, especially in roof and floor tiles.

A metamorphic facies (sets of mineral assemblages) is another concept used to classify metamorphic rocks. In general, metamorphic rocks do not drastically change their chemical composition during metamorphism (changes in pressure and temperature). The cases in which changes in chemical composition do occur are collectively called metasomatism (e.g., Harlov and Austrheim, 2013). Such changes include the replacement of one rock by another of a different mineralogical and chemical composition. Thus, the mineral assemblages that are observed must be an indication of certain tectonic settings under the temperature and pressure conditions that the rock was subjected to. This pressure and temperature environment is referred to as metamorphic facies (e.g., Harlov and Austrheim, 2013). Facies boundaries are defined by the appearance or disappearance of a mineral or group of minerals, and not a specific pressure and temperature.

The concept of metamorphic facies was first proposed by Eskola (1939). The main categories of metamorphic facies are granulite facies, epidote-amphibolite facies, amphibolite facies, glaucophane-schist facies, and eclogite facies. *Granulite facies* are rocks which form under an intense temperature of 650 to 1,100 degrees Celsius and pressures of 2 to 13 kilobars may be reached. The minerals found in the rocks of the granulite facies include pyroxene, biotite, garnet, calcium plagioclase, and quartz. *Amphibolite facies* are rocks which are formed under conditions of moderate to high temperatures of 500 to 750 degrees Celsius maximum and pressures of 3 to 12 kilobars may be reached. Amphibole, diopside, epidote, plagioclase, almandine, and wollastonite are minerals typically found in rocks of the amphibolite facies. The disappearance of epidote and an increase in calcium and in plagioclase are characteristics of chemical changes as metamorphic intensity increases through these facies. *Glaucophane Schist facies* are rocks which are formed at temperatures of 200 to about 500 degrees Celsius and pressures of 6 to 12 kilobars may be reached. Aragonite and white mica are minerals typically found in rocks of the glaucophane schist facies. *Epidote-amphibolite facies* are the rocks which form under moderate temperatures of 250 to 400 degrees Celsius and up to 4 kilobars. Calcite, quartz, tremolite, epidote, and diopside are minerals typically found in rocks of the Epidote-amphibolite facies. *Eclogite facies* are rocks which are formed under a temperature of about 400 to 1,000 degrees Celsius and pressure of 8 to 10 kilobars may be reached. Aragonite, quartz, plagioclase, diopside, and hypersthene are minerals typically found in rocks of the eclogite facies.

3.2.3. Sedimentary Rocks

Sedimentary rocks are formed from sediments normally inside water bodies (sea, oceans, etc.). Sediments refer to (1) all solid particles that are the result of the breakdown of pre-existing rocks (igneous, metamorphic, and sedimentary) and (2) minerals extracted from fossil plant matter and skeletal structures. Examples of sediments are shown in Figure 3.5. A sedimentary rock is any rock composed of sediments.

Figure 3.5. An illustration of physical (mechanical) weathering processes, which is the breaking down of rocks into smaller pieces, known as sediments. Weathering is part of a cycle that includes erosion (i.e., the movement of sediments from one place to another), transportation and deposition (i.e., the placement of sediment at a new location), burial (i.e., the accumulation of sediment layers), diagenesis (i.e., turning loose sediment into sedimentary rock), heat, pressure, and melting. Weathering, erosion, and transportation processes create an apron of debris at the margins of slopes.

Seventy-five percent of the Earth's surface (land) is composed of sediments and sedimentary rocks and most of the seafloor, except at spreading ridges. Igneous and metamorphic rocks represent 95 percent of the crust, and igneous rocks represent 100 percent of the mantle because they are formed beneath the Earth's surface, whereas sedimentary rocks are formed, in most part, at the Earth's surface. In most cases, sediments are turned into sedimentary rocks only when they are buried by other sediments to depths of several hundreds of meters.

The sedimentary rocks come in four main classes, depending on the origin of the sediments that comprise them: (1) clastic sedimentary rocks, which are made of sediments resulting in a physical and chemical breakdown of pre-existing rocks (Figure 3.6); (2) chemical sedimentary rocks, which are made of sediments, resulting in chemical reactions, especially minerals precipitated by or evaporated from water (Figure 3.7); (3) biochemical sedimentary rocks, which are made of sediments, resulting from the remains of shells, often

Figure 3.6. Examples of clastic sedimentary rocks. (a) shows a conglomerate and (b) shows a breccia. In a conglomerate, the coarse-grained clasts are well-rounded and are embedded within a fine-grained matrix. In a breccia, the coarse-grained clasts are very angular. (c) shows a sandstone rock. It is composed almost entirely of sand-sized grains (0.063 to 2 millimeters) cemented together through lithification. (d) Shale is a fine-grained rock formed by the compaction of well-rounded silt-and clay-sized grains (too small to see with the naked eye).

broken shells, of organisms; and (4) organic sedimentary rocks, which are organic sediments that come from the remains of plants (Figure 3.8). Examples of clastic sedimentary rocks are breccia, sandstone and shale. Conglomerates and breccia contain an abundance of coarse-grained clasts. In a breccia, the coarse-grained clasts are well-rounded and are embedded within a finer-grained matrix. In a breccia, the coarse-grained clasts are very angular. They are also embedded in a fine-grained matrix. Sandstone rock is composed almost entirely of sand-sized grains (0.063 to 2 millimeters) cemented together through lithification. Shale is a fine-grained rock formed by the compaction of well-rounded silt- and clay-sized grains (too small to see with the naked eye).

Examples of chemical sedimentary rocks are dolomite, chert, rock salt, and rock gypsum. Dolomite, also known as dolostone and dolomite rock, is a carbonate rock composed primarily of the mineral dolomite (calcium magnesium carbonate). It is formed by the post-depositional alteration of limestone by magnesium-rich groundwater. Chert is a hard rock made of micro- and cryptocrystalline silica. Chert can develop as nodules inside other rocks or as rock layers. Most types of chert are hypothesized to originate from silica derived from a solution in water, biochemical sediments, or lava flow and volcanic ash. Rock salt forms from the evaporation of ocean or saline lake waters. So it is composed of sodium chloride, also known as salt, which is an ionic compound with the formula NaCl. It is often mined for use in the chemical industry or for use as a winter highway treatment. Gypsum as a rock type is an aggregate of gypsum crystals.

Examples of biochemical sedimentary rocks are limestone and chalk. Fossiliferous limestone contains visible fossil-sized shells, shell fragments, and other fossils in a fine-grained matrix. Shells indicate past life. Non-fossiliferous limestone (also called crystalline limestone) is a very fine-grained limestone devoid of obvious fossils. Chalk, a very fine-grained limestone, contains microscopic fossils. It is a soft, friable, and porous variety of limestone. Chalk is composed of calcareous tests of marine microorganisms like coccoliths and foraminifera. Coal is a combustible rock containing more than 50 percent by weight of carbonaceous material. It is a major fossil fuel, although its economic importance is declining.

Examples of organic sedimentary rocks are peat and lignite coal, which are made of organic materials derived from plants and animals. Note that limestone ($CaCO_3$) is widespread on Earth—most with biologic origin. It contains as much carbon as the atmosphere of Venus.

Figure 3.7. Examples of chemical sedimentary rocks. (a) shows dolomite, also known as dolostone and dolomite rock. It is a carbonate rock that is formed by the post-depositional alteration of limestone by magnesium-rich groundwater. (b) shows chert. It is a hard rock made of silica. (c) shows rock salt. It forms from the evaporation of ocean or saline lake waters. So it is composed of sodium chloride, also known as salt. It is often mined for use in the winter highway treatment. (d) shows a rock of gypsum. Gypsum as a rock type is an aggregate of gypsum crystals.

All rocks are important to understand the Earth's history, but sedimentary rocks play a special role because they preserve evidence of surface processes responsible for them (Carlson et al., 2009). Oceans rise and fall, lakes come and go, streams run and disappear, and deserts become forests and forests become swamps. Sedimentary rocks contain this information. It is the only rock type that preserves fossils. In addition, some sediments and sedimentary rocks are themselves natural resources or they are the host for resources such as coal, gas, oil, and water.

As discussed earlier, most sedimentary rocks form on the continents. They contain critical clues to the history of the continents, Earth's climate, important clues to the history of life, and important and valuable natural resource deposits. Many sedimentary rocks have a single layer with a definable top and bottom.

Figure 3.8. Examples of biochemical and organic sedimentary rocks. (a) shows fossiliferous limestone, which contains visible fossil-sized shells, shell fragments, and other fossils in a fine-grained matrix. (b) shows a non-fossiliferous limestone (also called crystalline limestone). It is a very fine-grained limestone devoid of obvious fossils. (c) shows chalk, a very fine-grained limestone, containing microscopic fossils. Chalk is a soft, friable, and porous variety of limestone. (d) shows coal.

In summary, Table 3.5 contains information on how the major rock types are formed, their mineral composition, their textures, and their subcategories. Shale is composed primarily of clay minerals. Sandstone is composed primarily of quartz. Carbonate rocks are made up primarily of carbonate minerals, with the two main types of carbonate being limestone and dolomite. Evaporites are salts formed from the evaporation of sea water: gypsum, anhydrite, halite, etc. Before we can fully comprehend the fundamental concepts of porosity and permeability that we will develop in later chapters, we need to gain a thorough understanding of sedimentary rocks.

Table 3.5. A contrast between (I) igneous, (M) metamorphic, and (S) sedimentary rocks based on the ways they are formed [F], their composition [C], and their secondary classification [S]. Note that glassy means that no mineral grains or crystals are present; aphanitic means that mineral grains are present, but they are too small to distinguish with the naked eye; porphyritic means that there are large crystals of one or more minerals set within a groundmass of fine-grained material; and pyroclastic means that there are angular fragments of volcanic rock within a finer-grained matrix.

[F]:	(I)	Melting and crystallization of molten rocks.
	(S)	Accumulation of sediments (e.g., break down of pre-existing rocks, shell fragments, and organic matter) are compressed and cemented together.
	(M)	Pre-existing rocks that are substantial changes by heat and pressure; new minerals and textures are formed.
[S]:	(I)	• Intrusive igneous rocks: granite, diorite, and gabbro. Intrusives are generally crystalline because they have a long time to cool with large crystals which can be seen with the naked eye. They are coarse-grained. • Extrusive igneous rocks: rhyolite, andesite, and basalt. Extrusives can be glassy, aphanitic, porphyritic, and/or pyroclastic. They are fine-grained.
	(S)	• Clastic sedimentary rocks: sandstone and shale. • Biochemical sedimentary rocks: limestone and chalk. • Chemical sedimentary rocks: dolomite and chert. • Organic sedimentary rocks: peat and lignite coal
	(M)	• Foliated rocks: slate, schist, gneiss, mylonite • Non-foliated rocks: marble, quartzite, and hornfels.
[C]:	(I)	Largely of feldspar and quartz, some biotite, pyroxene, amphibole, and olivine, especially for basalt and gabbro.
	(S)	• Clastics: Quartz, feldspar, amphibole, clay, and silt. • Biochemicals include shells of organisms • Chemicals and organics: halite and gypsum.
	(M)	Biotite, garnet, quartz, feldspar, amphibole, and pyroxene.

Clastic sediments and clastic sedimentary rocks. Detrital (clastic) sediments are all solid particles of preexisting rocks yielded by mechanical and chemical weathering. As shown in Table 3.6, detrital sediments can be divided into four main classes: gravel, sand, silt, and clay (Selley and Morill, 1991). We basically move from coarse grains (visible grain size) to fine grains (microscopic grain size), as illustrated in Figure 3.10. Gravels include pebbles, cobbles, and boulders. Sands are sediment with grain size between 0.0625 and 2 millimeters. Mud is sediments with a grain size less than 0.0625 millimeters. It is important to keep in mind that clay (Figure 3.2) and silt are mud.

> **Quiz:** Determine the rock types in Figure 3.9.
> Select your answers among these three types: igneous, metamorphic, and sedimentary.

(a) (b) (c)

Figure 3.9. An illustration of the three major rock types.

Clastic (detrital) sedimentary rocks form the clastic sediments that are mostly clay minerals, quartz, feldspars, and micas. Clastic rocks with an average grain size greater than 2 millimeters are known as *rudites.* The common rudites are conglomerates and breccia. Clastic rocks with an average grain size between 2 millimeters and 0.5 millimeters are known as *arenites.* The common arenites are sandstone, quartzite, and arkose. Sandstone is composed principally of quartz. Clastic rocks with an average grain size smaller than 0.5 millimeters are known as lutites. The common lutites are shale and siltstone. Shale is composed principally of clay minerals.

As illustrated in Table 3.7, conglomerates contain pebbles (large grains) held together by a matrix of sand-, silt-, and clay-sized particles. These grains are poorly sorted. The combination of poorly sorted, predominantly coarse, rounded grains suggests that conglomerates form in high-energy environments, such as steep-gradient streams, and indicates that they spent considerable time in the transportation process and were ultimately deposited in a high-energy environment capable of carrying large clasts.

Table 3.6. Detrital sediments are generally divided into gravel, sand, silt, and clay along with associated rocks that form them.

Name of the Particle	Size Range (mm)	Example	Consolidated Rock
Boulder, cobbles, gravel, pebbles	> 2	> Pea	Conglomerate or Breccia
Medium sand	1/16 - 2	Table salt	Sandstone
Fine sand	0.075 - 0.42	Powdered sugar	Sandstone
Silt	1/256 - 1/16	Powdered sugar	Silstone
Clay	$< 1/256$	Powdered sugar	Claystone, Mudstone, and Shale

Sandstone (Figure 3.11) comprises about 20 percent of all sedimentary rocks and is formed in a variety of different environments, including fluvial (rivers), marine, coastal (oceans and beaches), aeolian (windblown), and glacial (ice). The differences in texture, sorting, and rounding help geologists decipher the environmental conditions that formed the sandstone. Quartz is by far the dominant mineral in sandstone (50-60 percent of the framework). The other dominant minerals are feldspars (10-20 percent of the framework) and clay (less than 5 percent of the matrix). Quartz sandstone is a single mineral rock. The only mineral it contains is quartz (SiO_2), but as many separate grains rather than a single crystal. If the clasts are so small you cannot see them, then your rock is shale, siltstone, or claystone, made of silt- or clay-sized particles: siltstone (greater than 66 percent silt), shale (about 66 percent silt and 33 percent clay), and claystone (greater than 66 percent clay).

Everyone has contact with products made from shale. If you

live in a brick house, drive on a brick road, live in a house with a tile roof, or keep plants in "terra cotta" pots, you have daily contact with items that were probably made from shale. Today, most of these items are made of synthetic (manufactured) clays that are mixtures of finely ground shale with water.

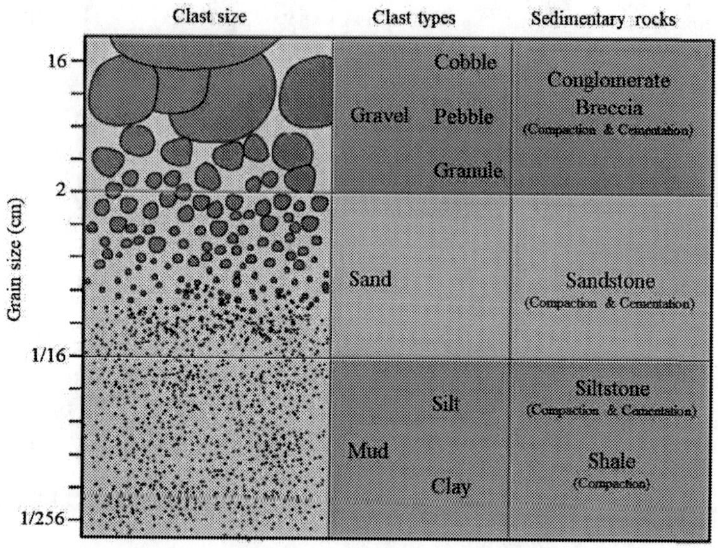

Figure 3.10. Important classifications of the detrital (clastic) sediment are as gravel, sand, and mud. Gravel is sediment with a grain size greater than 2 millimeters. Gravel includes pebbles, cobbles, and boulders. Sand is sediment with a grain size between 0.0625 and 2 millimeters. Mud is sediment with a grain size less than 0.0625 millimeters. Clay and silt are mud.

Cement is another common material that is often made from shale. To make cement, crushed limestone and shale are heated to a temperature that is high enough to evaporate all water and break down the limestone into calcium oxide and carbon dioxide. The carbon dioxide is lost as an emission, but the calcium oxide combined with the heated shale makes a powder that will harden if mixed with water and allowed to dry. Cement is used to make concrete and many other products for the construction industry.

Table 3.7. A description of clastic rocks based on their sediment composition. Column 2 (sediments), column 3 (composition) and column 4 (rocks).

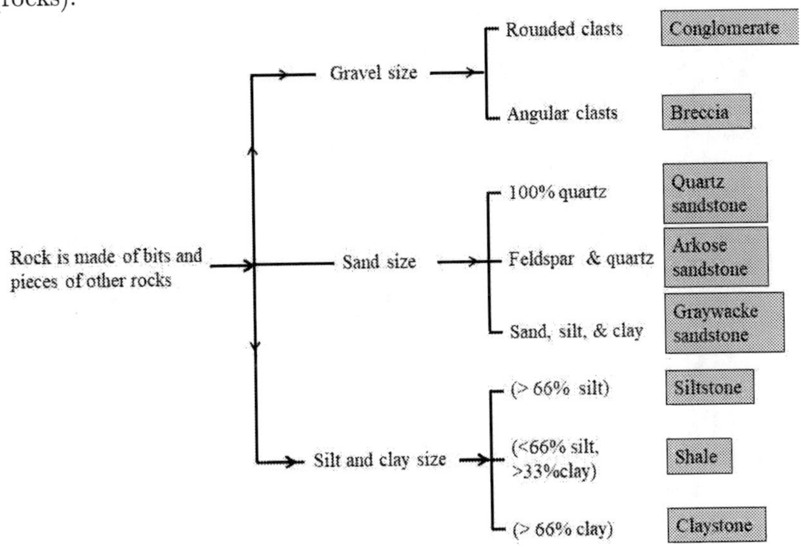

Chemical sedimentary rocks form when conditions favor a chemical reaction or process that causes chemicals dissolved in water to precipitate—precipitation is the separation of a solid from a solution like water. When water in a salty sea or lake evaporates, for example, it may leave behind halite (salt) and gypsum deposits, leading to rock salt and rock gypsum—rocks formed by the evaporation of sea or lake water are generally called evaporates and are the most common class of chemical sedimentary rock. In calcium-rich waters, changes in temperature or acidity may cause calcium carbonate to precipitate. The accumulation of calcium carbonate deposits can lead to the formation of crystalline limestone. Sometimes magnesium in water that enters the pores of a limestone rock can replace the calcium in the rock, turning limestone into another chemical sedimentary rock called dolostone. Groundwater flowing through rock can also precipitate quartz to replace minerals that were present. This precipitation produces non-biogenic chert. Also groundwater containing dissolved calcium and bicarbonate ions can precipitate calcite to form a chemically precipitated limestone called travertine. This precipitation can occur in lakes, hot springs, and caves.

Figure 3.11. (a) An illustration of sandstone and shale as part of a rock formation. (b) An illustration of sandstone, shale, and coal as part of a rock formation.

All these chemical sedimentary rocks, including their composition and texture, are shown in Table 3.8. Note that crystalline limestone and dolostone (also known as dolomite) are carbonate rocks because they are formed due to the precipitation of calcium carbonate or magnesium carbonate, respectively, from water. Note also that dolomite is a secondary, rather than a primary, sedimentary mineral, meaning that carbonate sediments do not start their lives as dolomite. Much dolomite is precipitated in the original calcium carbonate sediment.

Biochemical sediments and sedimentary rocks. The difference between chemical and biochemical sedimentary rocks is that in biochemical sedimentary rocks the shells of living organisms play a role in turning the ions into sediment. That means the presence and nature of biochemical sedimentary rocks are linked to the life requirements of the organisms that make them. In chemical sedimentary rocks, however, the process is inorganic, often resulting from a body of water evaporating and concentrating the ions. It is possible for one type of sedimentary rock to form from both chemical (inorganic) and biochemical (organically mediated) processes. That is the case with limestone and chert.

Table 3.8. Chemical sedimentary rocks are made from minerals that precipitate out of water solutions. Chemical Cherts are formed by replacing silica (SiO_2) with calcite, long after limestone was deposited. Rock salt and rock gypsum are rocks formed from evaporated sea and lake water. Rock salt is formed of halite (salt; NaCl) and rock gypsum is formed from gypsum (CaSO4H2O). Travertine is made calcium carbonate ($CaCO_3$) precipitated from groundwater. Dolostone is made of limestone altered by Mg-rich fluids.

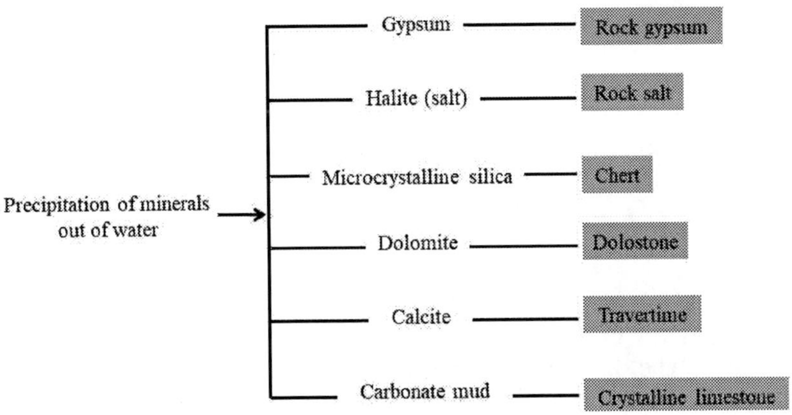

Lab: You can see evaporation and precipitation at work in a small-scale experiment. Stir a spoon of table salt into a glass of hot water until the salt crystals dissolve completely. Put the glass next to a window for several days, and the water will evaporate, leaving behind salt crystals. Table salt comes from the chemical sedimentary rock halite.

Limestone is made of calcite and aragonite, and aragonite reverts to calcite with time. It can occur as a chemical sedimentary rock, forming inorganically, due to precipitation, but most limestone is biochemical. In fact, limestone is by far the most common biochemical sedimentary rock, with sometimes lots of visible fossils. Other times, recrystallization during lithification has destroyed the remains, like in micrite and chalk. Micrite limestone is made of calcite mud and chalk, which is made of plankton shells. Biochemical chert is made of cryptocrystalline (microscopic crystals) quartz, formed from the shells of plankton that sat on the sea floor and dissolved into a silica-rich ooze. Silica in the pore fluid solidifies into a gel and the silica gel precipitates chert.

Lithification is here a process that transforms sediments into rock formation, by expelling much of the water contained in the sediment was deposited and forcing the rock grains to come much closer together to and re-crystallize.

About one-fifth of all sedimentary rocks are carbonate rocks (limestone, dolomite, and chalk). The two main kinds of carbonate rocks are limestone dolomite (dolostone) and chalk. Together with sandstone and shale, limestone and dolostone constitute reservoirs for most of the world's petroleum and gas reserves. Note that chalk is white and crumbly. It is formed from the shells of sea creatures and it does not weather, which is why there are chalk cliffs still standing by the sea.

Quiz: (1) What are the four types of sedimentary rocks? (2) How much of the Earth's crust is made up of sedimentary rock? (3) What is the difference between the terms *clastic* and *detrital*?

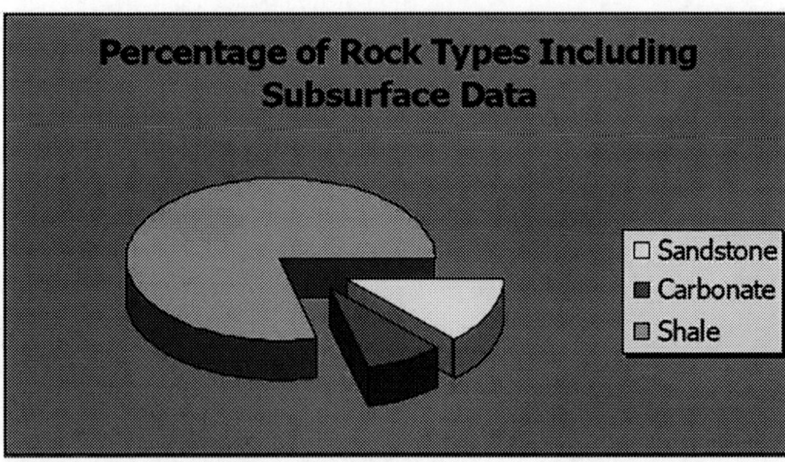

(a)

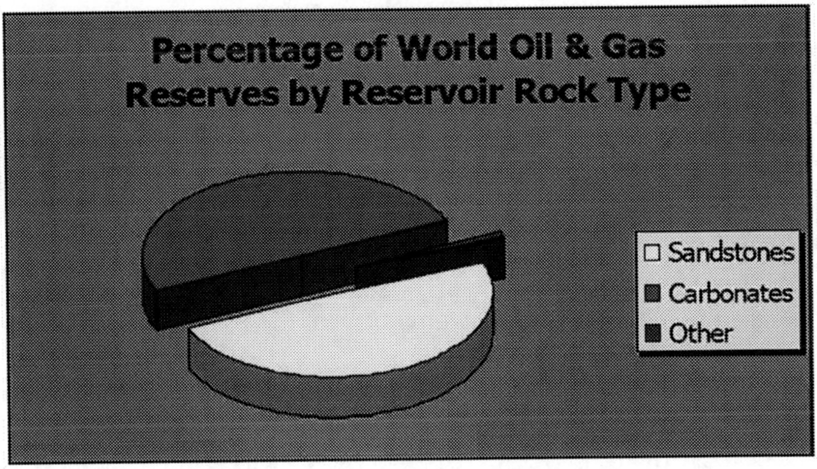

(b)

Figure 3.12. (a) Shale, sandstone, and carbonates make up about 99 percent of all sedimentary rocks, with shale being the most abundant and carbonates the least abundant. Shale is more abundant than sandstone and carbonates combined. Therefore, there is a likelihood that shale gas is more abundant in the solid earth than conventional oil and gas. (b) Most conventional petroleum reservoirs are either sandstone or carbonates. The amount in carbonate petroleum reservoirs is almost equivalent to the amount in sandstone petroleum reservoirs. Notice that more than 75 percent of sedimentary rocks do not contain conventional oil and gas.

Organic sediments and organic sedimentary rocks. Organic or biological sedimentary rocks are formed by living organisms, including corals, sometimes excrement, swamp vegetation, and bacterial activity, either directly or as a byproduct of living organisms, shown in Table 3.9. Typically, when the remains of living organisms build up and are compacted by sediment, these biologic components are intermixed with the other types of sedimentary material, but when prominent, they become organic rock. Coal, for example, is made from long-dead vegetation crushed by thick layers of sediment and chemically altered through heat and pressure. These plant remains accumulate to form peat with a carbon content of about 60 percent. Organic material deposited in water with low oxygen content lead to oil and natural gas reservoirs through the subsurface heat that can change organic solids to oil and natural gas in porous rocks.

Table 3.9. A limestone composed of $CaCO_3$ derived from the hard parts of once living organisms.

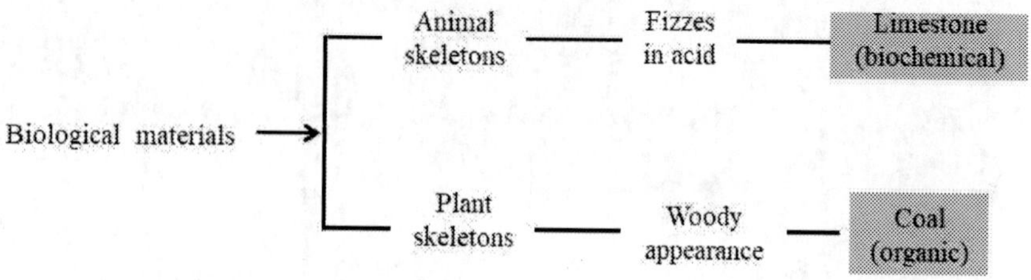

3.2.4. Source Rocks, Reservoirs and Seal Rocks of Conventional Hydrocarbons

Sedimentary rocks tend to contain voids (open space) between sediment particles. Thus, most conventional petroleum reservoirs are found in the coarser sediments—that is, porous rock (i.e., a rock capable of holding petroleum) and permeable (i.e., a rock that permits the flow of petroleum) in which oil accumulates; petroleum exists in the pore space. Shale, sandstone, and carbonates (limestone, dolomite, and chalk) make up about 99 percent of all sedimentary rocks, with shale being the most abundant and carbonates the least abundant (see Figure 3.12a). Yet the conventional amount of oil (accumulation of oil whose upward migration in the earth is impeded by impermeable rock and therefore can be produced by drilling wells through the impermeable rock) in carbonate petroleum reservoirs is almost equivalent to the amount of oil in sandstone petroleum reservoirs, as illustrated in Figure 3.12b. Again, this distribution is for conventional oil and gas. The distribution drastically changes if we consider unconventional oil and gas contained in shale (remember that unconventional hydrocarbon accumulations are so closely surrounded by impermeable rocks that they can only be recovered by fracking rocks). Because shale is more abundant than sandstone and carbonate combined, the likelihood of finding shale gas is greater than that of finding conventional oil and gas.

Petroleum (oil and gas) is a naturally occurring flammable liquid formed from organic matter: tiny plants, algae, and bacteria (plankton) that lived in the ocean millions of years ago. As living organisms, they absorbed energy from the sun-that was stored in their bodies as carbon molecules. When they died, they sank to the bottom of the sea. Vast quantities of these remains long ago settled on sea or lake bottoms, mixing with sediments and being buried under special anoxic conditions.[1] As further layers settled on sea or lake beds, intense heat and pressure built up in the lower regions. The burial and thus the heating of such organic matter induce reactions leading to the generation of gas, then oil and gas (140 degrees Celsius to 250 degrees Celsius), and finally gas alone (140 to 250 degrees Celsius), as the temperature and residence time at high temperatures increase. The rocks from which this generation of oil and gas takes place are known as source rocks. They are generally sedimentary rocks, typically shale or limestone, which are rich in organic matter.

Oil and gas that are formed deep in the subsurface tend to rise toward the earth's surface because they are less dense than water and most surrounding rocks. Their upward movement is known as migration. Some oil and natural gas have migrated all the way to the surface and escaped. Other migrations of oil and gas are halted by impermeable rocks, known as seals or cap rocks. Seals tend to be fine-grained or crystalline low-permeability rocks such as mudstone/shale, cemented limestone, chert, anhydrite, and salt (see Table 3.1). Seals can also develop along fault planes, faulted zones, and fractures.

The presence of seals is critical for the accumulation of petroleum in commercial quantities in the subsurface. In the absence of seals, petroleum will rise to the Earth's surface and be destroyed. Although seals are critical for the accumulation of petroleum in commercial quantities, they are not leak-proof. However, their large thicknesses and lateral extent minimize the leakages. The most common lithology that forms a petroleum seal is mudrock. Mudrocks are composed of either carbonate or siliciclastic minerals (or both), and mudrock sequences are often thick ($> 50\,\mathrm{m}$) and laterally continuous ($> 10\,\mathrm{km}^2$). Examples of mudrock seals are found in the Gulf of

[1] Anoxic conditions will occur if the rate of oxidation of organic matter by bacteria is greater than the supply of dissolved oxygen.

Mexico, the Niger Delta, the northwest Australian shelf, northwestern Europe, the North Sea, and Southeast Asia. Halite forms part of the sealing lithology in many of the large Middle Eastern petroleum accumulations.

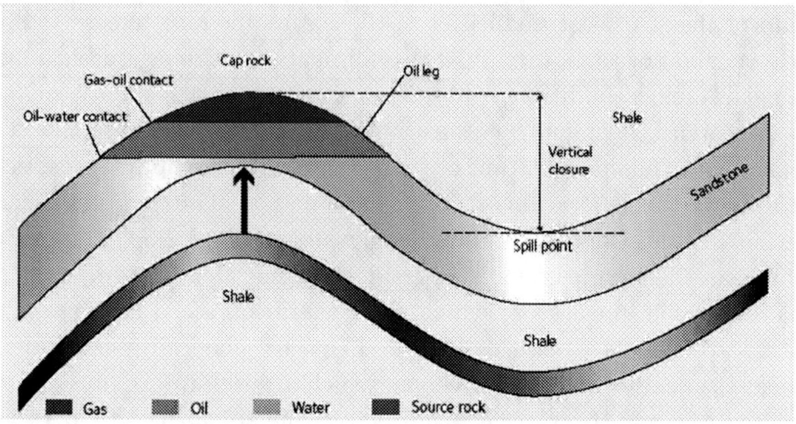

Figure 3.13. An illustration of the prerequisites for a hydrocarbon trap. The trap configuration shown here is an anticlinal fold. The single-headed arrow depicts the primary migration of oil and gas from a maturing source rock into an overlying sandstone reservoir (courtesy Statoil).

In addition to seals, a petroleum reservoir is needed in order to accumulate petroleum in commercial quantities—that is, porous rock (i.e., a rock capable of holding petroleum through open spaces in the rock that stores fluid) and permeable rock (i.e., a rock that permits the flow of petroleum—the ease of fluid movement through a pore space) in which oil accumulates. These reservoirs are typically sandstones and carbonates. The goal of petroleum exploration is to find present-day traps whose seals are intact and whose reservoir rocks are (i) sufficiently porous to contain commercial quantities of petroleum and (ii) sufficiently permeable to allow them to flow fast enough toward production wells.

Trap is the term that describes the body, bounded by seals and containing a reservoir, in which petroleum can accumulate as it migrates from a source rock to the Earth's surface, an illustration of the prerequisites for a hydrocarbon trap. Although there are many trap configurations, the one shown in Figure 3.13—an anticlinal fold—is perhaps the simplest. The single-headed arrow depicts the primary migration of oil and gas from a maturing source rock into an overlying sandstone reservoir.

Notice that petroleum migrations (i.e., the process whereby petroleum moves from its place of origin, the source rock, toward the Earth's surface) are generally divided into three stages: (i) primary migration: expulsion of petroleum from the source rock, (ii) secondary migration: the journey from the source rock to the trap, and (iii) tertiary migration: leakage and dissipation of petroleum at the Earth's surface. Primary migration is understood to occur when the petroleum volume in the source rock matches the volume of the pore space in the source rock—that is, when the source rock is fully saturated with petroleum. Secondary migration takes petroleum from the source rock to the trap through carrier beds. On its escape from the source rock, petroleum is dispersed over a large area. By the time petroleum reaches the relatively restricted area of a trap, it can occupy more than 90 percent of the pore volume in the reservoir. Secondary migration is arrested once migrating petroleum enters a trap. Disruption or overfilling of the trap can lead to remigration of petroleum to upper structures under the same secondary-migration process. Such secondary migration ends when petroleum approaches the Earth's surface. Tertiary migration includes leakage, seepage, dissipation, and alteration of petroleum as it reaches the Earth's surface. Note that the endpoint of primary migration, which is also the starting point of secondary migration, is still poorly understood.

A petroleum geology system is a dynamic system that functions in restricted geological space and on a restricted time scale. Figure 3.13 depicts a petroleum system that requires a timely convergence of geologic events that are essential to the formation of petroleum deposits. These events involve mature source rock, hydrocarbon expulsion, hydrocarbon migration, hydrocarbon accumulation (reservoir), and hydrocarbon retention (seal). To have a commercial petroleum prospect, several requirements must be met. There must be (1) a petroleum source rock, (2) a reservoir rock, and (3) a seal. First, petroleum migrates from the source rock to the porous and permeable reservoir, where petroleum is trapped by the seal rock. Note that natural gas can also exist in shale deposits, which formed about

350 million years ago. Shale, which consists mainly of consolidated clay-sized particles, is the Earth's most common sedimentary rock. Shale not only forms the geologic seal that retains the conventional oil and gas within producing reservoirs, preventing petroleum from migrating to the surface (secondary migration), but it also contains large amounts of natural gas.

3.2.5. Black Shale: Unconventional Hydrocarbons

Again shale is a fine-grained sedimentary rock that forms from the compaction of silt- and clay-sized mineral particles that we commonly call mud. Other constituents might include organic particles, carbonate minerals, iron-oxide minerals, sulfide minerals, and heavy mineral grains. Texturally, shale is made up of many thin layers and readily splits into thin pieces along the laminations. Only some types of shale are rich in organic materials that can generate oil and/or natural gas.

Shale has a very small particle size, so the interstitial spaces are very small. In fact, they are so small that oil, natural gas, and water have difficulty moving through the rock. Shale can therefore serve as a cap rock for water, oil, and/or natural gas traps, as described in the previous subsection. However, in some cases, the interstitial spaces in shale can hold significant amounts of water, gas, or oil, but not be able to effectively transmit them because of the low permeability. In other words, the pore spaces in shale are so tiny that gas, oil, and/or water have difficulty moving through the shale and into the well. The oil and gas industry has overcome these limitations of shale by using horizontal drilling and hydraulic fracturing, to create artificial permeability within the rock.

Let us elaborate a little bit more on the genesis of hydrocarbons in shale. Salt rocks were originally deposited in low-energy depositional environments, such as tidal flats, swamps, lakes, and deep-ocean basins, where the clay particles dropped out of suspension from the water column. During the deposition of the muds that form shale, organic matter is also deposited, particularly in swamps and lakes. Plant and algal remains and the animal forms that consume them (organic matter) are buried with the sediment. As the sediment load increases due to continued burial, heat and pressure convert some of this organic matter into hydrocarbons. Conversion of this initial organic matter into petroleum increases rock pressure, which expels gas, oil, and/or water from the shale into adjacent strata. Continued pressure increases from burial may allow much of the gas, oil, and/or water to migrate from the organic shale into the overlying reservoir rocks. The gas and oil that remain are tightly bound in the shale and are known as the unconventional shale gas and the unconventional shale oil.

Figure 3.14. (a) Black shale, (b) gray shale, (c) red shale, and (d) green shale.

The amount of organic matter in a rock is described by a parameter called total organic carbon (TOC), which is measured in percentage or ppm (parts per million) with respect to rock volume. Shale oil and shale gas are produced from organic-rich shale, also known as black shale (see Figure 3.14a), when their TOCs are greater than 1 percent. The average organic content of black shale is about 2 percent (Pratt et al., 1988) but can be as high as 8 percent.

Like most rocks, the color of shale is often determined by the presence of specific materials in small amounts. A few percentages of organic materials or iron can significantly alter the color of a rock.

A black color in sedimentary rocks generally indicates the presence of organic materials (Swanson, 1961). Just one or two percentages of organic materials can impart a dark gray or black color to the rock. More importantly, this black color generally implies that the shale was formed from sediment deposited in an oxygen-poor environment (anoxic conditions). An oxygen-poor environment also provides the proper conditions for the formation of sulfide minerals, such as pyrite, another important mineral found in most black shale. The presence of organic debris in black shale makes them candidates for oil and gas generation. The Barnett shale, Marcellus shale, Haynesville shale, Fayetteville shale, and other gas-producing rocks are all dark gray or black shale that yields natural gas. The Bakken shale of North Dakota and the Eagle Ford shale of Texas are examples of types of shale that yield oil.

Note that back shale is a rock that may or may not be a source of hydrocarbons. The propensity depends on the type of organic matter within the rock, what pressure and temperature exposure it has had, and what minerals are mixed in with the clay and organics. The organics may be oil or gas prone, the rock may be immature—not yet cooked enough or it may be baked beyond generation and all the movable hydrocarbons are already gone. If it is all clay and organics, it will never produce, as it needs a majority of either clastic or carbonate content to provide pathways out of the framework.

Gray shale (see Figure 3.14b) sometimes contains a small amount of organic matter. However, gray shale can also be rocks that contain calcareous materials or simply clay minerals that result in a gray color. Shales that are deposited in oxygen-rich environments often contain tiny particles of iron oxide or iron hydroxide minerals, such as hematite, goethite, or limonite. Just a few percent of these minerals distributed through the rock can produce the red (due to hematite), brown, or yellow (due to limonite or goethite) colors exhibited by many types of shale (see Figure 3.14c). Green shale (see Figure 3.14d) is occasionally found because some of the clay minerals and micas that make up much of the volume of these rocks typically have a greenish color.

Box 3.1. Geological Time Scale

The geological time scale is the age names of major Earth events as recorded in rocks and fossils, including the oxygenation of the atmosphere, the evolution of life forms, Pangea, and the mass extinctions. We here summarize this convention.

The **Hadean Eon** is the oldest eon and dates from 4.5 to 4.0 billion years ago. Earth then was characterized by a partially molten surface, volcanism, and asteroid impacts. It was mostly made of molten material.

The differentiation of the Earth from a homogeneous planet into a heterogeneous one with layers of felsic and mafic crust, ultramafic mantle, and iron and nickel core occurred in this eon. The **Archean Eon**, which lasted from 4.0 to 2.5 billion years ago, represents the beginning of much more robust rock and fossil records. It includes the so-called *late heavy bombardment* that occured after the planets formed, about 4.1-3.8 billion years ago. Meteorites and comets started impacting objects throughout the solar system. Earth, the Moon, and all planets in the solar system were pummeled by materials from the asteroid and Kuiper belts. The Moon's craters are from this event. The **Proterozoic Eon** ranges from 2.5 billion to 541 million years ago. During this time, most of the central parts of the continents had formed and plate tectonic processes had started.

Photosynthesis by microbes and plants completely transformed the oceans and the atmosphere by adding massive amounts of free oxygen gas (O_2) and initiating what is called the Great Oxygenation Event (GOE). The **Phanerozoic Eon** is the most recent, 541 million years ago to today. The Phanerozoic rock records have revealed an abundance of fossils that preserve body parts of Phanerozoic organisms like claws, scales, shells, teeth, and bones.

The Phanerozoic is subdivided into three eras: Paleozoic, Mesozoic, and Cenozoic. During the **Paleozoic Era**, which lasted from 541 to 252 million years ago, sea-levels rose and fell four times. The assembly of Pangea was completed in the late Paleozoic Era. The beginning of the Paleozoic Era is marked by the sudden appearance of biological diversity called the *Cambrian Explosion* (541 to 485 million years ago). One of the five documented large mass extinctions occurred at the end of the Paleozoic Era. It is known as the *Permian Mass Extinction.* It is estimated that up to 96 percent of marine species and 70 percent of land-dwelling vertebrates went extinct. During the **Mesozoic Era**, which lasted from 252 to 66 million years ago, Pangea started to break apart, mammals (most notably reptiles such as dinosaurs), birds, and flowering plants developed. The Atlantic Ocean opened as the young Mid-Atlantic Ridge, with South America separating from central and southern Africa and Greenland and Scandinavia parting ways. Similar to the end of the Paleozoic era, the Mesozoic Era ended with the so-called K-Pg Mass Extinction 66 million years ago. Ninety percent of plankton species, 75 percent of plant species, and all the dinosaurs went extinct. The impact that created the Chicxulub Crater (Yucatan Peninsula) is considered the cause of this extinction. The **Cenozoic Era** is the period when mammals became a dominant and large life form, including human ancestors.

Drilling and well development methods, in the late 1990s, such as hydraulic fracturing and horizontal drilling, have allowed us to produce the oil and natural gas trapped within the tight matrix of organic shale. Wells can now be drilled and steered horizontally in a targeted layer of rock (even if it is relatively thin) for thousands of meters, and thereby dramatically increase the number of natural fractures encountered. Additional fractures are created through fracking. Water is pumped down the well under pressures high enough to locally fracture the shale, significantly extending and adding to the natural fracture system and thereby allowing the trapped shale fluid to flow to the well. To keep the natural and artificial open fractures from closing due to the pressure of the overlying rock layers, sand or other materials of various sizes (called proppants) is also pumped into the fracture zones to provide pathways for fluid flow. The Barnett Shale of Texas was the first major natural gas field developed in a shale reservoir rock.

In summary, black shale, like sandstone, contains pores capable of storing hydrocarbons. However, these pores can be extremely small and very poorly connected to each other, making it almost impossible for fluid to flow through the shale. The word *tight* is often used to describe this characteristic. The world's most prolific known source rocks are black marine shale (Degens et al., 1981). They have high organic contents relative to similar rocks and they are generally deposited in relatively quiet shallow seas and on continental margins.

3.3. The Rock Cycle

Rocks on and below the surface of the Earth are constantly changing, albeit slowly, on the scale of millions of years, except when volcanic eruptions or earthquakes cause quick changes. The changes happen through a cycle of creation, transformation, and destruction, called the primary rock cycle or igneous-sedimentary-metamorphic cycle, as illustrated in Figure 3.15. In other words, rocks in the subsurface and at the Earth's surface are constantly being recycled through the processes of weathering (that is, the breaking down of rocks into smaller pieces, known as sediments), erosion (that is, the movement of sediments from one place to another), transportation and deposition (that is, the placement of sediment at a new location), burial (that is, the accumulation of sediment layers), diagenesis (that is, turning loose sediment into a sedimentary rock), heat, pressure, and melting. There is no particular order in which the changes happen. In fact, the rock cycle is different from other cycles, such as the water cycle (see Ikelle, 2018), because there is no particular order in which the changes occur. We will elaborate on the processes and loops associated with the rock cycle in this section.

3.3.1. Some Fundamentals of the Rock Cycle

In its simplest presentation (see Figure 3.15a), the rock cycle can be described as a cyclic picture with two inside loops as follows:

If igneous and metamorphic rocks are exposed at the surface, they weather. Weathered rock fragments (sediments) are transported, deposited, and lithified into sedimentary rocks. If igneous and/or sedimentary rocks are subjected to temperatures and pressures that exceed those under which they have been solidified, they may undergo changes to form metamorphic rocks—these changes are globally called metamorphism. The melting of metamorphic rocks provides the magmas that form igneous rocks.

Actually, things are more open-ended. For example, sedimentary rocks are commonly eroded to provide sediments that form new sedimentary rocks. That is the outside loop on the bottom left in Figure 3.15b, and the other outside loops are likewise transformations within the other two major types of rock. The four inside loops are also added to the no-longer cyclic picture. In the end, it is easier to say that nature can transform rocks from any one of the three fundamental groups to any one of those groups, with two exceptions. One exception is that matter cannot go directly from the sedimentary realm to the igneous realm because the transition from the low

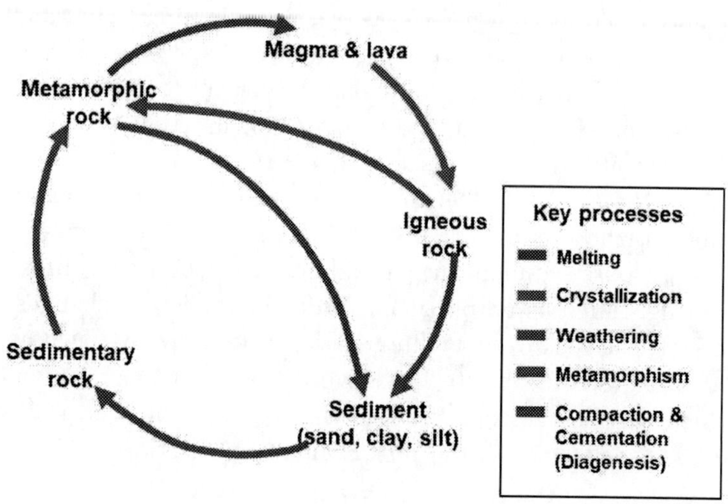

(a)

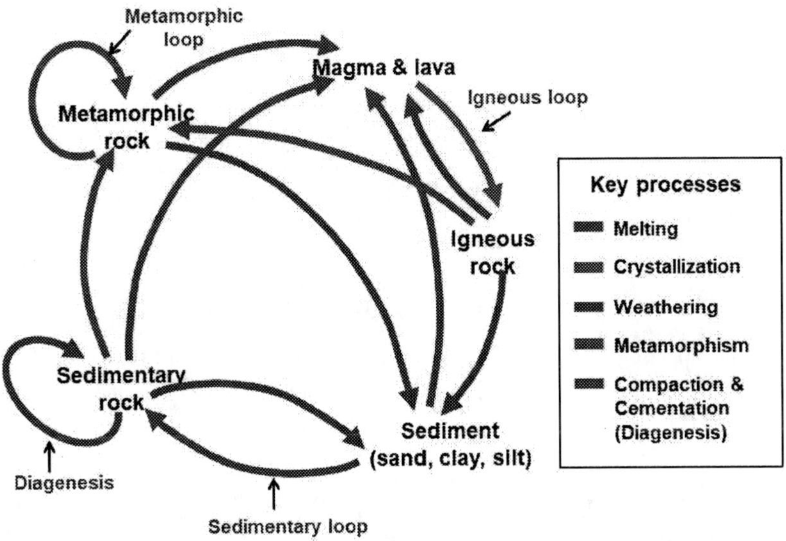

(b)

Figure 3.15. The rock cycle, in this graphic representation, shows the interrelationships of rocks and the processes that form them in the Earth's crust. (a) This version highlights how each type of rock can be "altered" into other rocks. Igneous rock can change into sedimentary or metamorphic rock. Sedimentary rock can change into metamorphic or igneous rock. Metamorphic rock can change into igneous or sedimentary rock. Notice that the rock cycle is not a one-way road. (b) Another more complete version of the rock cycle. It looks a lot more complex, but it clearly shows that the rock cycle contains many loops. For example, a sedimentary rock can be weathered, and eventually the sediments are deposited and become another sedimentary rock. An igneous rock can be melted and cooled to become another igneous rock. Diagenesis is the term used for all of the changes that a sediment undergoes after deposition and before the transition to metamorphism. These changes include compaction, deformation, dissolution, cementation, recrystallization, hydration, and bacterial action.

temperatures at which sedimentary rocks form and exist to the high temperatures at which minerals melt to give magmas would have to go through the temperatures at which any sedimentary rock would be metamorphosed. The other exception is that the process of lithification, which transforms sediments into rock formation, is a unique process in sedimentary rocks. Most other rocks (with the exception of pyroclastic igneous rocks) form initially as rocks (e.g., igneous rocks form as magma or lava cools and solidifies; metamorphic rocks start off as rocks and are altered by high pressure or temperature). We are not saying that all sedimentary rocks form through the lithification of sediment. Some, especially chemical sedimentary rocks, initially form as solid masses. However, most sedimentary rocks start off as unconsolidated sediment.

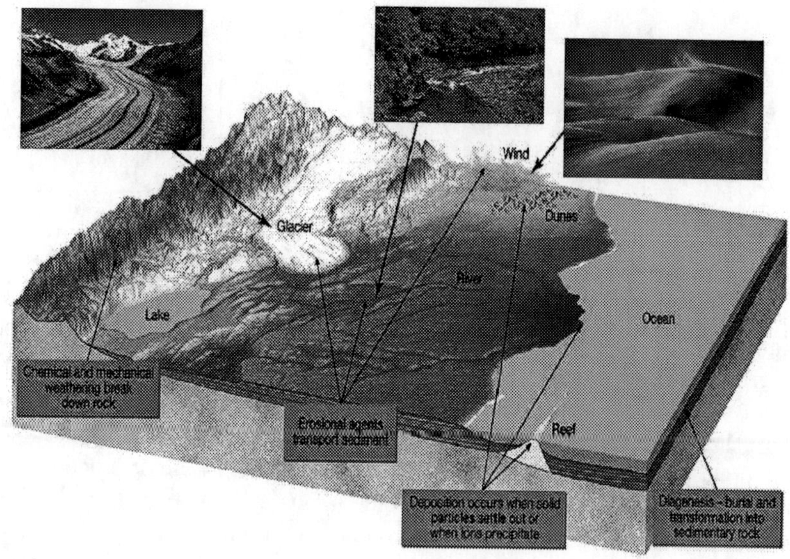

Figure 3.16. The sedimentary stages of the rock cycle. Weathering breaks down rocks. Erosion carries away particles. Erosion and transportation agents include landslides, rivers, sea, lake, and wind. Deposition occurs when particles settle or precipitate. Burial occurs as layers of sediment accumulate. Diagenesis lithifies the sediment to make sedimentary rocks.

3.3.2. Rock Weathering

Sedimentary rocks are formed through the weathering and transporting of sediments, followed by their deposition, compaction, and cementation into rocks, as illustrated in Figure 3.16. Because some of these geological processes are not self-explanatory, let us recall them.

Weathering is the forces and conditions that exist at the Earth's surface. These forces and conditions can cause solid rocks at or near the Earth's surface to split, as you can see when crossing and working in alleys to your home or office, and even to break down into smaller fragments and/or different minerals from those originally present; that is weathering. In other words, pre-existing rocks can be broken up into smaller pieces and/or their original minerals can be changed into new minerals during the weathering process. Mountains of igneous, metamorphic, and sedimentary rocks exhibit clearly the weathering. Once the overlying materials (e.g., vegetation) of the mountains have been eroded away and the rocks are exposed, weathering can begin (see Figure 3.5).

Figure 3.17. An example of an acidic pond. Rainwater is oxidizing and acidic. When rainwater enters rock and sediments, it has considerable potential for causing chemical reactions, as illustrated here.

Chemical weathering: There are three types of weathering processes: chemical, physical, and biological. Chemical weathering is the process of breaking down rocks by chemical reactions through rocks' exposure to water, oxygen, carbon dioxide, and acid. In other words, the most common agent of chemical weathering is rainwater. Chemical weathering, such as acid rain, eats away at certain types of rocks such as limestone, creating cracks and holes. So the exposures to water can alter minerals in the rock. Because we are dealing with chemical changes, it is important to keep in mind that chemical weathering not only breaks down the rocks, but it also alters the minerals' composition. Actually, no two rocks look exactly the same because they have been subjected to chemical weathering.

Rainwater is oxidizing and acidic. Acidity is acquired through nitrous acid formed during thunderstorms, carbonic acid from a solution of carbon dioxide in the atmosphere, and sulfuric acid from volcanic gases (a reminder: The solution is freshwater in lakes, groundwater, or seawater). As rainwater seeps through soil, it also acquires humic acids from decaying organic matter. Thus, when rainwater enters rocks and sediments, it has considerable potential for causing chemical reactions, as illustrated in Figure 3.17.

The reactions associated with chemical weathering are generally classified as hydrolysis, oxidation, carbonation, and absorption. (1) Hydrolysis refers to water acting to "loosen" the chemical bonds of rocks to break down minerals. It works faster in slightly acidic water. Groundwater, including water from rainfall, is normally acidic and is capable of breaking down rocks naturally, given sufficient time. (2) Oxidation refers to the chemical reaction of rock minerals with oxygen (water includes oxygen), which changes the mineral composition of the rock. This process is also capable of breaking rocks down naturally. An easily recognizable example of oxidation is rust (see Figure 3.18a) caused by prolonged contact with water. This reaction is responsible for the red surface of Mars. (3) Carbonation refers to the chemical reaction of porous rock minerals or soft minerals with carbon acid. Carbonation occurs when rain water, which is naturally slightly acidic, due to atmospheric carbon dioxide, combines with carbon dioxide in the atmosphere (or in the soil) to form a weak acid (carbonic acid). Carbonic acid is water combined with carbon dioxide. It dissolves and/or breaks down the minerals of rocks, as illustrated in Figure 3.18b. Carbonation is sometimes referred to as dissolution. It is the driving force behind the sinkholes, caverns, and underground rivers of karst topography. It also frequently occurs on gravestones, stone statues, and building blocks made from porous rocks. (4)

Finally, the hydration (or absorption) of water into some minerals (mainly clays) causes them to expand, resulting in the breakdown of rocks by fracturing. Hydrolysis is the opposite of hydration in the sense that water breaks down the chemical bonds of a mineral instead of creating a new mineral. It is a decomposition reaction.

Mechanical (physical) weathering is the process of physically breaking down rocks into small pieces (sediments) without changing their chemical composition. It always involves fractures. It can be caused by freezing and thawing of water in cracks in the rock (frost wedging), formation of salt crystals within the rock (salt wedging), roots and plants pushing into the rocks (root wedging), animals burrowing and digging through the ground (burrowing animals), and removal of overburden (exhumation) through erosion or mining. (1) Frost wedging (also known as freeze-thaw) is the process by which water seeps into cracks in a rock, expands on freezing, and thus enlarges the cracks and eventually breaks the rock apart, as illustrated in Figure 3.18c. It is most effective in mountainous regions where water from rainfall and/or snow-melt is collected in the cracks of rocks. At night, temperatures drop and the water freezes and expands. The increase in volume of the ice exerts pressure on the cracks in the rock, causing them to split the rock further open and create new cracks. In general terms, repeated thermal expansion and contraction can break rocks, as one can observe with wildfires or in deserts with extreme daily heating and cooling. (2) Salt wedging is the process by which salty water (rain water usually has salts dissolved inside) seeps into rocks, and then the water is evaporated on a hot day, as salt crystals grow within cracks in the rock. These crystals exert pressure on the rock and cause it to weaken and eventually break apart. (3) Root wedging is the process by which the growth of roots can force their way into even the tiniest cracks, and then widen those cracks, as they grow and break them apart, as illustrated in Figure 3.18d. In other words, root wedging is the same as frost wedging, except that roots pry open the cracks, instead of ice. (4) Although animals do not normally burrow through solid rock, they excavate and remove huge volumes of soil, and thus expose the rock to weathering by other mechanisms. (5) A mass of rock is exposed by weathering and by removal of the overlying rock. This unloading promotes cracking of the rock, known as exfoliation, and the development of cracks leads to other kinds of weathering. (6) Rocks rubbing or crashing into each other through streams, glaciers, wind, landslides, and flowing debris can also break down rocks. This process is known as abrasion.

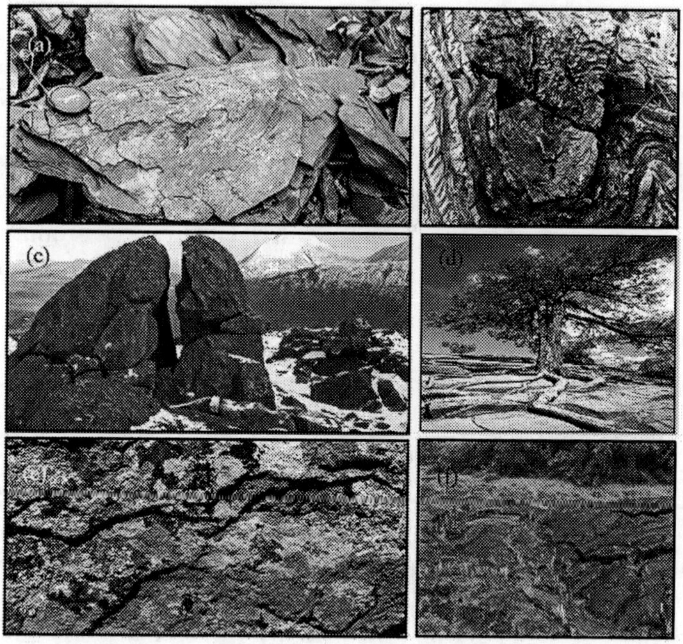

Figure 3.18. Illustrations of biological, chemical, and physical weathering. (a) Chemical weathering. Oxidation takes place on shale with the familiar rusting (red, orange, and yellow stains), which characterizes the oxidation reaction. (Photograph by Stan Johnson). (b) Chemical weathering. The limestone has been carbonized to different degrees in different areas because of compositional differences. The buff-colored bands are volcanic rock. (c) Physical weathering. Frost wedging is caused by water that fills cracks, freezes, expands, and forces cracks to open, causing them to grow. (d) Physical weathering. Roots pry open the cracks. (e) Biological weathering. Once rock is weathered into smaller particles, microorganisms and small plants begin to establish themselves there. The weathering process creates finer particles of new minerals. (f) Igneous rock resulting from lava flow that is being broken down through biological processes.

Biological weathering is the process of weakening rocks and eventually breaking them down by the actual molecular breakdown of minerals. It is caused by organisms—plants, animals, and bacteria—that release acid-forming chemicals that cause weathering and also contribute to the breaking down of rocks and landforms. Organic material is added to the soil from the decomposition of animals and plants. These remains provide nutrients for other organisms. Soil fertility is increased as more organisms inhabit the soil and as decayed material (humus) is added to mineral particles. Humus also increases the soil's capacity for holding water and air, both necessary for plant growth.

Typical biological weathering examples are (1) the activity of boring algae on shells, (2) marine organisms, like sea urchins, can burrow into rock, and (3) lichen and other organisms can create acids that attack minerals, as illustrated in Figures 3.18e and 3.18f. Growing plant roots can exert stress or pressure on rock. Although the process is physical, the pressure is exerted by a biological process (i.e., growing roots). Biological processes can also produce chemical weathering, for example, where plant roots or microorganisms produce organic acids, which help to dissolve minerals.

Although physical, chemical, and biological weathering are separate processes, two or all three processes can act together in nature.

Lab: Consider a pottery container with water in wintertime. Verify that when water is frozen and becomes ice, it can have about 10 percent relative volume change, i.e.,

$$\frac{dV}{V} = 0.1 \ .$$

This quiz illustrates frost wedge.

3.3.3. Sediment Erosion, Transportation and Deposition

In terms of producing sediment, erosion involves removal and beginning the transportation process of the weathered products from their original location. Sediment can then be transported by sliding down slopes, by being picked up by the wind, by being carried by running water in streams, rivers, or ocean currents, or by moving ice. The distance the sediment is transported and the energy of the transporting medium all leave clues in the final sediment that tell us something about the mode of transportation. Sediment is deposited when the energy of the transporting medium becomes too low to continue the transporting process. In other words, if the velocity of the transporting medium becomes too low to transport sediment (e.g., winds/water currents die down), the sediment will fall out and become deposited. The final sediment thus reflects the energy of the transport medium.

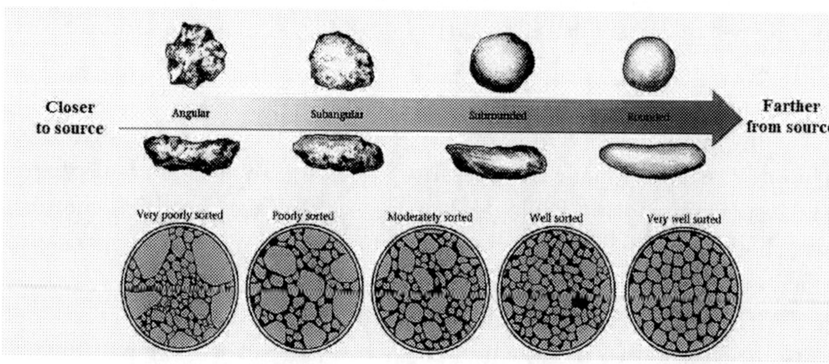

Figure 3.19. Classifications of the detrital (clastic) sediment are also based on sphericity (rounding) of grains and distribution of grain sizes. (Top row) Rounding describes the relative shape of the grains, including from deviations in rounded or spheroidal grain shapes. (Medium row) Angularity indicates the absence of rounding. (Bottom row) Sorting describes the grain-size distribution or range of grain sizes in a rock. Poorly sorted rocks contain a wide range of grain sizes, including fine, medium, and coarse. Well-sorted rocks contain almost all grains of the same size.

The rate of erosion varies widely across the Earth. It is normally higher on steep mountain slopes and lower on horizontal areas. Biological processes, such as the development of vegetation, can slow it down. Erosion occurs on land and along some coasts, but is largely absent under the sea.

Water is the most important agent of erosion and transportation of geological materials. Water removes loose particles from the surface and washes them into channels and eventually into streams. Creeks and rivers transport both large and small particles, and continue the breaking down of large pieces of rock into smaller pieces. Most of the material transported by a river is suspended in the water. These materials include clay and fine silt, but in fast-flowing rivers or during floods, sand, gravel, and even boulders can be carried in the water (see Figures 3.10 and 3.19 for classified grains). The size of particles that will be moved by water is directly proportional to the velocity of the water.

Ice in the form of glaciers can erode, transport, and deposit sediment, especially during ice ages such as the one in which we live. The resultant deposits are normally very poorly sorted and contain a heterogeneous range of sediment particles, ranging from clay to boulders. These deposits are thus sometimes referred to as "boulder clay" or "tillite."

Gravity is another powerful agent of local transportation. It is central to the occurrences of landslides or avalanches. The resultant sediment deposits are normally angular and poorly sorted.

Wind or eolian action is an important process of transportation and deposition in deserts. A distinction is normally made between the sand-bed load that is transported close to the ground and the fine-grained clay and silt that are carried into the atmosphere in suspension and then settle out under gravity many days after a sandstorm has ended. Sand, on the other hand, is blown along the ground in migrating ripples and mega-ripples (dunes). The resultant sediment is often extremely well-sorted.

3.3.4. Sedimentary Rock Formation: Lithification

We now have to consider lithification, the process by which unconsolidated sediments are cohesively bound to form solid sedimentary rock. Compaction and cementation are generally the stages of the lithification process, as illustrated in Figure 3.20. Compaction is a physical process and cementation is a chemical process.

Compaction (a physical process). As more sediment is being deposited, there is an increase in weight or pressure that usually expels much of the connate water—the original (normally marine)

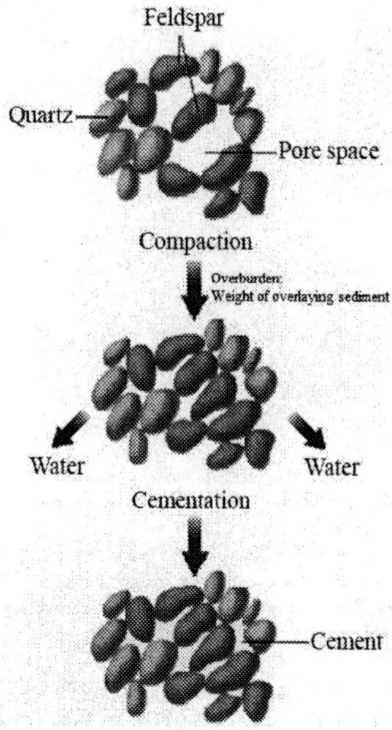

Figure 3.20. An illustration of the processes of converting sediments into sedimentary rocks by lithification, which include compaction (first stage) and cementation (second stage). Compaction reduces pore space due to the weight of accumulating sediment that forces the grains together, and then by reducing the pore space and forcing water out of the sediment. Upon compaction, the expelled water has to go somewhere. The typical direction of flow is upward. In the second step, pore water expelled from deeply buried sediments migrates upward toward the Earth's surface. As the water rises and cools, ions dissolved in the water precipitate, forming minerals that cement the grains together.

water with which a sediment was deposited—and forces the rock grains to come much closer together. As the grains are forced against each other, their outer surfaces usually dissolve and recrystallize, thus welding the grains together. The rearranging of sediment is so that there is much less open space between the grains. Compaction alone is sufficient for lithification of fine-grained sediments (e.g., clay and silt), but for coarse-grained sediments (e.g., sand and gravel), cementation, involving precipitation[2] of minerals in pore spaces, is necessary.

So compaction is the rearrangement of sediments so that there is less open space between grains. Note that, in the initial stages, compaction involves dewatering and a closer packing of grains. The typical direction of the flowing of the expelled water is upward (not necessarily vertically upward, but often updip). Further compaction through overburden pressure can result in local fracturing and bending of weak grains and even the solution of grains at points

Note also that there are three varieties of water infill in sediment pores: meteoric, connate, and juvenile. Meteoric water originates from the rain and snow from the atmosphere. Connate water has evolved from the original water associated with deposition and bacterial activity, and juvenile waters are of magmatic origin (water-derived magma). Connate waters are commonly different from meteoric waters in terms of their temperature, salinity, and chemistry because of various reactions between the sediment and the original fluid that take place immediately after deposition. The water table is defined as the point where the pore water pressure and the atmospheric pressure are equal. It is important to realize that compactional water, which is evolved connate water moving through the sediments as a result of compaction, can also be different from connate water.

Lab: Try to lithify pure quartz sand by squeezing it with a piston. Verify that you will fail, as no matter how hard you squeeze, the sand stays loose.
(2) Add now some soft rock fragments, like calcite, or interstitial mud. Verify that you would end up with at least partly lithified material: it would not fall apart readily in your hands. In other words, to arrive at lithified rock, you would need more than just compaction. You need cementation-substances that glue sediment.

Cementation (a chemical process). Pore water expelled from deeply buried sediments migrates upward toward Earth's surface. As the water rises and cools, ions dissolved in the water precipitate and

[2]Precipitation is the separation of a solid from a solution like water.

form new minerals like calcite, hematite, quartz, chert, or any combination of minerals. These minerals are cements. So cementation is the process by which dissolved substances in the water precipitate out and act as glue (cement) to hold the grains together and consequently make a solid rock. Calcite and silica (quartz) are common cements.

The cementation process can be relatively fast (weeks) or incredibly slow (millions of years). It can completely fill the pore space (pervasive cementation) or only be sporadic. Partially cemented pore space results in friable sedimentary rocks that easily fall apart when handled (see Figure 3.21). Most non-carbonate petroleum reservoirs occur in sandstones that have only been partially cemented and so retain much of their depositional porosity.

Recrystallize (a chemical process). Note that, in some rare lithification processes, a recrystallization occurs when unstable minerals recrystallize to form more stable minerals. Recrystallization most often occurs during the formation of biochemical sedimentary limestone rocks that previously contained aragonite, a chemically unstable form of calcium ($CaCO_3$)

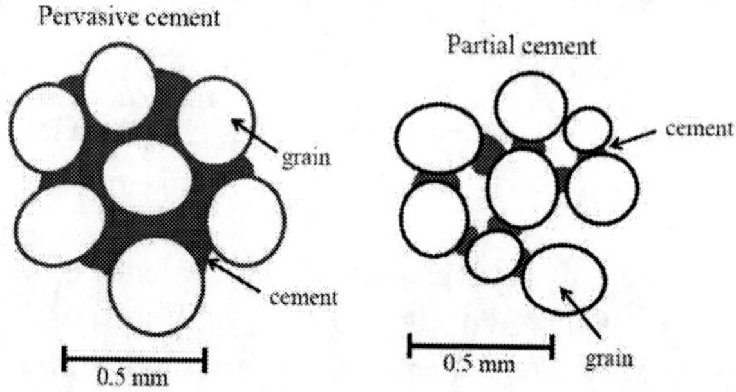

Figure 3.21. Coarse-grained sediment (gravel and sand) is most likely to be lithified through the action of cementation. The cementation can completely fill the pore space (pervasive cementation) or only be sporadic. Partially cemented pore space results in friable sedimentary rocks that easily fall apart when handled. Calcite, hematite, quartz, chert, or any combination of minerals, are typical cements.

Box 3.2. Measuring the Size of an Earthquake

Here is the general formula for all the earthquake scales:

$$M = \log_{10}(A/T) + F(z_0, d) + C \ , \tag{3.1}$$

where A is the amplitude of the signal, T is the dominant period of the earthquake recordings, F is a correction for the variations of amplitudes with the earthquake depth, z_0, and distance, d, from the recording station, and C is a regional correction. We deduce the Richter scale by taking $T = 0.8$, z_0 constant, and C, for Southern California; i.e.,

$$M_L = \log_{10}(A_{max}) + 2.76\log_{10}(\Delta) - 2.48 \ , \tag{3.2}$$

where A_{max} is the maximum amplitude recorded at 100 km from the epicenter, Δ is the distance from the recording station to the epicenter. A more-consistent measurement of earthquakes nowadays is the moment magnitude (M_W), which is the scaled version of seismic moment (M_0), because fault geometry and displacement are a part of the seismic moment; i.e., (Hanks and Kanamori, 1979)

$$M_W = (2/3)\log_{10}(M_0) - 10.73 \ ,$$

with $M_0 = \mu\, S\, L$ in dyne-cm (dyne = gcm/s^2), μ is the rigidity in dyne/cm^2, S is the fault area in cm^2, and L is the average displacement in cm (Aki, 1966), as depicted in Figure 3.22. An earthquake with a fault surface of 100 km times 12 km, an average slip of 6 m, and an average rigidity of 3×10^{11} dyne/cm^2 yield

$$\begin{aligned} M_0 = \mu S L &= 3 \times 10^{11} \times 6 \times 10^2 \times 100 \times 10^5 \times 12 \times 10^5 \\ &= 2.16 \times 10^{27} \text{ dyne} - \text{cm} \end{aligned}$$

and $M_W = \frac{2}{3}\log_{10}(M_0) - 10.73 = 7.5$. Below are magnitudes versus rupture areas and slip.

M_W	Recording range (km)	Slip (m)	Rupture area (km^2)	M_L
-4	< 0.03	0.01	0.003 micro	-4
-2	< 0.75	0.1	0.3 micro	-2
0	< 3	1	30 micro	0
2	< 30	10	4 micro	2
4	< 150	20	0.002	4
6	< 1500	30	1.36	6
8	everywhere	40	1,000	6.9
9.5	everywhere	40	176,000	6.9
10	everywhere	40	1 million	6.9

Dissolution (a chemical process). The basic natural processes of converting sediment to clastic and chemical sedimentary rocks are valid for all sedimentary rocks, including organic sedimentary rocks.

However, the depositional mechanisms, diagenesis, and response to structural deformation of organic sedimentary rocks can differ significantly from inorganic sedimentary rocks because organic deposition includes biogenic materials such as allochems. Allochems are carbonate particles (fossils, molds, oolites, or intraclasts) that have been transported a short distance and deposited as sediment. The open spaces between allochems after deposition decrease with burial through compaction and cementation to form carbonate rocks. However, these rocks can still include open space within the allochems produced by dissolution.

3.4. The Tectonic Plate Theory

An earthquake is a vibration of the Earth's crust produced by sudden deformation of a portion of the Earth's crust that rapidly releases energy. The energy released by an earthquake is in the form of waves which radiate in all directions from its source (also known as the focus or hypocenter). The epicenter is the position on the surface of Earth directly above the focus. Earthquakes occur in or beneath the continental crust as well as in or beneath the oceanic crust. Earthquakes in or beneath the ocean can cause upward or downward movements of the seafloor, which in turn displace huge amounts of seawater and create powerful waves known as tsunamis. These earthquakes can also trigger tsunamis by unleashing underwater landslides that displace huge amounts of seawater.

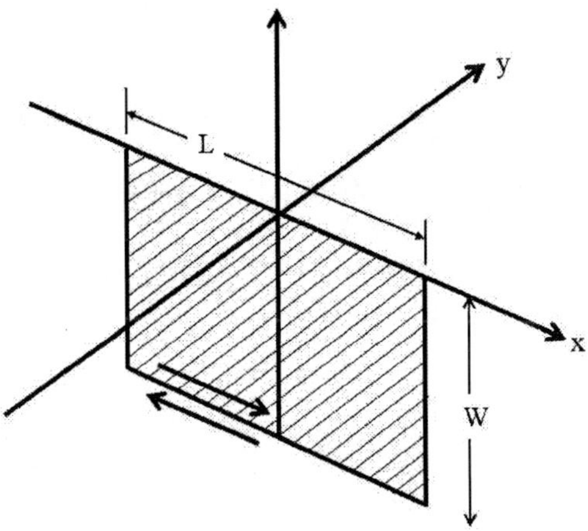

Figure 3.22. Cross-section showing the fault area and the parameters that are used in the calculation of seismic moment.

Between 1960 and 1970, the theory of plate tectonics was developed to explain the correlations between earthquake epicenters and Earth's surface locations of volcanic eruptions, continental drift, mid-ocean ridge systems, deep oceanic trenches, and seamount chains, among other phenomena. The 1968 paper of three seismologists of Lamont (Columbia University), Brian Issacks, Jack Oliver, and Lynn Sykes in the *Journal of Geophysical Research* was a significant turning point in the acceptance of the tectonic plate theory by the scientific community. Basically, these three scientists organized their seismic observations into coherent tables which incontrovertibly showed the partition of the globe into moving tectonic plates (see Figure 3.23). Earth's lithosphere was divided into 14 tectonic plates: the North American Plate, the Juan de Fuca Plate, the Pacific Plate, the Cocos Plate, the Caribbean Plate, the Nazca Plate (named after the Nazca region of southern Peru), the Scotia Plate, the Eurasian Plate, the Arabian Plate, the Australian-Indian Plate, the Philippine Plate, the South American Plate, the African Plate, and the Antarctic Plate. Note that the tectonic map does not coin-

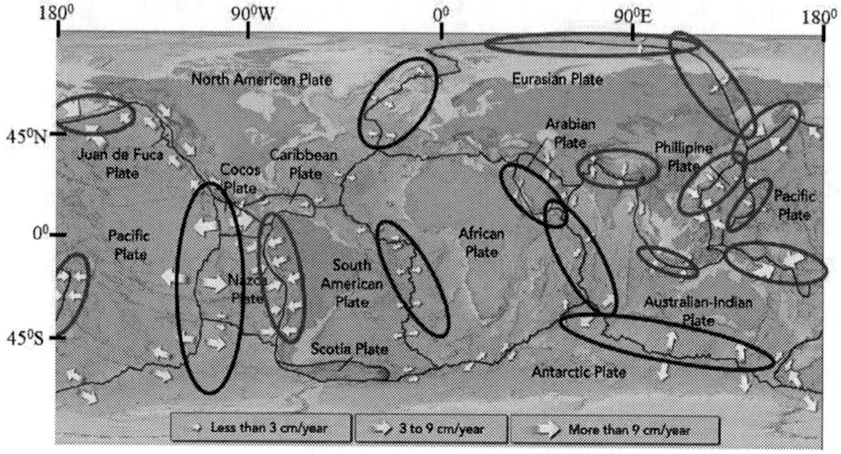

(a)

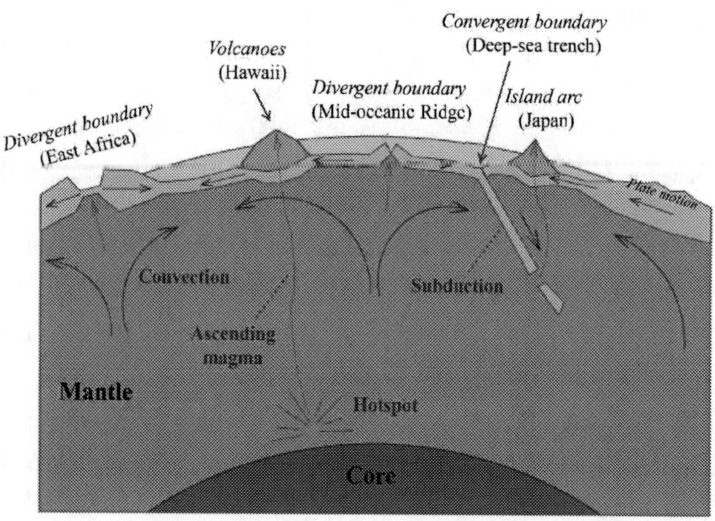

(b)

Figure 3.23. (a) A partition of the globe into 14 moving plates which move relative to each other at rates of 1 to 15 cm/year, resulting in continental drift and convergent and divergent plate boundaries. The arrows indicate the speed and direction of the plates. The tectonic map does not coincide with the geographic map. (b) At the Mid-oceanic Ridge, new sea-floor crust is formed out of magma rising from the mantle.

cide with the geographic map. The plates move relative to each other at rates of 1 to 15 cm/year, resulting in continental drift. Deformation occurs at the plate boundaries, while internally the plates remain largely intact. The map of the Earth's surface constantly changes because of plate tectonics. We have a dynamic planet!

We here briefly review basic facts about earthquake and volcanoes description and characterization along with their relationships with plate tectonics. This background is essential for the description of monitoring and prediction of earthquakes and volcanoes in the subsequent chapters.

Box 3.3. Measurements of Volcanic Explosiveness

Volcanoes at the Earth's surface are the results of ascending magma at hotspots (e.g., Hawaii), mid-oceanic ridges (e.g., Mid-Atlantic Ridge), continental rifts (e.g., East Africa), and subduction volcanic island arcs (e.g., Japan). Volcanic eruptions are characterized as non-explosive (mild) or explosive (violent) eruptions. A volcanic eruption is a non-explosive eruption if it has low-viscosity magma and low-dissolved gas. A volcanic eruption is explosive if it has high-viscosity magma and high-dissolved gas. The quantity, known as the volcanic explosivity index (VEI), is generally used to measure the explosiveness of volcanic eruptions (Newhall and Self, 1982). Volume of ejecta (i.e., volcanic materials thrown out of solid Earth into the atmosphere), eruption cloud height, and the duration of the eruption are the determinant factors of VEI. Ash, lava, lava bombs, and ignimbrite are all treated alike by the VEI. The scale of VEI generally varies from 0 to 8. For non-explosive eruptions and very small eruptions, defined as less than 10000 m^3 of ejecta, VEI=0 (or VEI-0). Eruptions rated at VEI-1 produce between 1.0×10^4 and 1.0×10^6 m^3 of ejecta. Above VEI=1, the scale becomes logarithmic, meaning that each step on the scale represents a tenfold increase in ejecta. VEI-2 eruptions produce between 1.0×10^6 and 1.0×10^7 m^3 of ejecta. The largest volcanoes in history have VEI=8. They produce more than 1.0×10^{12} m^3 of ejecta and have cloud-column heights of over 40 km.

The last VEI-8 eruption was in New Zealand (Oruanui Ignimbrite), about 27,000 years, which means that there have not been any Holocene (last 10,000 years) eruptions with VEI=8. Pintatubo in 1991 was a VEI-6 eruption and Tambora in 1815 was a VEI-7 eruption. Below are the most explosive volcanic eruptions in the Japanese Islands in the last 200 or so years. All of them are Subduction-Zone volcanoes and are all monitored 24 hours except Oshima-Oshima which is located in the sea of Japan. [Data sources: Tilling, 2009; GVPc (2009)].

Location	Date	VEI	Heigh (m)	[Lat (°), Lon (°)]
Asamayama	May 9, 1783	4	1783	[+36.4, +138.5]
Usuzan	Mar. 9, 1822	4	725	[+42.5, +140.8]
Hokkaido	Sep. 25, 1856	4	1140	[+42.0, +140.6]
Nasudake	Jul. 1, 1881	4	1917	[+37.1, +139.9]
Bandaisan	Jul. 15, 1888	4	1819	[+37.6, +140.0]
Suwanosejima	Oct. 2, 1889	4	799	[+29.5, +129.7]
Tarumaesan	Apr. 12, 1909	4	1024	[+42.6, +141.3]
Sakurajima	Jun. 12, 1914	4	1118	[+31.5, +130.6]
Hokkaido	Jun. 17, 1929	4	1140	[+42.0, +140.6]

3.4.1. Plate Boundaries: Mid-Ocean Ridges, Oceanic Trenches and Transform Faults

Plate boundaries have tectonic activity because the plates are moving relative to one another (relative motion). Two plates can move away from each other, toward each other, or sideways relative to one another—that is, there are three types of plate boundaries (Table 3.10): divergent boundaries, convergent boundaries, and transform boundaries. At a divergent boundary, two plates move apart relative to one another and magma fills the space between the plates in most cases. At a convergent boundary, two plates move toward one another and sometimes one plate slides under the other. At a transform boundary, two plates move horizontally past one another. There is about a 70,000-km long chain of divergent boundaries and about as long a chain of convergent boundaries.

Both chains are segmented and connected by transform boundaries, as illustrated in Figure 3.23. This network results from the patterns of plate motions and mantle convection that we review here.

Table 3.10. The three types of plate boundaries with their relative motions and their subcategories.

Plate boundaries	Motions	Subcategories
Divergent (crust is recycled)	⇐ | ⇒	Oceanic-Oceanic Continental-Continental
Convergent (crust is destroyed)	⇒ | ⇐	Oceanic-Oceanic Oceanic-Continental Continental-Continental
Transform	↑ | ↓	Transform faults link the pieces of mid-ocean ridges and trenches as they progressively change their orientations or their expansion rates.

Divergent boundaries. There can only be divergent oceanic-oceanic boundaries and continental-continental boundaries but no oceanic-continental boundary or continental-ocean boundary. Mid-ocean ridges are the result of divergent oceanic-oceanic boundaries, where new crust forms as two plates move apart. Figure 3.24 shows the distribution of ocean ridges. The Mid-Atlantic Ridge is the longest mountain chain underwater in the world. Some of these mountains are so high that they stick out of the water and form volcanic islands like Iceland. The Mid-Atlantic Ridge is created by underwater volcanic eruptions and it is more than 55,000 km long, more than 800 km across, and spread at about 2 cm per year. The East Pacific Rise, separating the Pacific and the Nazca Plates, in Figure 3.24b is the fastest-spreading ridge in the world. It spreads at about 7 cm per year. Note that Africa is surrounded by mid-ocean ridges and has moved relatively little since the breakup of the Pangaea supercontinent.

(a)

(b)

Figure 3.24. (a) is a map of the mid-Atlantic ridge and its fractures. It is the longest mountain range in the world. (Adapted from http://images.suite101.com/files/articles/102000/102057). (b) shows the distribution of oceanic ridges. (Adapted from Reynolds et al., 2010).

On the continents, divergent motion can split a continent into two pieces, with a new ocean basin forming as the pieces move apart. This process, called continental rifting, creates a continental rift, such as Africa's Great Rift Valley. East Africa will progressively move away from the African continent, as illustrated in Figure 3.25, with a large new ocean basin between them. The path of this splitting extends from the Red Sea in the Afar region of Ethiopia (about 10° N in latitude) to beyond the Zambezi River (about 15° S in latitude); a distance of about 4,000 km. It provides us with a great opportunity to understand how the Atlantic Ocean may have begun to form about 200 million years ago. The African rift is only spreading a few millimeters per year; it is still a long way to go before we see the Earth's next major ocean.

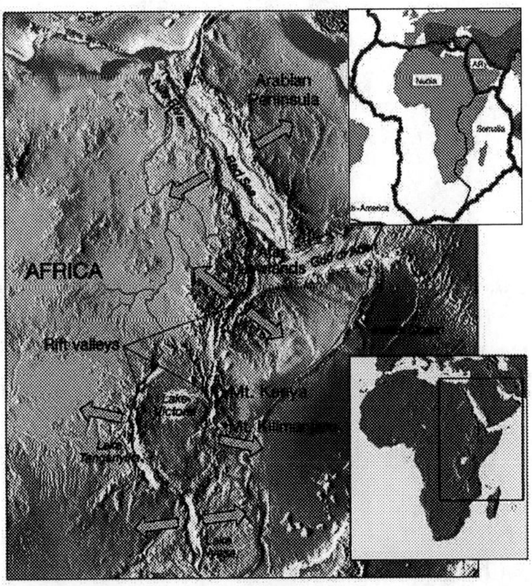

Figure 3.25. Divergent motion is splitting Africa into two continents, forming a new ocean basin as the new continents move apart. This process is called continental rifting.

Some plate tectonic models already divide the African Plate into two plates at the East African Rift: the Somalian Plate, which is the breakaway East Africa, and the Nubian Plate, which is the western part of the African Plate (e.g., Bird, 2003; DeMets et al., 2010). Actually, the model of DeMets et al. (2010) decomposes the eastern part of Africa into two sub-plates (Lwandle and Somalia).

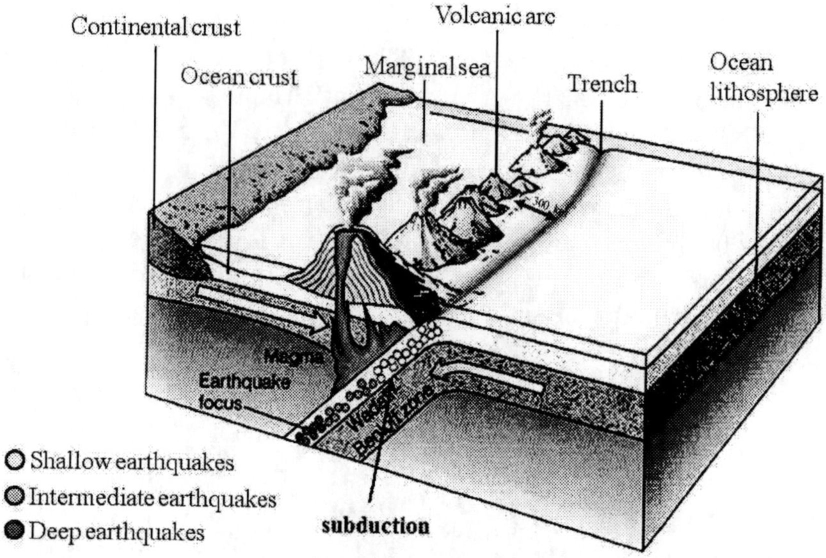

Figure 3.26. An illustration of a subduction zone at an oceanic-oceanic convergent boundary. At this boundary, one plate bends and slides beneath the other plate along an inclined zone. Often the subduction forms volcanoes on the ocean floor. If the volcanoes emerge as islands, a volcanic island arc is formed (Japan, Aleutian Islands, Tonga Islands).

Convergent boundaries. As we mentioned earlier, convergent boundaries form when two plates move toward each other. A convergent boundary can be oceanic-oceanic, oceanic-continental, and continental-continental boundaries. Ocean trenches, island arcs, and Earth's largest mountain belt have been formed at convergent boundaries.Many of Earth's most dangerous volcanoes and largest earthquakes also occur along these boundaries.

At the oceanic-oceanic convergent boundary (Figures 3.23, 3.26, and 3.27) and at the oceanic-continental convergent boundary (Figures 3.23, 3.27, and 3.28), one plate bends and slides beneath the other plate. In such a case there is a subduction, and the zone around the downward-moving plate is known as a subduction zone. In the case of oceanic-continental convergent boundary, the ocean plate which is less dense and less heavy always goes under a heavy continental plate. At the edges of the oceans are found deep-sea trenches, under which the oceanic crust dives down into the mantle. In other words, the deep sea trenches indicate subductions. The deepest trench in the world is the 11-km deep Mariana Trench, which is located in the western Pacific Ocean. Mariana Trench marks where the fast-moving Pacific Plate subducts under the slower moving Philippine Plate (oceanic-oceanic convergent boundary). It is deeper into the Earth's interior than Mount Everest, the world's tallest mountain, rises above sea level (about 9 km).

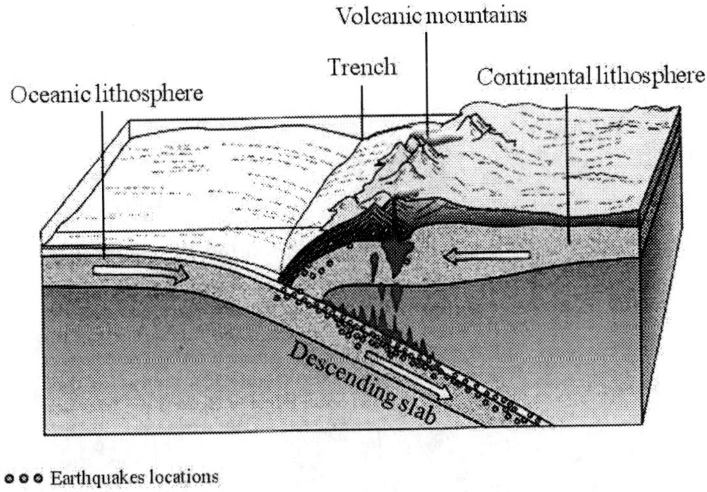

Figure 3.27. An illustration of a subduction zone at an ocean-continental convergent boundary. The denser oceanic plate subducts beneath the more-buoyant continental plate. The resulting volcanic mountain chain is called a continental volcanic arc (the Andes and the Cascades).

When the downward-moving oceanic plate reaches a depth greater than 100 km, the water it contains lowers the fusion point of the surrounding rocks, which creates magma. This magma is hotter and less dense than the rocks around and therefore starts to migrate toward the surface, where it eventually generates volcanism typical of subduction. The chain of volcanoes, known as the volcanic arc, is more or less always at the same distance (around 300 km) from the trench, which is the surface limit between the two plates. It generally runs parallel to the trench. The volcanic island arc exists for both oceanic-oceanic and oceanic-continental convergent boundaries. Often the oceanic-oceanic subduction process forms these volcanoes on the ocean floor. Such volcanoes sometimes emerge from the water to form volcanic island arcs (examples include Japan, the Aleutian Islands, the Tonga Islands, the Andes, the Caribbean Islands, and the Cascades). In other words, the ocean-ocean convergence helps us understand the formation of the Japanese Islands and Caribbean Islands, as examples. In fact, new islands are born around Japan every few years. After some million years, Japan will be a single landmass because continental crust formation is constantly replacing the oceanic crust.

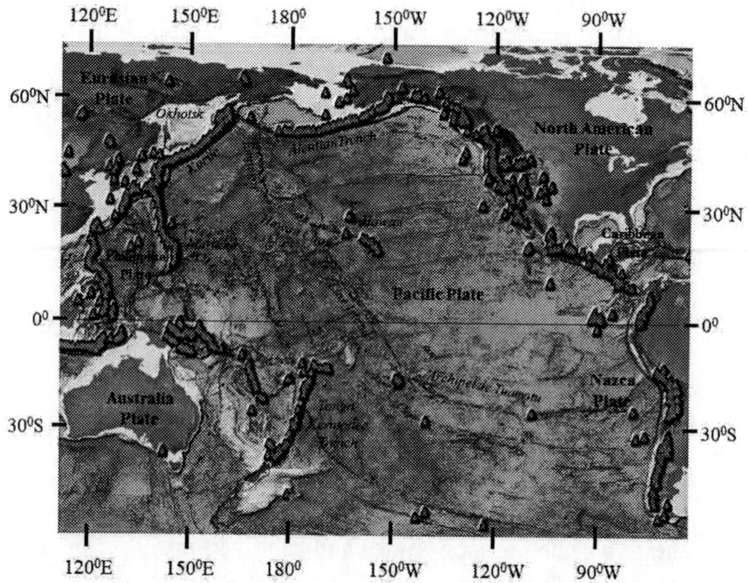

Figure 3.28. Here is the map of the Pacific ring of fire and its volcanoes. (Adapted from Reynolds et al., 2010).

An important geology notion associated with complex subductions is terrane. Figures 3.26 and 3.27 suggest that subductions continued, once begun, unabated to the present, or sporadically ceased. Inability to satisfy these conditions would indicate a failure of subduction and Figure 3.29 illustrates failure of a subduction zone. The

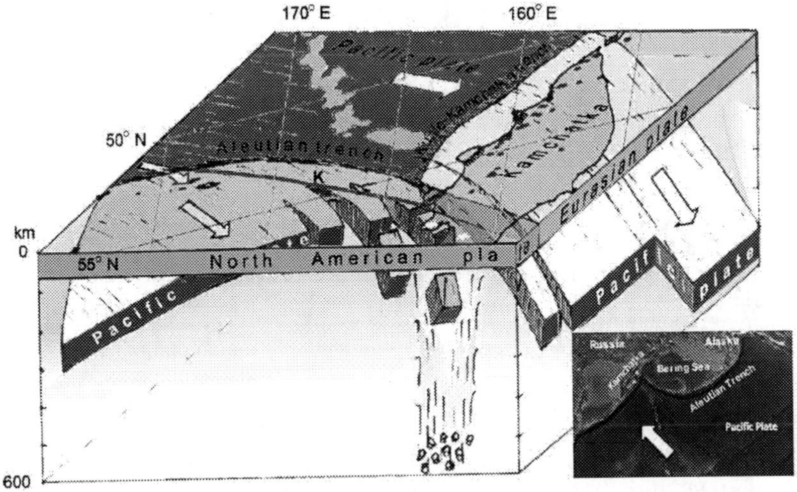

Figure 3.29. 3D-model of the geodynamics of the Kamchatka-Aleutian junction as a view from North West. Black stars are active volcanoes, white stars – Nachikinsky (N) and Khailula (K) extinct volcanoes located off the edge of the Pacific Plate, K-Komandorskaya microplate, which separates from the North American Plate. Some lithosphere blocks are torn off the Pacific Plate edge as a result of interaction (breaking) with the North American Plate. These blocks, having negative buoyancy, sink into the mantle, are heated, and can form a small mantle plume. We also appended a 2-D map of the region for clarity of the location of this junction. (Adapted from Avdeiko et al., 2007).

density and thickness largely of slab determine whether the subduction zone operates normally or fails. Contrasts in lithospheric bulk density (crust plus mantle) of < 0.1 cc determine whether normal subduction or collision occurs and/or only bodies of continental and intraoceanic arc crust greater than > 15 km thick cause subduction zone failure. This collision is called terrane accretion. So a terrane in geology is a fragment of crustal material formed on, or broken off from, one tectonic plate and accreted to the crust lying on another plate. The crustal block or fragment preserves its own distinctive geologic history, which is different from that of the surrounding area, hence, the exotic term *terrane.* A terrane is not necessarily an independent microplate in origin, since it may not contain the full thickness of the lithosphere. As illustrated in Figure 3.29, Alaska and Siberia have many geological terranes, which resulted from the subduction of oceanic plates that carried several continental fragments (e.g. Jones et al. 1987), and the edges of these terranes are often major geological faults (see Ikelle, 2022d).

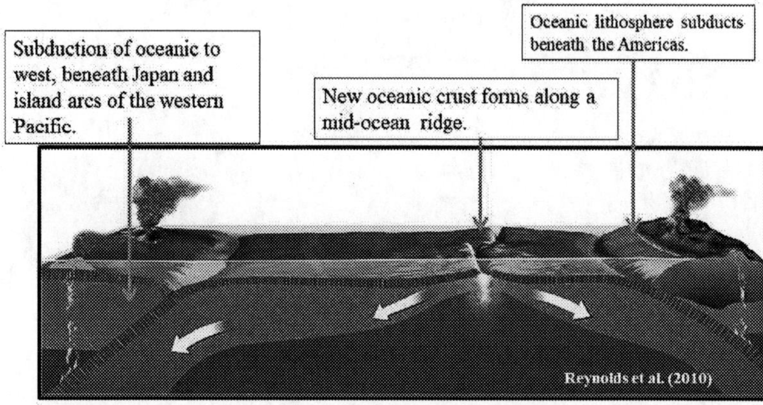

Figure 3.30. Here is a schematic of the tectonic plate of the Ring of Fire. It is a combination of two convergent plate boundaries and one divergent plate boundary.

Notice the concentrated seismicity along the subduction interface between the two convergent plates extends (Figures 3.26 and 3.27). This surface is known as the Wadati-Benioff zone or simply the Benioff zone, in honor of the seismologists who first recognized these zones, Kiyoo Wadati of Japan and Hugo Benioff of the United States. It marks the trace of the actively subducting lithosphere. The velocities of underflow along Benioff zones are roughly 100 km per million year (i.e., 10 cm/year).

Figure 3.30 illustrates the major tectonic-boundary mechanism along the circum-Pacific region, which is known as the *Pacific Ring of Fire.* It is a belt of oceanic trenches, island arcs, volcanic mountain ranges and plate movements and collisions that encircles the Pacific Ocean. The volcanic line around this ring extends from the southwestern Pacific, then through the Philippine Islands, Japan and Alaska, and down the western coasts of the Americas. Notice that the ring of fire is a combination of two convergent plate boundaries, with subduction on both sides of the Pacific, and one divergent plate boundary. The northwestward moving Pacific plate subducts beneath the Aleutian Islands arc in the north, along the Kamchatka peninsula and Japan in the west. To the south a number of smaller tectonic plates are in collision with the Pacific Plate from the Mariana Islands, the Philippines, Tonga, and New Zealand.

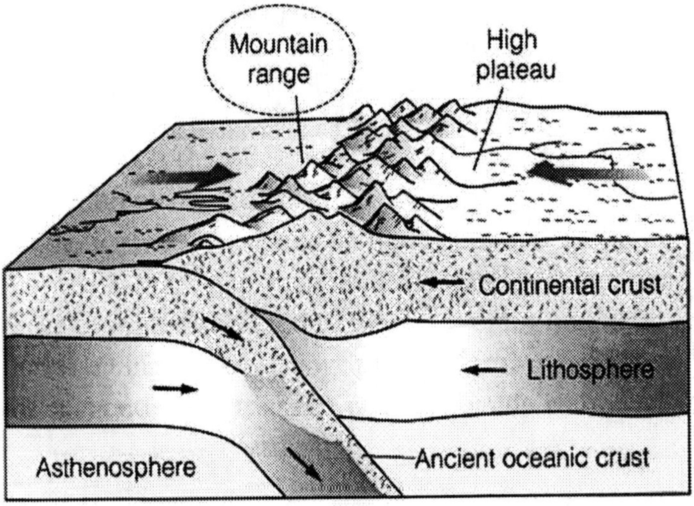

Figure 3.31. An illustration of a continental-continental convergent boundary. This type of boundary is commonly called a continental collision. The resulting collision between two continental blocks produces mountains (the Himalayas, the Alps, the Appalachians). In particular, the collision between the Indian and Eurasian Plates has pushed up the Himalayas and the Tibetan Plateau.

Note that the slab geometries of the subducting Pacific Plate of the Pacific Ring of Fire vary with regions. For example, the Marianas subduction zone is at an almost vertical angle whereas the Peru-Chile subduction beneath the South American Plate is less steep, and includes almost horizontal slab segments (e.g., Barazangi and Isacks, 1976; Cahill and Isacks, 1992; Ramos and Folguera, 2009).

Continental-continental convergent boundaries (Figure 3.31) are commonly called continental collisions. Neither plate is subducted. Instead, the crust tends to buckle and be pushed upward or sideways.

They produce large mountain ranges (e.g., the Himalayas, the Alps, and the Appalachians).

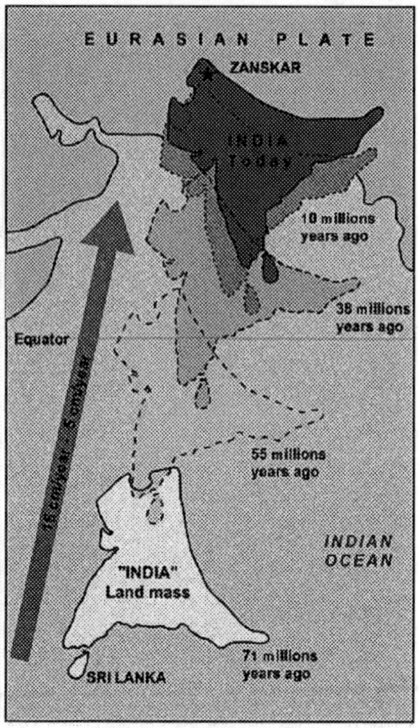

Figure 3.32. About 220 million years ago, India was a large island still situated off the Australian coast. When Pangaea broke apart about 200 million years ago, India began to forge northward at a rate of about 9 m a century. It rammed into Asia about 40 million years ago, creating the Himalayas, which stretch 2,900 km along the border between India and Tibet. (Adapted from http://upload.wikimedia.org/wikipedia/kn/thumb/2/2f/Himalaya-formation.gif).

Formation of the Himalayas (Figure 3.32) is one of the most dramatic and visible creations of tectonic-plate forces, in particular of a continental-continental convergent boundary. The Himalayas stretch 2,900 km and are as high as 9 km above sea level, along the border between India and Tibet. This immense mountain range began to form about 40 million years ago when two large landmasses, India and Eurasia, driven by plate movement, collided (a continental-

continental convergence). About 220 million years ago, India was

Table 3.11. The strongest earthquakes on record in 2019 with direct effects on people and land and the most explosive volcanic eruptions in the last 220 years on record in 2019 with direct effects on people and land. All these eruptions are subduction volcanism and no eruptions with VEI greater than 7 occurs in post-Holocene so far. (Sources: Tilling,2009; Naranjo et al., 1993; De Silva and Zielinksi, 1998; Hildreth and Drake, 1992). BCE= before the birth of Jesus Christ.

Location	**Date**	Magnitude
Chile (Valdivia)	May 22, 1960	9.5
Alaska	March 28, 1964	9.2
Japan (Tohoku-oki)	Mar. 11, 2011	9.1
Indonesia (Northern Sumatra)	Dec. 26, 2004	9.1
Kamchatka	Apr. 11, 1952	9.0
Ecuador	Jan. 31, 1906	8.8
Chile (Maule)	Feb. 27, 2010	8.8
Alaska	Apr. 4, 1965	8.7
Indonesia (Nias-Simeulue Sumatra)	Mar. 28, 2005	8.6
Assam-Tibet	Aug. 15, 1950	8.6
Alaska	Sept. 3, 1957	8.6

Location	Date	VEI	Subduction
Oruanui Ignimbrite (New Zealand)	~ 26,500 BCE	8	Taupo volcano
Mount Tambora (Indonesia)	Apr. 10, 1815	7	Sunda arc
Mount Pinatubo (Philippine)	Jun. 16, 1991	6	Luzon volcanic arc
Cerro Azul (Chile)	Apr. 10, 1932	6	Andean Volcanic arc
Novarupta (Alaska, USA)	Jun. 6, 1912	6	Aleutian Ramge
Santa Maria Volcano (Guatemala)	Oct. 24, 1902	6	Central America volcanic arc
Krakatoa (Indonesia)	Aug. 26, 1883	6	Sunda arc

still a large island situated off the Australian coast. When Pangaea broke apart about 200 million years ago, India began to forge northward at a rate of about 9 meters a century. When India rammed into Asia about 40 million years ago, the Himalayas, the highest continental mountains in the world, were formed. Moreover, the neighboring Tibetan Plateau, at an average elevation of about 4.5 km, is higher than all the peaks in the Alps except Mont Blanc and Monte Rosa.

In summary, Table 3.11 shows the strongest earthquakes and most explosive volcanic eruptions on recent record until 2019. All these earthquakes and volcanoes are associated with convergent plate boundaries. For example, the 1950 Assam-Tibet earthquake is related to continental-continental convergent boundaries of the northward motion of India. This motion has deformed all of the region (Pakistan, Tibet, China, Indochina, Mongolia, Indonesia, and Siberia) and creates many large earthquakes. All the other earthquakes are related to subductions along the Pacific Rim, and these subduction zones have the most faults capable of magnitude 9 or greater earthquakes. In fact, subduction zones account for more than 95 percent of the energy released by earthquakes globally. Tsunami hazards are associated with subduction zones. Subduction zones are also geodynamic systems that build island arcs. The sinking oceanic lithosphere interacts with water and other materials from the sinking lithosphere to cause the mantle to melt. These melts ascend and erupt at arc volcanoes. The most explosive volcanic eruptions of the last 220 years have also occurred in the Pacific Rim. While the forces powering plates is still an open issue in geosciences—we will revisit this issue in Ikelle (2022d)—one point that is clear and widely accepted is that the subduction process is one of the dominant driving forces of the plate tectonics phenomena.

Mantle convection refers to the mechanism of divergent and convergent plate boundaries. Plates of the lithosphere are moved around by the underlying hot mantle convection cells.

Convection is how heat passes through fluids (see Figure 3.33a). One important property of fluid is that it rises when heated. The hot fluid becomes less dense and rises up. Cooler fluid is less dense and so it sinks down. In other words, the fluid density differences

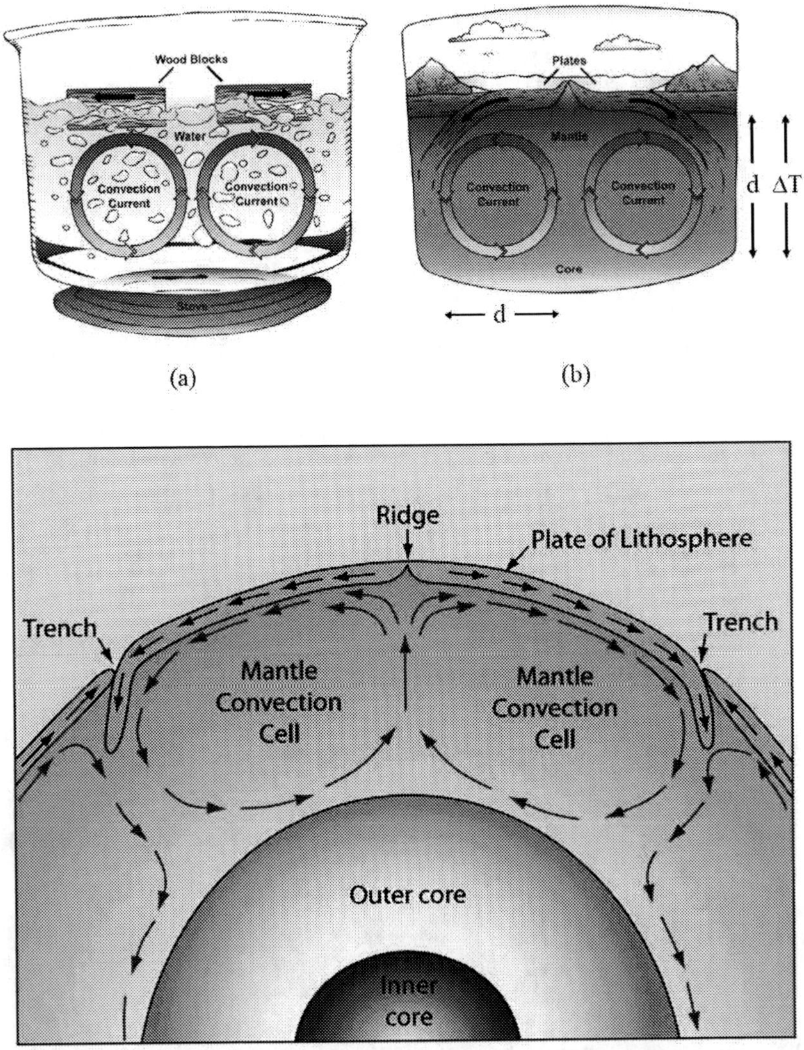

(c)

Figure 3.33. (a) The hot fluid becomes less dense and rises. Cooler fluid is less dense and so it sinks down. This up-and-down motion creates what are convection currents. (b) In the interior of Earth, we have convection between the cooler crust and the hotter core. The mantle is considered as a layer of fluid heated below and cooled above with $d \approx 700$ km and ΔT is the temperature difference across the layer. (c) Another illustration of the mantle convection. (Adapted from Kious and Tilling, 2001).

are a direct result of temperature differences between the fluid and the solid wall surface. This up-and-down motion creates what are called convection currents. Convection currents are circular movements of heated fluid that help spread the heat. So, as illustrated in Figure 3.33a, we need a fluid layer bounded by two surfaces and a heater to produce convection. In the interior solid wall surface. This up-and-down motion creates what are called convection currents. Convection currents are circular movements of heated fluid that help spread the heat. So, as illustrated in Figure 3.33a, we need a fluid layer bounded by two surfaces and a heater to produce convection. In the interior of solid Earth, we have convection between the cooler crust (cool rocks at shallow depth) and the hotter core (hot rocks at great depth). In other words, the mantle can be described as a layer of fluid heated from below and cooled from above (see Figure 3.33b). The "fluids" (mantle plume) here are the sublithospheric mantle's rocks, which heat up and partially or completely melt in the deeper and hotter mantle. These so-called fluids are viscous (viscosity is the rock's resistance to flow; it decreases with increasing temperature). In any event, we have mantle convection as first pointed out by Bull (1921) and Holmes (1931).

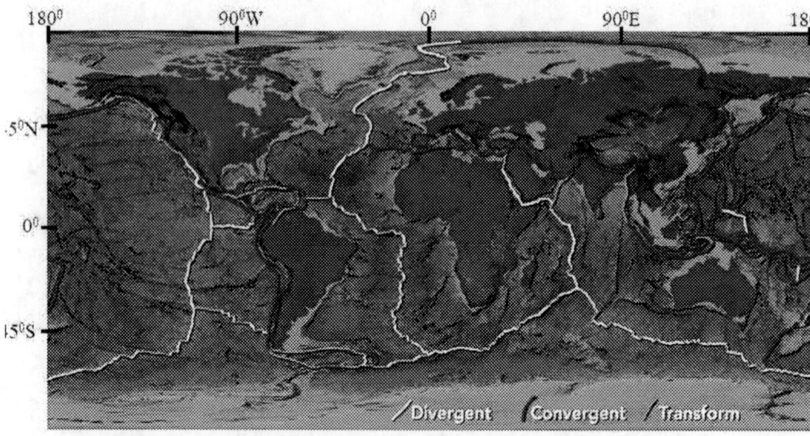

Figure 3.34. Another world tectonic map showing the three types of plate boundaries. (Adapted from Reynolds et al., 2010).

Convection is how heat passes through fluids (see Figure 3.33a).

One important property of fluid is that it rises when heated. The hot fluid becomes less dense and rises up. Cooler fluid is less dense and so it sinks down. In other words, fluid density differences are a direct result of temperature differences between the fluid and the solid wall surface. This up-and-down motion creates what are called convection currents. Convection currents are circular movements of heated fluid that help spread the heat. So, as illustrated in Figure 3.33a, we need a fluid layer bounded by two surfaces and a heater to produce convection. In the interior of solid Earth, we have convection between the cooler crust (cool rocks at shallow depth) and the hotter core (hot rocks at great depth). In other words, the mantle can be described as a layer of fluid heated from below and cooled from above (see Figure 3.33b). The "fluids" (mantle plume) here are the sublithospheric mantle's rocks, which heat up and partially or completely melt in the deeper and hotter mantle. These so-called fluids are viscous (viscosity is the rock's resistance to flow; it decreases with increasing temperature). In any event, we have mantle convection as first pointed out by Bull (1921) and Holmes (1931).

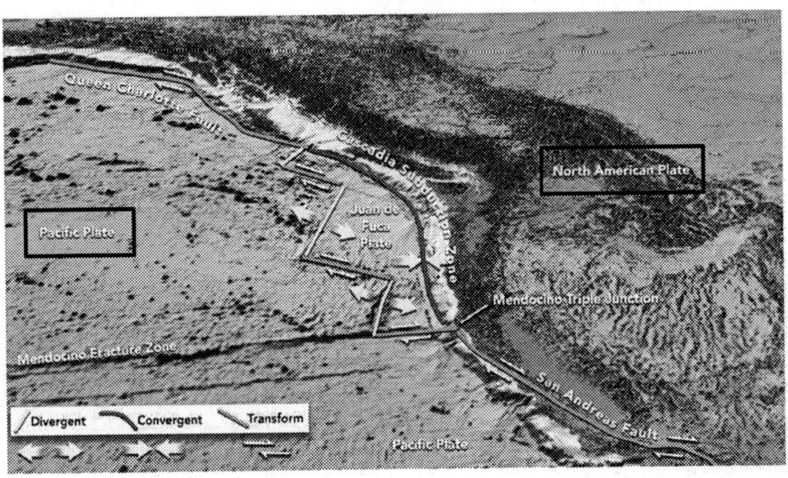

Figure 3.35. A map of the boundary between the Pacific Plate and the North American Plate. Note that this boundary is mostly a transform boundary, with the Pacific Plate moving northwest relative to the main part of the North American continent. (Adapted from Reynolds et al., 2010).

The mantle starts at 10 to 70 km depth and goes down to the outer core at 2,900 km depth. In the mantle, hot material rises toward the lithosphere, and then reaches the base of the lithosphere, where it cools and sinks back down through the mantle (Figure 3.33c). The cool material is replaced by additional hot material, and so on, forming a large convection cell. This slow but incessant movement in the mantle causes the rigid tectonic plates to move (float) around the earth's surface at an equally slow rate. Figure 3.33c is a very simplified version of mantle convection. We will revisit the convention model in later chapters of this book.

Transform Boundaries. At transform-fault boundaries, or simply transform boundaries (Figure 3.34), plates slip horizontally past each other along transform faults.[3] The concept of transform faults was first presented by Canadian geophysicist J. Tuzo Wilson. In the oceans, transform faults are associated with mid-ocean ridges. These faults combine with spreading centers to form a zigzag pattern on the seafloor (see Figure 3.24a). Transform faults can link different types of plate boundaries as illustrated in Figure 3.34. Actually, most transform faults are found in the oceans. Some faults occur beside or within a continent, sliding one large crustal block past another, such as the Alpine Fault in New Zealand and the San Andreas Fault in California. The San Andreas Fault zone is about 1,300 km long and tens of kilometers wide. Along it, the Pacific Plate has been grinding horizontally past the North American Plate for 10 million years, at an average rate of about 5 cm/year. This transform fault connects the East Pacific Rise, a divergent boundary to the south, with Juan de Fuca, another divergent boundary to the north. The boundary between the Pacific Plate and the North American Plate (Figure 3.35) is a transform boundary in large segments, with the Pacific Plate moving northwest relative to the main North American land mass.

In this subsection we have explained the mechanism of con-

[3]Faults are surfaces, usually planar, where the two sides move relative to each other. Direction of motion is known as slip direction. The faults can be classified as thrust fault, normal fault and strike-slip fault based on the nature of relative movement along the fault.

tinental drift by using transform plate boundaries in addition to mid-ocean ridges (oceanic-oceanic divergent boundaries), the great rift valley (continental-continental divergent boundaries), ocean trenches (oceanic-oceanic and oceanic-continental convergent boundaries), the subduction of oceanic plates (oceanic-oceanic and oceanic-continental convergent boundaries), mantle convention (divergent and convergent boundaries), and formations of large mountain ranges (continental-continental convergent boundaries). In summary, the creation, expansion, and subduction described here take place on a geological time scale. For example, it took 100 million years to create the seafloor of the Atlantic Ocean. The speed of creation, expansion, and subduction is measured in centimeters per year. Obviously, this speed cannot directly endanger humans. Note also that, among the terrestrial planets (Mercury, Venus, Mars, and Earth), only Earth appears to have subduction zones. Mars has immense volcanoes of the Tharsis region, the solar systems largest "hotspot" volcanoes. The Insight mission on Mars, equipped with seismometers, has not yet recorded a significant volcano or tectonic earthquake (Ikelle, 2020).

Note that the origin of the Earth's magnetic field is due to convection of an electrically conducting fluid outer core. This convection is an example of a homogeneous dynamo, or a dynamo without wires, consisting of currents in a continuum. The resulting Earth's magnetic field is essentially a dipole field, exhibiting some small non-dipole components. It varies in both space and time, consistent with an origin due to convection in a conducting, fluid outer core. We will revisit this issue in later chapters.

3.4.2. The Hotspot Theory

As described above, the plate tectonics theory can effectively explain solid Earth's magnetism, deformation, and earthquakes along plate boundaries, but cannot explain some within-plate geological phenomena, especially seamount chains. In other words, there is one observation about the seafloor that is not yet captured by the tectonic plate theory—the seamount chains (i.e., isolated submarine mountains) such as those of Hawaii, the Louisville, Azores, the Canaries, the Réunion, and Tahiti. None of these locations are at plate boundaries. The tectonic plate theory uses a complementary theory, the hot-spot theory (also known as the mantle-plume theory), to explain seamount chains; i.e., the hot-spot theory explains seamount chains such as those of Hawaii, the Azores, the Canaries, Reunion, and Tahiti. Tuzo Wilson postulated in 1963 that volcanic chains like the Hawaiian Islands result from the slow movement of a tectonic plate across a fixed hotspot deep beneath the surface of the planet. Richards et al. (1989) have identified some 40 to 50 hotspots around the globe (see Figure 3.36). Hotspots are thought to be caused by a narrow stream of hot mantle, called a mantle plume, convecting up

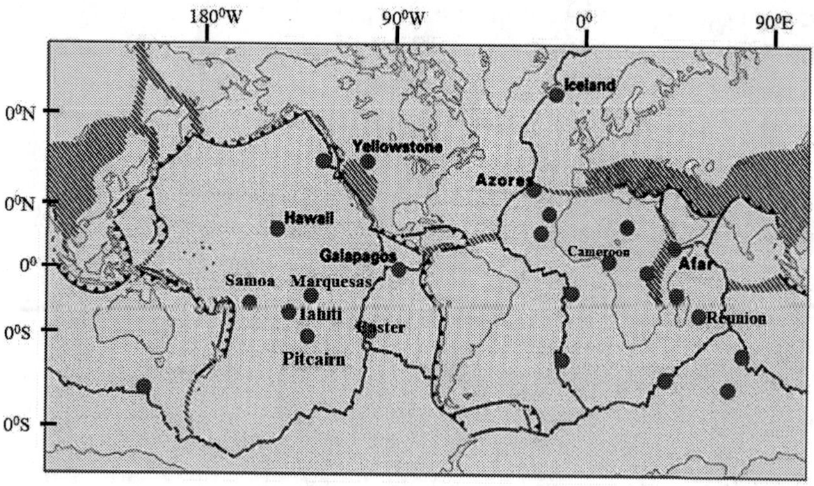

Figure 3.36. The red dots represent some 40 to 50 hot spots that geologists have identified. (Adapted from pubs.usgs.gov/publications).

from the earth's core-mantle boundary. Mantle plumes are mushroom-shaped trails of hot rock that rise from deep inside the mantle, melt as they rise, and erupt from volcanoes at hotspots at the surface. When a tectonic plate moves over a fixed hotspot, a chain of volcanoes is formed (Figure 3.37). The volcanoes get younger from one end to the other. Hotspots can be further divided into three types: (1) midplate continental hotspots like Yellowstone in the United States of America, (2) midplate oceanic hotspots like Hawaii and Bermuda, and (3) near ridge hotspots like Iceland, the

Galapagos Islands, and the Azores. Each of these volcanic tracks has a mantle source at a different depth.

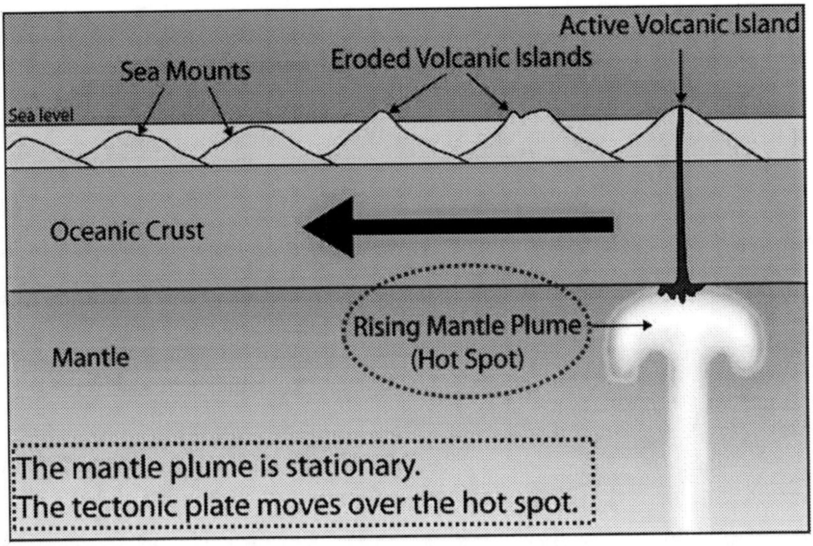

Figure 3.37. An illustration of a hotspot in the context of tectonic plate theory. The heat, which is concentrated in the mantle, can produce magma (called a mantle plume), which rises to Earth's surface. The tectonic plate moving over a fixed (stationary) hot spot forms a chain of volcanoes.

Let us use the following analogy to improve our understanding of seamount formation. Suppose that your shoes capture some mud. As you continue walking, mud will continue to drop along your path. Now think about the plate at a hotspot. At this hotspot, which is a stationary point, the moving plate captures magma, and the plate spreads this magma as it moves. That is how seamount chains are formed. For example, the Pacific Plate has moved over a hotspot, producing the Hawaiian Islands.

According to Wilson's hotspot theory, the volcanoes of the Hawaiian chain should get progressively older and become more eroded as they travel beyond the hotspot. The oldest volcanic rocks on Kauai (Figure 3.38), the northwesternmost inhabited Hawaiian island, are about 5 million years old and are deeply eroded. By comparison, on the "Big Island" of Hawaii, the southeasternmost in the chain and presumably still positioned over the hotspot, the oldest exposed rocks are less than 1 million years old, and new volcanic rock is continually being formed. The hotspot in the Central Pacific created the linear chains of the Hawaiian Islands, thanks to the rapid movement of the Pacific Plate (9 cm/year). When plate movement is slow (1 or 2 cm/year), there are large gaps in islands and seamount chains, as in the Azores or the Canaries. Note that as one island volcano becomes extinct, another develops over the hotspot, and the cycle is repeated. This process of volcano growth and death, over many millions of years, has left a long trail of volcanic islands and seamounts across the Pacific Ocean floor.

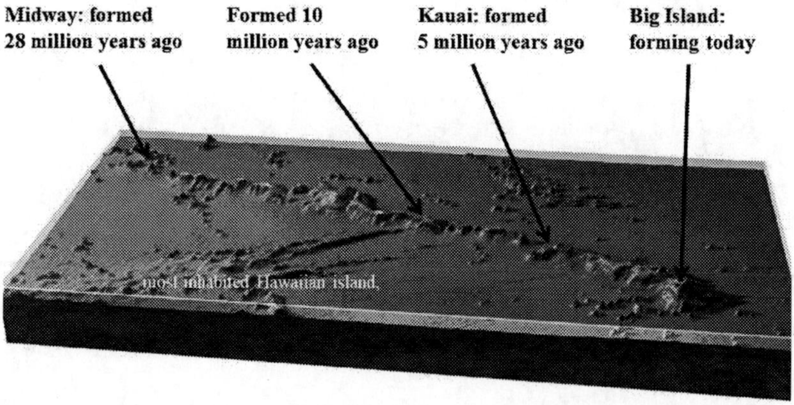

Figure 3.38. The Hawaiian chain is a seamount chain created by the Pacific Plate moving over a hotspot. Notice that the volcanoes get progressively younger from one end to the other. The volcanic rocks on Kauai, the most-inhabited Hawaiian island, are about 5 million years old and are deeply eroded. By comparison, on the "Big Island" of Hawaii, which is presumably still positioned over the hot spot, the oldest exposed rocks are less than 1 million years old, and new volcanic rock is continually being formed.

Hotspots can also create terrestrial volcanoes. The Yellowstone Supervolcano, for instance, sits over a hotspot in the middle of the North American Plate, with a series of ancient calderas stretching across southern Idaho. The Yellowstone hotspot fuels the geysers, hot springs, and other geologic activity in Yellowstone National Park,

Wyoming.

A fundamental question about hotspots that is still largely unresolved and that we will revisit in later chapters is the motion of hotspots. Some hotspots, such as Iceland, are located on plate boundaries and cannot be used as a fixed-hotspot reference frame. Therefore they move with plate boundaries to which they are associated with relative to others and even relative to the underlying mantle. Other hotspots are located within plates such as Hawaii, Tahiti, and Cameroon. Are the second group of hotspots fixed relative to the mantle and/or fixed relative to one another? Because a rigid body motion on a sphere is a rotation about a geocentric axis, the answer to this question can lead us to the realization of a global fixed reference frame of the Earth that coincides with Earth's geocentric axis and geocenter. Moreover, all plates can then be defined relative to such a frame. We will return to this question in the next chapter.

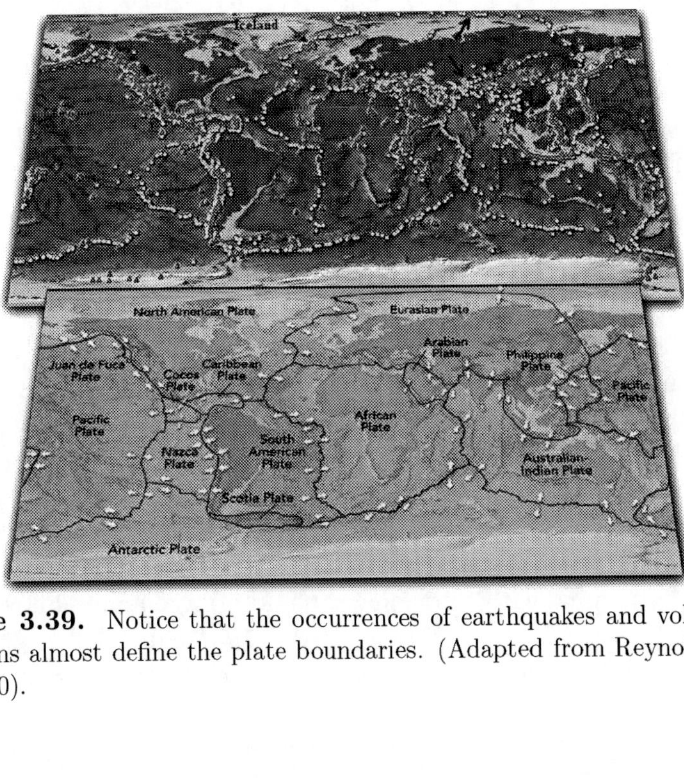

Figure 3.39. Notice that the occurrences of earthquakes and volcanic eruptions almost define the plate boundaries. (Adapted from Reynolds et al., 2010).

3.4.3. The Earthquake Map versus the Plate Map

Let us start by looking at the distribution of earthquakes and volcanic eruptions. Figure 3.39 shows a map of the epicenters of earthquakes of a magnitude greater than 5. These epicenters are clearly not incoherently distributed on Earth. They are much more common when located beside trenches. Similarly, volcanic eruptions are much more common than trenches, just as earthquakes are. It is obvious that there is a significant correlation between earthquakes and volcanic eruptions. However, there are also some exceptions. Let us look at these correlations and non-correlations in more detail. Ecuador and its capital, Quito, for example, are under the threat of both volcanic eruptions and earthquakes. The whole of Central America is under the twin threat of earthquakes and volcanic eruptions. An earthquake in Guatemala in 1976 caused 22,000 deaths, an earthquake in Nicaragua in 1972 led to 5,000 deaths, and an earthquake in El Salvador in 1986 caused 10,000 deaths. Volcanic eruptions occurred in Irazu in 1964 and Arenal in 1970, both in Costa Rica. On the west coast of South America, an earthquake in Chile in 1960 caused 5,700 deaths, and one in Peru in 1970 left 66,000 dead. The volcanic eruption of Nevado del Ruiz, Colombia, in 1986 left 23,000 dead. In the Pacific Rim (i.e., cities located around the edge of the Pacific Ocean), the situation is quite similar. Volcanic eruptions and earthquakes have occurred in the New Hebrides (Vanuatu), New Zealand, Papua New Guinea, and the Philippines. Japan is well known for its volcanic eruptions and earthquakes, as are Indonesia and the less-densely-populated areas of Kamchatka (a 1,250-km-long peninsula in the Russian Far East, between the Pacific Ocean to the east and the Sea of Okhotsk to the west, with a 10,500-meter-deep trench along its Pacific coast) and Alaska. These are clear examples of the correlation between earthquake occurrences and volcanic eruptions.

Let us now look at the cases in which there is no correlation between earthquake occurrences and volcanic eruptions. Figure 3.39 shows an area with earthquakes, but no volcanoes between the Eurasian and North American Plates. On July 26, 1963, a 6.1-magnitude earthquake occurred in Skopje (in the present-day Republic of Macedonia, then part of Yugoslavia), killing more than

1,070 people. About 80 percent of the city was destroyed. Yet there are no active volcanoes in the region. China has experienced numerous earthquakes but has no active volcanoes (Figure 3.39). On May 12, 2008, more than 67,000 people died when a 7.9-magnitude earthquake struck Sichuan, Gansu, and Yunnan provinces in western China. On April 14, 2010, an earthquake with a magnitude of 6.9 struck Yushu, Qinghai, China; 2,698 people were killed. On Réunion Island there are spectacular volcanic eruptions, including the Piton de la Fournaise volcano, which is one of the most active volcanoes in the world and erupts almost every year, spewing fluid basalt fountains and flows, but there is no serious threat of earthquakes in that area (Figure 3.40). Although the noncorrelations are not as common as the correlations, they do exist. In theory, regions with earthquakes and no volcanoes can be simpler to model and therefore to monitor than those with correlations because of the isolated tectonic setting.

(a)

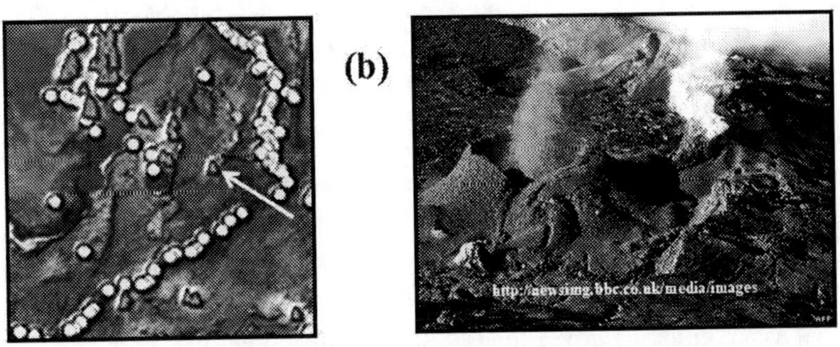

Figure 3.40. (a) An example of a region with volcanoes but no earthquakes. The arrow indicates this region, which is Reunion Island. (b) shows lava spewing from the Piton de la Fournaise volcano on the island of La Reunion in the Indian Ocean. It is one of the most active volcanoes in the world, erupting almost every year with fluid basalt fountains and flows. (Adapted from http://newsimg.bbc.co.uk/media/images).

Notice also that the three types of earthquake faulting mechanisms (Figure 3.41) can be connected to the three types of plate boundaries, as follows. Divergent boundaries correspond to normal faulting, transform boundaries correspond to strike-slip faulting, and convergent boundaries correspond to thrust faulting. Most earthquakes occur at plate boundaries. Ridge earthquakes are relatively small and almost nobody is killed because they take place kilometers under the seafloor. Earthquakes in trenches are violent, very deep, and very deadly (as in Mexico, Guatemala, Chile, and Japan). Earthquakes are generally located no deeper than the first 50 km of the crust, except for those in trenches. The trench earthquakes can occur at depths of 300, 400, and 700 km. The tectonic plate theory allows us to explain these observations as follows: in the regions in which plates are cold, rigid, and subducting, the potential earthquake locations could also be subducted in the Earth's interior. The earthquakes along transform faults are very violent. Typical examples are California (San Francisco, 1906), the long fault of Anatolie (Erzincan, 1939), and Jordan (Jericho, 1250). In summary, most large earthquakes occur at places at the boundaries where the plates rub against each other, collide, or subduct into the Earth's interior.

3.4.4. The Active Volcanic Map versus the Plate Map

A comparison of the map of active volcanoes and plate boundaries also shows remarkable coincidences. Zones of oceanic trenches, or subductions, are also places where there are active volcanoes. Such zones exist in Japan, Indonesia, the Philippines, New Guinea, Alaska, the western United States, Mexico, Ecuador, Peru, Chile, Italy, and Greece. Zones of oceanic ridges are also regions where there are active volcanoes. They are less spectacular because they are located 1 or 2 km below the seafloor. Their effects rarely show up at the sea surface. However, transformation faults do not produce volcanic eruptions. This is a major difference between earthquake distribution and active volcano distribution; therefore there is no correlation between earthquakes and volcanic eruptions at the transformation boundaries. There are no active volcanoes in the former Yugoslavia, in northern Turkey, along the San Andreas Fault, or in the Jordan River delta. At these places, the plates rub against each other. There is no vertical transfer of materials and no pop-up of materials from the interior of the earth. Thus there are no active volcanoes, but these are active seismic areas because of the transformation boundaries.

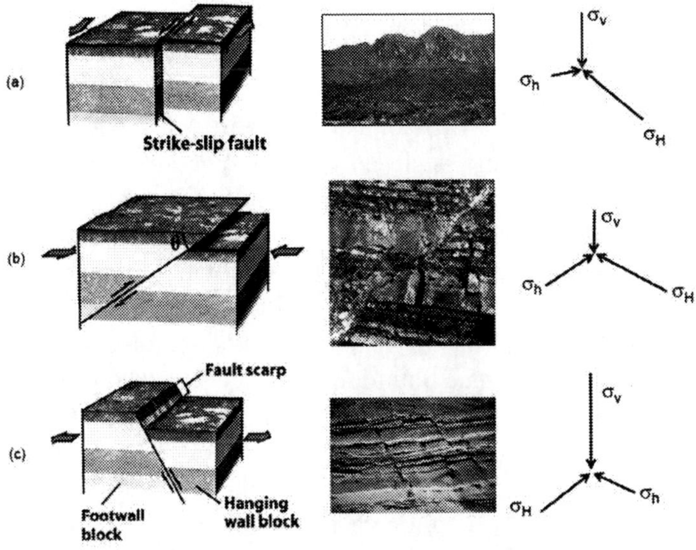

Figure 3.41. An illustration of the three fault types. These three types of earthquake faults are essentially controlled by the relative magnitude of the three principal stresses, σ_h, σ_H, and σ_v. (a) Strike-slip faults (also known as the transform fault) with $\sigma_H > \sigma_v > \sigma_h$. In this case one block slides sideways (lateral shearing) past the other. The strike-slip fault can be the result of one tectonic plate movement relative to an adjacent plate with no vertical motion, and the fault surface is nearly vertical. The Haiti earthquake of magnitude 7 on January 12, 2010, was caused by a strike-slip fault. (b) The thrust fault (also known as the reverse fault), with $\sigma_H > \sigma_h > \sigma_v$. In this case, a hanging wall went up relative to a footwall. It is caused by crustal shortening (e.g., convergent plate boundaries). The slope (dip) of the fault is not generally steep. The Chile earthquake of magnitude 8.8 on February 27, 2010, was caused by a thrust fault. (c) The normal fault (or reverse fault) with $\sigma_v > \sigma_H > \sigma_h$. In this case, a hanging wall went down relative to a footwall. It is caused by crustal stretching (e.g., divergent plate boundaries). The Nevada earthquake of magnitude 7.1 on December 16, 1954, was caused by a normal fault. The orange arrows indicate the pressure that causes faulting.

In addition to the distribution of volcanoes based on plate boundaries, we must add the volcanism of hotspots located inside the plates. This volcanism is not correlated with tectonic earthquakes because there are no plate boundaries. Iceland, which was in the news in the spring and early summer of 2010, constitutes a unique case (Figure 3.42). Its volcanic eruptions caused an enormous disruption of air travel across western and northern Europe in April and May 2010. From April 14 to 20, ash covered large areas of northern Europe. About 20 countries closed their airspaces to commercial jet traffic and this move affected about 10 million travelers. Volcanic ash can damage jet engines by causing excessive abrasion and even complete in-flight failure. The area has a hotspot that has formed on the oceanic ridge and remains (see Figure 3.42). It is a kind of hybrid ridge-hotspot point—a very complex case. Iceland is located right on top of a divergent boundary. In fact, the island exists only because of this feature. As the North American and Eurasian Plates were pulled apart, volcanic activity occurred along the cracks and fissures. After many eruptions over time, the island grew out of the sea. Iceland widens by 19 mm/year due to the continuous spreading between the North American and Eurasian Plates. Also Iceland is one of the few places of the Mid-Atlantic Ridge elevated enough above the sea level.

In theory, regions with volcanoes and no earthquakes can be simpler to model and therefore to monitor because of the isolated tectonic setting.

So the tectonic plate theory provides explanations for Earth's major surface processes—the distribution of earthquakes, volcanoes, and mountains; and the distribution of ancient organisms and mineral deposits (Table 3.12). The theory is a well-tested and widely accepted view that best explains geological, geophysical, and geochemical observations. Actually, the tectonic plate theory is more than a theory; it has become a paradigm because it can explain a large number of interrelated phenomena.

Earthquakes can also occur within plates, although plate-boundary earthquakes are much more common. Less than 10 percent of all earthquakes occur within the plate's interior. As plates continue to move and plate boundaries change over geologic time, weakened boundary regions become part of the interiors of the plates. These zones of weakness within the continents can cause earthquakes in response to stresses that originate at the edges of the plate or in the deeper crust. The New Madrid earthquakes of 1811-1812 and the

1886 Charleston earthquake occurred within the North American Plate.

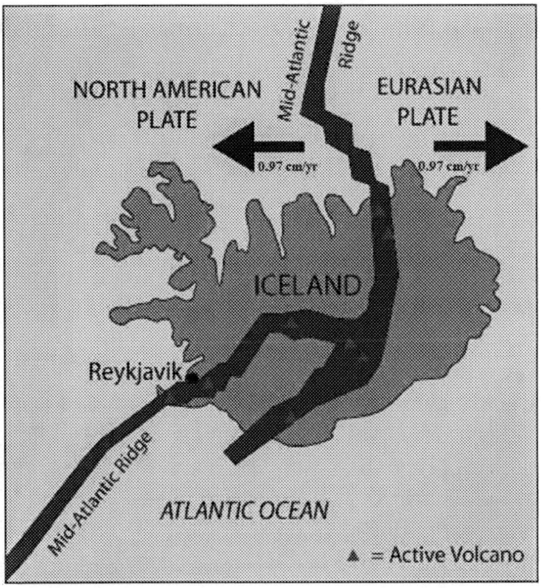

Figure 3.42. Iceland (in green) is located on a ridge on top of a divergent boundary; the Mid-Atlantic Ridge splits Iceland and separates the North American and Eurasian Plates. Black arrows indicate the spreading direction between the Eurasian and North America plates. The map also shows the locations of some of Iceland's active volcanoes (red triangles). (Adapted from http://pubs.usgs.gov/dynamic/understanding.html).

To summarize, plate tectonics is the kinematic theory that describes the large-scale motions and features of the crust in terms of the relative motions and interactions of large, rigid, interlocking plates. Our description of plate tectonics here was essentially along the lines of the studies of McKenzie and Parker (1967), Morgan (1968), Le Pichon (1968), and Isacks et al. (1968). These studies are mostly kinematic in the sense that they predominantly rely on the travel times and speeds of plates while almost ignoring the forces that cause these plate motions (i.e., the driving mechanism of plates or the dynamics of plate tectonics). Some of these studies recognize this fact. One might have expected that subsequent studies would include comprehensive dynamic descriptions of plate tectonics to complement the current kinematic description. More than 60 years later, the description of causes of the plate motions in a consistent way with the current understanding of plate tectonics still remains elusive to geoscientists and physicists. We will return to these issues in later chapters. Note also that plate tectonics are also useful in exploration for mineral and hydrocarbon resources and modeling climate change and sea-level change over geological time-periods (Ikelle, 2017b and 2020).

Table 3.12. An explanatory summary of earthquakes and volcanic eruptions based on plate boundaries and hot-spot locations. Regions where there is a twin effect of earthquakes and volcanic eruptions are indicated by "yes" for earthquakes, volcanoes and correlations. Regions where there are earthquakes but no volcanoes are indicated by "yes" for earthquakes and "no" for volcanoes and correlations. Regions where there are volcanoes but no earthquakes are indicated by "no" for earthquakes, "yes" for volcanoes, and "no" for correlations.

	Divergent		Convergent				
	O-O	C-C	O-O	O-C	C-C	Transform	Hotspot
Earthquakes	Yes	Yes	Yes	Yes	Yes	Yes	No
Volcanoes	Yes	Yes	Yes	Yes	No	No	Yes
Correlations	Yes	Yes	Yes	Yes	No	No	No

O-O: Ocean-Ocean; O-C: Ocean-continent; C-C= Continent-continent

3.4.5. Where else are Earthquakes and/or Volcanoes in our Solar System?

As described in Chapter 1, the four terrestrial planets have roughly the same structure: a central metallic core, mostly iron, with a surrounding silicate mantle underlying a lithosphere layer. Three of the four solar terrestrial planets (Venus, the Earth, and Mars) have even substantial atmospheres. Yet, so far, plate tectonics do not appear to be happening on Venus, Mars, and Mercury. So Earth may be today unique in our Solar System because of its plate tectonics and volcanism that have made it habitable by us.

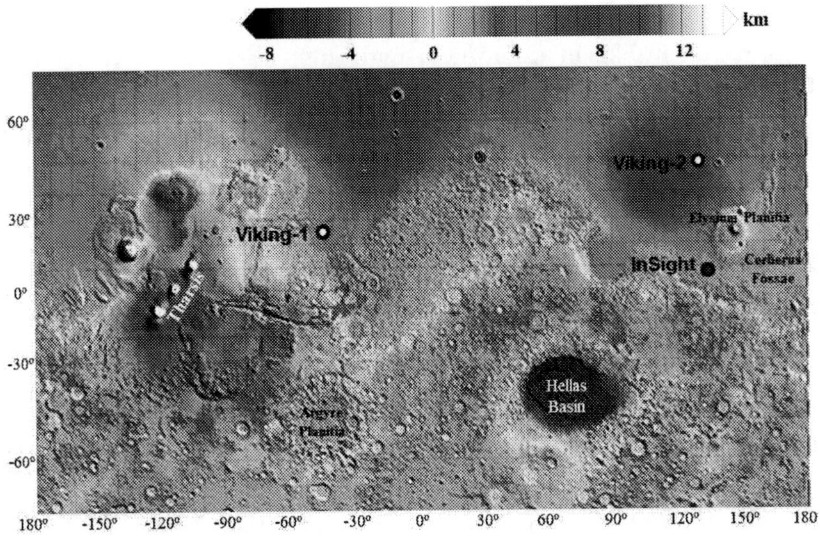

Figure 3.43. Topographic map of Mars showing the region around the InSight lander, Viking-1 lander, and Viking-2 lander. Two of the strongest marsquakes detected so far have been in the Cerberus Fossae region.

As we discussed above, the Earth's crust is fragmented into pieces of variable shape and size, called plates. The motions of these plates induce continent drifting, formation of mountains, volcanic eruptions, and earthquakes. Unlike the Earth, the crust of each of the other terrestrial planets (Mars, Venus, and Mercury) is so far thought to be one piece; i.e., their crusts may not be fragmented into plates. In other words, Mars, Venus, and Mercury may be described as single-plate planets and therefore may have no tectonic plates. However, in order to definitely determine that the crusts of these planets do not have tectonic plates and no volcanism, we still have to eliminate the possibilities of transforming faults within them and/or quakes related to moving magma and volcanic fluids.

Before we further discuss these possible quakes, let us note that moving magma and volcanic fluids cause earthquakes. Earthquake activity always increases before an eruption because liquid magma, hot fluids, and gases are less dense than the surrounding rocks, so they force their way through shallow underground fractures and passageways to reach the surface. When magma and volcanic gases or fluids move, they will either cause rocks to break or cracks to vibrate, just like in man-made hydraulic fracturing. When rocks break, high-frequency earthquakes are triggered. However, when cracks vibrate, either low-frequency earthquakes or a continuous shaking volcanic tremor is triggered. Because the signals generated by the ascending magma ascending toward the surface is the result of elastic wave propagation through the Earth just like earthquakes, these signals and physics behind them is termed volcano seismology. The appearance of continuous seismic signals at active volcanoes demonstrates the most profound difference between tectonic earthquakes and volcano earthquakes; tectonic earthquake signals are transient whereas volcano earthquake signals are long-lasting and continuous.

Quakes on Mars. The seismology on one of the terrestrial planets, namely Mars, has also been ongoing since 1975 (see Figure 3.43). The first attempts to detect seismic activity on Mars were with the Viking-1 program in 1975 (Anderson et al., 1977a and 1977b). Seismometers were mounted on top of the Viking landers, and the seismometers on the Viking-1 lander failed to unlock.

Viking-2 landed on Mars in September 1976 immediately following the first successful spacecraft landing on Mars by Viking-1. This time the seismometers were operational. However, no events were convincingly detected during 19 months of nearly continuous operation, with the possible exception of one event (Anderson et al., 1977a). This event occurred when no wind data was recorded, but recent analysis (Lorenz et al., 2013) suggests that this event was wind-induced landing noise. The severe blowing Martian wind was considered the reason for the failure of Viking-2's seismic components of its operations on Mars. Remember that sensors must be coupled to the ground (i.e., correctly attached to the ground) to measure ground motion. In the Mars seismic experiments, the coupling must be able to sustain the severe Martian meteorological conditions. The Viking-2 operational system and sensor design may have failed to meet this requirement.

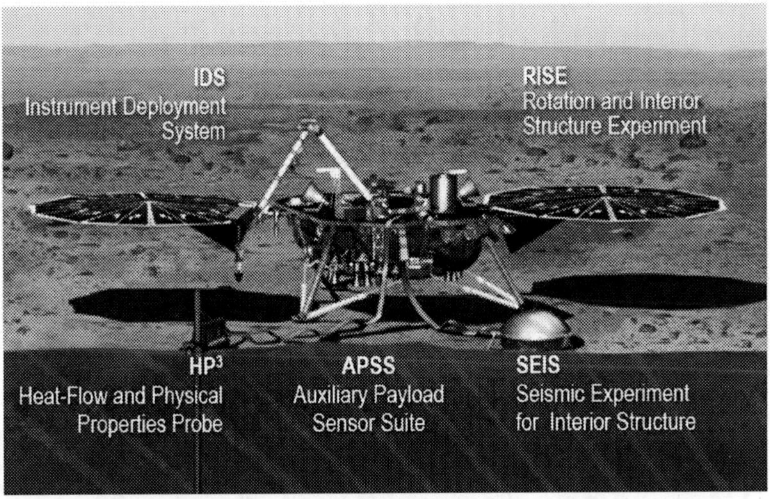

Figure 3.44. NASA's InSight lander on Mars. Its prime mission is to record marsquakes that can then be used to investigate the interior of Mars. Expectations are that significant marsquakes will only occur as a result of meteorite impacts. Note that the seismometer is covered with wind and thermal shield. InSight also includes a heat flow and physical properties package (HP3) for conducting the first geotechnical survey of Martian soil. Rotation and Interior Structure Experiment (RISE) uses InSight's radio connection with Earth to collect surface data on the planet's surface.

Forty-two years (in 2018) after the landing of the Viking seismometers on Mars, the NASA-ESA InSight mission took off on May 5, 2018, and landed on Mars's surface on November 26, 2018, with its multiple-component seismometer, as illustrated in Figure 3.44. It landed in Elysium Planitia, a large region of ancient lava flows. All the lessons of the Viking missions' failure with respect to seismic recording were assumed to be learned, and seismometers were carefully designed and made (see Lognonne et al., 2019). On April 1, 2021, NASA's InSight lander detected two strong, clear quakes originating at a location on Mars called Cerberus Fossae. The new quakes have magnitudes of 3.3 and 3.1 (see Figure 3.45); the previous quakes were magnitudes 3.6 and 3.5. These events resemble those related to moving magma and volcanic fluids and therefore they are likely related to volcanism (see Ikelle and Amundsen, 2005 and 2018) rather than any plate motions. The availability of these data is still limited hence data interpretation is still an ongoing process. To add a little bit of perspective to marsquakes, InSight has recorded about four or so moderate quakes (i.e., M_W between 3 and 4) so far in more than two years. On the Moon, more than 2,000 moderate moonquakes were recorded in six years by the Apollo program, which is more than 160 moonquakes in six months of a magnitude greater than 3 every year. On Earth, we record on average more than 13000 moderate earthquakes of a magnitude greater than 3 every year.

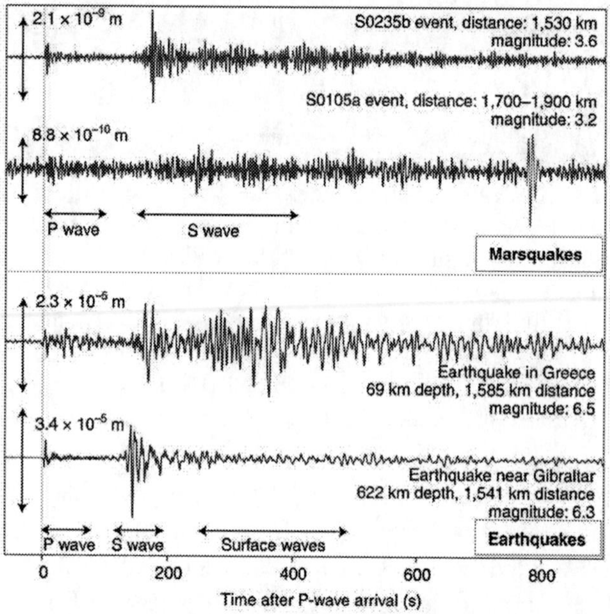

Figure 3.45. Comparison of seismograms of earthquakes and marsquakes. Note how small are marsquakes. (Source: Knapmeyer-Endrun and Kawamura, 2020).

No available data suggest Earth's uniqueness compared to the other terrestrial planets, in terms of plate tectonics and volcanism, 4.6 billion years ago. The accretion of the four terrestrial planets is almost identical. The fact that evolution of the plate tectonics and volcanism these four planets turn out differently is still largely a mystery in science. All of the four planets have been subjected to heavy bombardments, widespread volcanism, and possible plate tectonics in the past. All geologic and tectonic activities have since then reduced or stopped. Recent studies suggests that major volcanism on Mercury stopped at around 3.5 billion years ago, in stark contrast volcanism was active a few hundred million years ago on Venus, a few million years ago on Mars (McSween, 1994; Nyquist et al., 2001), and it still takes place on Earth today. The surface of these planets is covered by rocks resulting from lava flows which can provide their ages and therefore confirm the occurrences of these tectonic events. Because outgassing from volcanoes (gas blown out of volcanoes), along with evaporation of surface liquid, sublimation of surface ice, impacts by the solar wind and comets, and/or the planet's early magnetic field trapping gases in the solar system are key factors in the formation of the planet's atmosphere.

It is possible that Mercury had an atmosphere 3.5 billion or so years ago and this atmosphere disappeared with the end of volcanism. Just like the Earth, Mars has volcanoes, craters, hematite spheres, and water features, including dried-up riverbeds, past flows leading into great oceanic basins. The Martian prominent volcanic region is Tharsis and now Cerberus Fossae, where, there are several very large volcanoes that resemble some terrestrial volcanoes, such as those in Hawaii, except that those on Mars are many times larger. Yet, Mars today displays almost no evidence of plate motion. Long linear mountain chains and subduction zones are absent, transcurrent faults and compression features of any kind are rare, and recycling of crustal materials is extremely slow compared to the relatively rapid rate of surface materials recycling by erosional processes, and subduction on Earth. On Mars, there is densely cratered terrain that probably dates back approximately four billion years.

Unlike other terrestrial planets, the surface of Venus is not heavily cratered. Most of the surface has been covered by lava flows in the last billion years and more recently in the past hundreds of millions of years. This implies that the surface of Venus today has very little signs of past volcanic or tectonic activities. Volcanoes probably continue to erupt on Venus today. Many features on Venus (folded mountain belts and rift zones) are like Earth, but, again, no Earth-like plate tectonic indications. Venus seems to have a single lithospheric plate. In other words, Venus today is not like Earth, nor like the Moon, Mars, or Mercury.

It is likely that all terrestrial planets in our solar system are on the same evolutionary path with different time-limes in their evolution. This path can be summarized by the following key stages:

Accretion, heating, differentiation → formation of solid crust heavy cratering → widespread effusive volcanism → reduced volcanism, possible plate tectonics → mantle solidification, end of tectonic activity → cool interior, no activity .

These stages represent major geological, tectonics, even magnetic changes. Therefore if they occur at different times on different planets, huge geological differences between Mercury, Earth, Mars, or Venus can be expected. The 40 or so percent increase in solar luminosity in the last 4 billions, the planets relative positions with respect to the Sun and Jovian planets, and the early bombardment of planets by numerous asteroids and comets that brought some of the materials from which atmospheres and oceans formed are examples of factors that may explain why planets have different geological, tectonic, atmospheric, and even magnetic evolutions.

Quakes on the Earth's Moon. All these planets are cooling and contracting. Mercury may have lost its heat much earlier than the other planets. As a result, Mercury began to contract, and the crust essentially sealed off any conduits by which magma could reach the surface and/or pushed most hot materials from the planet's interior to the surface by contraction. Because a volcano is the result of hot material escapes from deep interior of the planets to a region that is less hot, sealing off any conduits by which magma means the end of volcanism. The same process of cooling and contraction shutting off volcanism may have happened recently to Venus and Mars. Also, as the interior of the planet cooled and its magnetic dynamo shut down, Mars, Venus, and Mercury no longer retained significant global magnetic fields.

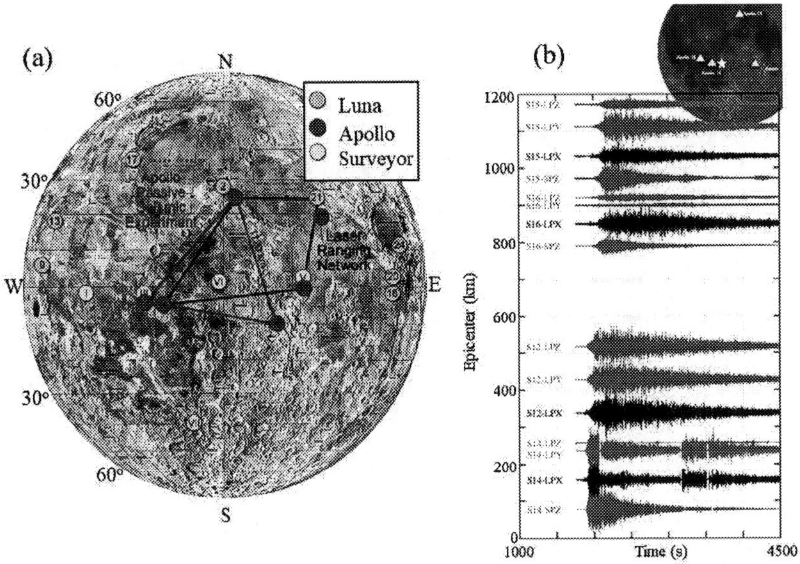

Figure 3.46. (a) The locations of the Apollo, Surveyor and Lunar landing sites. The red lines denote the Apollo-based passive seismic network (seismometers at Apollo 12, 14, 15 and 16 sites), the black lines denote the laser retro-reflector network (based on [NASA, 2009], courtesy of C. Neal). (b) Representative lunar seismograms with their epicenters. (Source: Lognonné and Kawamura, 2014).

The cooling and contracting of terrestrial planets extends to terrestrial planets like Earth's Moon. The moon is much smaller than Earth and therefore largely cooled off long ago, so one might not expect much, if any, tectonic activity. Yet, the moon is still tectonically active, like Earth, generating moonquakes as our planet creates earthquakes. The moonquakes likely happen because the moon is quivering as it shrinks.

Actually space exploration based on seismology has been underway since 1969. NASA's Apollo astronauts installed five seismometers at their landing sites, which measured thousands of quakes while operating on the Earth's moon between 1969 and 1977 (e.g., Latham et al., 1969, 1970; Bates et al., 1979). The Apollo 12, 14, 15, and 16 seismometers radioed data back to Earth until they were switched off in 1977. More than 12,000 seismic events were recorded.

However, in 2010, high-resolution images from NASA's Lunar Reconnaissance Orbiter revealed a network of thousands of faults on the moon that appeared to be less than 50 million years old. Researchers suggested these faults were evidence of lunar tectonics, although it was unclear how recent this activity was.

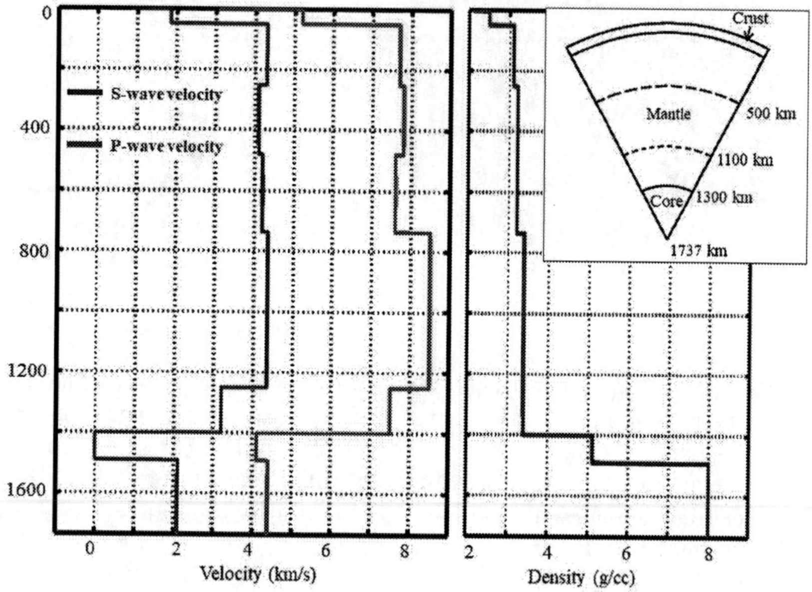

Figure 3.47. A model of the internal lunar structure was obtained using a seismic array processing technique. The moon appears to be differentiated into a crust, mantle, and metallic core. Appended here is a seismic model of the interior of the moon. (Adapted from Jolliff et al., 2006).

Seismometers at four Apollo landing sites on the moon recorded 28 shallow moonquakes between 1969 and 1977, ranging from magnitude 1.5 to 5 on the Richter scale. Some of these shallow quakes might in theory result from activity on lunar faults, but the locations and depths of the sources of these quakes were uncertain.

Moonquakes can have origins other than tectonic activity in the lunar crust. For example, the Apollo missions detected about 11,000 moonquakes happening about 500 to 680 miles (800 to 1100 kilometers) beneath the lunar surface. Previous research suggested that these deep moonquakes resulted from Earth's gravitational pull on the lunar interior, much as how the moon's gravitational pull on Earth's waters results in the tides, said study co-author Nicholas Schmerr, a planetary seismologist at the University of Maryland at College Park.

By using the methods for locating the epicenter and focus of earthquakes, described in Chapter 5 of Ikelle (2017b and 2020), these data have revealed that there are at least four different kinds of moonquakes:[4] (1) Deep moonquakes originated from depths between 700 and 1,200 km below the surface, much deeper than any quake on Earth, probably caused by tides; (2) vibrations from the impact of meteorites as the Moon does not have an atmosphere of any significance, most meteorites do not burn up due to the frictional heat as they do on their way through the Earth's atmosphere, but instead make it all the way to the lunar surface; (3) thermal quakes caused by the expansion of the rigid crust when first illuminated by the morning Sun after two weeks of a deep-freeze lunar night with a temperature difference of more than 200 degrees Celsius between day and night on the lunar surface; and (4) shallow moonquakes only 20 or 30 km below the surface. Examples of deep, shallow, and meteoroid-caused earthquakes are shown in Figure 3.46.

Most deep moonquakes have magnitudes smaller than 3 on the Richter scale. The other groups of moonquakes have even smaller magnitudes. However, 28 sizable moonquakes with magnitudes of up to 5 were recorded. They occur at depths of less than 200 km. It turns out that these moonquakes have a fixed time period of occurrence of 27 days. As the time it takes the Moon to circle the Earth once is 27 days, these quakes are again likely caused by the tidal forces that the Earth exerts on the moon. These moonquakes are challenge to NASA's current plans to return astronauts to the Moon by 2024, to lay the foundation that will eventually enable human exploration of Mars, because astronauts, visitors, and settlers on the Moon will certainly be felt by them. Choosing less moonquake hazardous locations is the issue as we are still learning to develop a quake hazard map on a planet or moon without plate tectonics.

---

[4]Moonquake is a quake which, much like an earthquake, would be a shaking of the surface or interior of the Moon as a result of the sudden deformation in the Moon's interior or at the Moon's surface.

Small-range moonquakes can also have an impact on scientific experiments on the Moon because shallow moonquakes last a remarkably long time, more than 10 minutes. On Earth, vibrations from quakes usually die away in less than a minute because of the high attenuation of seismic waves throughout the Earth, especially as they pass through fluids. On the Moon, seismic waves attenuate less rapidly because the Moon's crust is dry.

Although moonquake data are limited by their spatial distribution (five sensors only), Toksöz et al. (1972, 1974), Nakamura et al. (1982), and Wieczorek and Phillips (1997) were able to determine the structure of the Moon's internal layers from the travel times of moonquake seismograms. They found that the Moon could be divided into a crust, mantle, and core, similar to the Earth and other terrestrial planets. The resulting Moon's internal and internal velocities calculated from the moonquake data are shown in Figure 3.47.

Quiz: We can reliably and precisely define the time cycle of moonquakes and not the time cycle of earthquakes. Why?

The thickness of the crust varies between 65 km (Toksöz et al., 1972) and 55 km (Toksöz et al., 1974). By using the travel times of the P- and S-waves in deep moonquake data with foci greater than 1,000 km, Nakamura et al. (1982) and Goins et al. (1981) found that the interface between the upper and lower mantle is about 500 km deep. Hood et al. (1982), Weber et al. (2011), and Garcia et al. (2011) use the fact that the Apollo 12, 14, 15, and 16 landers were also all equipped with magnetometers to measure the Moon's weak magnetic field. Most of the strength of the magnetic field detected at the surface is core-induced magnetic fields, as described in Chapters 1 and 2 of Ikelle (2022d). By combining travel times from moonquakes, related to core-reflected waves (ScS), with the Moon's internal conductivity distribution, they infer the existence of a solid metal core with a radius less than 360 km. Current estimates are that the core radius of the Earth's Moon is between 150 and 360 km. A large-aperture network with a large number of sensors is required to further narrow this range.

Table 3.13. An explanatory summary of the current state of earthquakes and volcanic eruptions on terrestrial planets.

	Divergent		Convergent				
	O-O	C-C	O-O	O-C	C-C	Transform	Hotspot
Mars	N/A	N/A	N/A	N/A	N/A	?	Very likely
Venus	N/A	N/A	N/A	N/A	No	?	Very likely
Mercury	N/A	N/A	N/A	N/A	No	?	No
Earth	Yes	Yes	Yes	Yes	Yes	Yes	Yes

O-O: Ocean-Ocean; O-C: Ocean-continent; C-C= Continent-continent

Table 3.13 summarizes our current understanding of the state of plate tectonics and volcanism on terrestrial planets. We need to include the active source (i.e., the manmade source), described in (see Ikelle, 2022b), in these missions. The obvious solution is a drop weight as a seismic source. We can even implement a drop-weight source by landing spacecraft. The landing of the InSight follow-up mission can be used as a seismic source. Hopefully, more data acquisition on Jovian planets will follow soon. Strangely enough, the seismology of giant (Jovian) planets can turn out to be more effective than that of terrestrial planets just like marine seismology has turned out to be more effective than land seismology because sensor coupling and large variations in planet surface topography are not issues. Moreover, we are dealing with a relatively homogeneous environment like in a marine environment on Earth. In addition, our data are essentially fluid waves, which can be recorded from sensors mounted to the spacecraft, inside the spacecraft, or inside multiple satellites, which communicate with the spacecraft or the Earth in order to keep track of the locations of their recordings. Seismometers in this case record pressure variations instead of ground motion. One can also include electromagnetic sensors to record the electromagnetic field. Modern seismology shows that the combined analysis of seismic and electromagnetic data improves the description and characterization of the medium under consideration. The passive sources in this case are atmospheric blasts and the multiple collisions in the solar system. One can also consider generating fluid waves with a charged-particle gun fired into space from a satellite. Because most of the wave propagation takes place in the air, one can even detect giant-planet responses on the Earth's surface.

Quiz: Use the artificial intelligence description in the previous chapter to propose a classification of the three key rocks in oil and gas exploration: sandstone, shale, and coal.
Hint: Define input vector and features of classification.

3.5. Essay: Monitoring Subduction Mega and Large Earthquakes

Japan is a very active seismic region with earthquakes famous for their societal impacts and international reach such as the 2011 Tohoku-oki earthquake and the nuclear disaster associated with its tsunami. Likewise, Chile is the site of some of the largest earthquakes in the world. On average, a magnitude-8 earthquake occurs there every 10 years and probably at least one magnitude-9 earthquake occurs there every 100 years. Lessons learned from the 2011 Tohoku-oki 9.1 earthquake in Japan and the 2010 Maule magnitude-8.8 earthquake in Chile suggest that there are ways of monitoring and even predicting the next large subduction earthquakes, assuming that large repeated earthquakes generally exhibit similar characteristics, including the magnitude, distribution, and style of slip on the fault. We here describe these lessons, which are essentially based on the behavior of small earthquakes. The differences between the physics of small earthquakes and that of large earthquakes can be used to monitor the next subduction mega-earthquakes, including those projected in Chile and Japan.

So far, the monitoring and prediction of earthquakes have been primarily based on the notion of *seismic gap* (Kelleher, 1972). The basic idea is that geological units that have ruptured in the past, but not ruptured for some time, are accumulating elastic strain energy. Sooner or later, they will rupture again and generate earthquakes; new gaps may start. For example, the last large earthquake in the Concepción area in Chile, before the 2010 Maule earthquake, was the

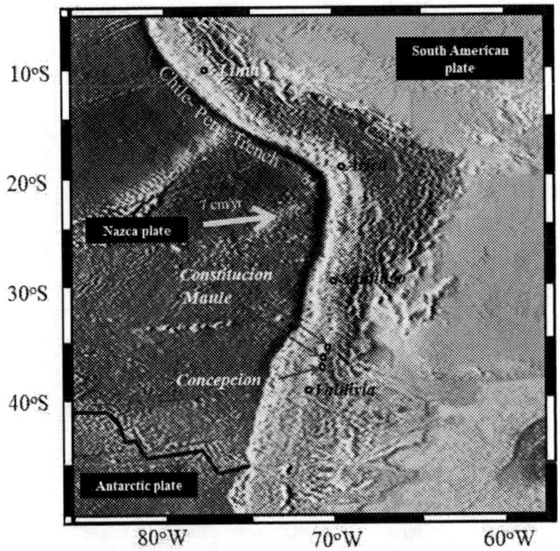
South American plate
Lima
Chile-Peru Trench
Arica
7 cm/yr
Nazca plate
Constitucion
Maule
Santiago
Concepcion
Valdivia
Antarctic plate
10°S
20°S
30°S
40°S
80°W
70°W
60°W

(a)

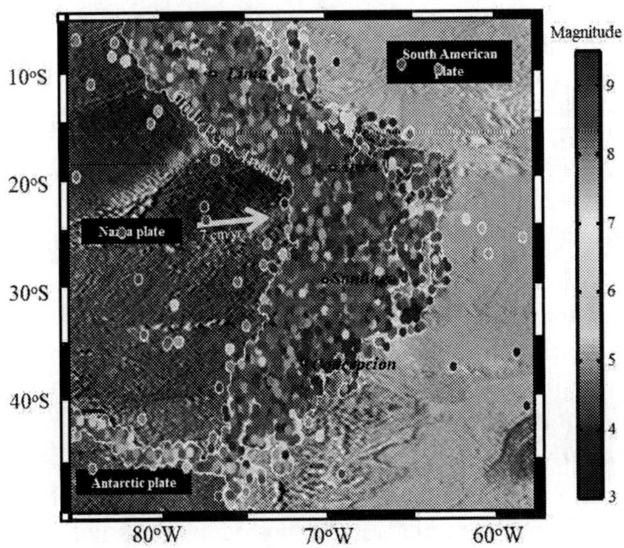
Magnitude
South American plate
Lima
Arica
Nazca plate
7 cm/yr
Santiago
Concepcion
Antarctic plate
9
8
7
6
5
4
3
10°S
20°S
30°S
40°S
80°W
70°W
60°W

(b)

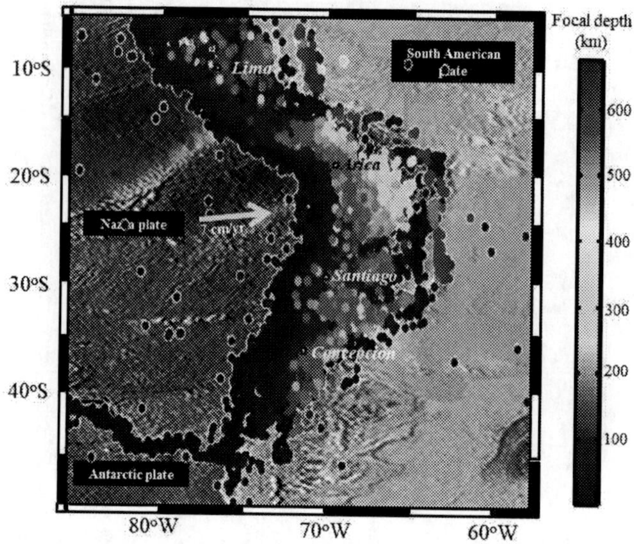

(c)

Figure 3.48. (a) The Chile-Peru region is under consideration in this study. The yellow arrow indicates the subduction of the Nazca Plate under the South American Plate. (b) Scatterplot of epicenters versus earthquake magnitudes of 3 or greater between 1900 and 2019. (c) Scatterplot of epicenters versus hypocenter depths of 3 or greater between 1900 and 2019. Note that the phase transition, from low density phase to high density phase, are at depths of about 300 km and 500 km.

large rupture of the 1960 Valdivia earthquake. The 2010 Maule earthquake broke at least some of the elastic strain energy accumulation since 1960 (and even before) in the Concepción area. The longer duration of a seismic gap, the higher likelihood of a magnitude-8 or greater earthquake; in other words, the gap is just based on the historical seismic recurrence in the region using past earthquakes. Unfortunately, the identification of these gaps does not solve the short-term prediction problem. It suggests areas for potential ruptures but it does not properly account for the complex tectonic changes underground.

Since the 1970s, we have now measured ground motion so precisely that even seismic events with magnitudes as small as 1 can be recorded. The network of sensors making these measurements on land and ocean floors is growing, although not rapidly enough for our needs to monitor the state of solid Earth stress. Moreover, these measurements are made in real time and are shared worldwide. By using these data, we here revisit the problem of monitoring subduction mega-earthquakes. Note that, through satellite Global Positioning Systems (GPS) and satellite interferometry (InSAR), we are also recording the degree to which plates move and deform the Earth's surface, including elastic strain accumulation, and even the crustal deformation produced by large earthquakes. These data are described in Ikelle (2022c) and are included in our studies in Ikelle (2022d).

Our description and analysis of the monitoring subduction mega-earthquakes uses Chile-Peru and Japan earthquakes as examples before drawing larger conclusions in Ikelle (2022d). We used data recorded by USGS (U.S. Geological Survey) that they made available through earthquakes.usgs.org. We will use basic statistic tools, described in Ikelle (2009 and 2017a), for our data analysis and basic results of fracture mechanics, as they relate to the study of rupture, described in Ikelle and Amundsen (2018), for the interpretation of some results of this analysis. Note that our data here are limited to the coordinates of the focus points (hypocenters) and earthquake magnitudes. An advanced analysis using seismogram recordings of the ground motions will be described in Ikelle (2022d).

3.5.1. Chile-Peru Seismicity

Plate tectonics. As illustrated in Figure 3.48, the seismicity of the Chile-Peru region is quite simple. The Nazca Plate converges toward the South American Plate at a rate of about 7 cm/year (DeMets et al., 1990 and 1994) and is among the highest plate motion rates on Earth and subducts under it. The Chile-Peru trench is the boundary between these two plates. Most earthquakes in South America are located around this boundary, that is, they are interplate earthquakes, as we can see in Figure 3.48b, obtained by using the USGS earthquake database. The most significant events are the 1960 Valdivia earthquake, with $M_W = 9.5$, and the more recent Maule earthquake, on February 27, with $M_W = 8.8$. There are also some intraplate earthquakes in the region (e.g., earthquakes within the interior of the South American Plate). The 1939 Chile intraplate earthquake with $M_W = 7.4$, is a typical example (Campos and Kausel, 1990; and Beck et al., 1998). Notice how the earthquake occurrences clearly define the boundary between the Nazca and Antarctica Plates, in Figure 3.48b. This result confirms the data quality of the USGS database. Note also that, in Figure 3.48c, there are almost no earthquakes with a focal depth of between 300 km and 500 km. These two depths limits phase transitions in the mantle in this region. Between these depths, the density of a new mineral assemblage becomes denser with depth. In other words, in phase transitions, we go from a low-density phase to a high-density phase.

Table 3.14. The largest earthquakes on record in the Chile-Peru region in 2018.

Date	M_W	Depth (km)	[lat (° S), lon (° W)]
1960-05-22	9.5	25	[38.1, 73.4]
2010-02-27	8.8	23	[36.1, 72.9]
1922-11-11	8.5	70	[28.3, 69.8]
2001-06-23	8.4	33	[16.3, 73.6]
2015-09-16	8.3	23	[31.6, 71.7]
2014-04-01	8.2	25	[19.6, 70.8]
1994-06-09	8.2	631	[13.8, 67.5]
1940-05-24	8.2	45	[11.1, 77.5]
1950-12-09	8.2	114	[24.0, 67.9]
1906-08-17	8.2	35	[32.4, 71.4]
1950-12-09	8.2	114	[24.0, 67.9]
1906-08-17	8.2	35	[32.4, 71.4]
1942-08-24	8.1	30	[14.8, 74.7]
1943-04-06	8.1	35	[31.4, 71.5]
1960-05-21	8.1	25	[37.8, 73.3]
1966-10-17	8.1	40	[10.7, 78.2]
1995-07-30	8	46	[23.3, 70.3]
2007-08-15	8	39	[13.4, 76.6]
1985-03-03	8	33	[33.1, 71.9]

In the Chile-Peru area, depicted in Figure 3.48a, there are more than 50,000 earthquakes of a magnitude-3.0 and greater between 1900 and 2019 in the USGS database, more than 33,000 earthquakes of a magnitude-4.0 and greater in the same period, and 52 earthquakes of a magnitude greater than 7.5. Table 3.14 captures the major earthquakes of this region. The Maule (about 60 km south of Constitución) earthquake of February 27, 2010, the seventh largest earthquake of the past century, affected 650 km of the Nazca-South American Plate boundary in south Central Chile and produced spectacular crustal deformation. Note that the rupture length of a fault is directly related to quake magnitude although nonlinearly (Madariaga, 1977; Kanamori and Brodsky, 2004); a longer rupture length implies

Box 3.4. A Mirror of Modern Science

"Only fools and charlatans predict earthquakes", Charles Richter (1900-1985). "Japan must admit it can't predict quakes", Robert J. Geller, Nature 545, 2017. Talking about California, "We are predicting another massive earthquake certainly within the next 30 years and most likely in the next decade or so," W. Pecora, U.S. Geological Survey Director, 1969. "Scientists will be able to predict earthquakes in five years," Louis Pakiser , U.S. Geological Survey, 1971. "We have the technology to develop a reliable prediction system already in hand," Alan Cranston, U.S. senator, 1973. "The age of earthquake prediction is upon us," U.S. Geological Survey, 1975. Several billion dollars of public money are invested each year to support investigation of earthquake prediction (e.g., USGS and higher-education institutions)—to save millions of lives in Tokyo, Bombay, Jakarta, Tangshan, Seattle, Los Angeles, Mexico City, Port-au-Prince, Istanbul, Tehran, Lima, Santiago, Bogota, Wellington ... . Yet in modern science, "the word *prediction* must be avoided in any manuscript because that is likely to upset geophysics establishment."

a higher magnitude. A relatively small tsunami hit coastal cities near the earthquake epicenter less than an hour after the main shock of the 2010 Maule earthquake. The biggest earthquake ever recorded was the Valdivia earthquake that occurred on May 22, 1960. It affected more than 1000 km of the Nazca-South American Plate boundary. Because it was far more powerful than the 2010 Maule earthquake, it produced a large tsunami that engulfed the whole Pacific rim, despite that its focal depth is about the same as that of the 2010 Maule

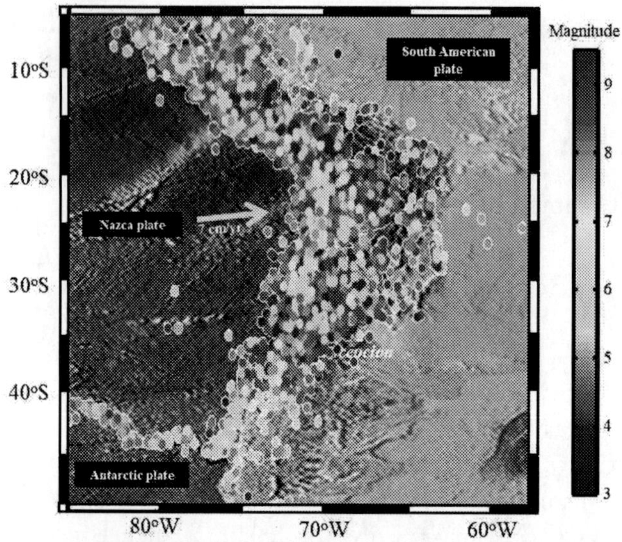

(a)

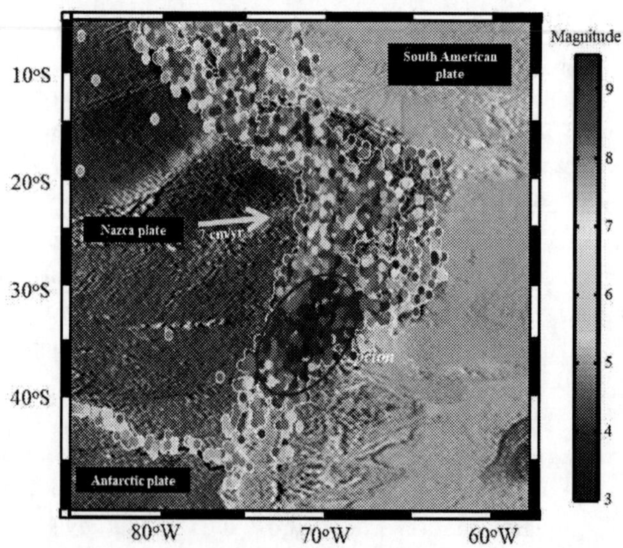

(b)

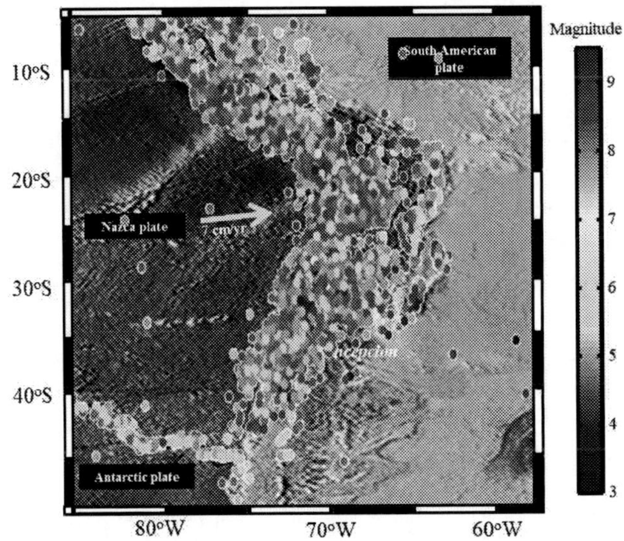

(c)

Figure 3.49. (a) Chile-Peru seismicity from 1900 to 31 December 1996 for earthquakes of magnitude 3 and greater. There is an abundance of light blue and yellow dots because recordings of small earthquakes in the early part of the twentieth century are rare. (b) Chile-Peru seismicity from May 1, 2000, to January 31, 2010. The circle shows the area with large occurrences of small earthquakes. (c) Chile-Peru seismicity from March 01, 2010, to November 21, 2019.

earthquake. Waves as high as 11 meters, traveling at an average of 325 kilometers per hour, were recorded 10,000 kilometers from the epicenter and as far away as Japan and the Philippines. It also affected Hawaii, eastern New Zealand, southeast Australia, and the Aleutian Islands.

For some time, it has been suggested that the Arca and Concepción regions are seismic gaps with large elastic strain energy accumulations and that they can break at any moment (e.g., Klotz et al., 2001; Ruegg et al., 2002 and 2009; Brooks et al., 2003; Moreno et al., 2008, and Vigny et al., 2009). The Concepción gap recently broke through the Maule earthquake and the Arca gap is still to break.

Data analysis. The key point of the seismicity studies is to capture the tectonic changes with time and space before a major earthquake and for lesson-learning exercises after the earthquake. Figure 3.49 shows magnitude-3 and greater earthquake distributions from May 01, 2000, to January 31, 2010 (i.e., a duration of 3,563 days, more than 10 years before the 2010 Maule earthquake, until 28 days before the earthquake), and the same distribution from March 1, 2010, to November 21, 2019 (i.e., a duration of 3,553 days, about 10 years after the 2010 Maule earthquake, until three days after the earthquake). Notice how the small earthquake occurrences dominate the seismicity of the entire region before the 2010 Maule earthquake and the particularly large cluster of small earthquakes in the Concepción area. Actually, the cluster of small earthquakes around the Concepción area captures the size and length of the ruptured area. It also explains why more aftershocks are located north of Maule than in the south. Moreover, the few intraplate earthquakes on the South American Plate are now reduced to almost zero. In contrast to the Concepción area, the earthquake distribution in the Arca area is quite similar to other areas of the region, even 10 years after the Maule earthquake, as we can see in Figure 3.48b. Maybe it is an indication that the major earthquake predicted in Arca is not yet imminent.

Table 3.15 provides even more details on the above observations. Year nine before the 2010 Maule earthquake, we had more than 2,700 earthquakes, with about 1,700 earthquakes of a magnitude of between 3 and 4. A year before the earthquake, we only had 593 earthquakes, with only 71 earthquakes of a magnitude of between 3 and 4 for the entire 12-month period in one of the most seismically active regions of the world, and a very large one. It is clear that solid Earth's evolution toward a major earthquake affects areas far away from the eventual epicenter, hypocenter, and ruptured area.

After the 2010 Maule earthquake, the number of earthquakes started going up again. Even after nine years, the number of earthquakes has not yet reached the high of nine years before the 2010 Maule earthquake. Moreover, the number of earthquakes with a magnitude between 3 and 4 is still very small, about 200, nine years after the 2010 Maule earthquake, compared to 1,700 earthquakes for the same duration of nine years before the 2010 Maule earthquake.

Short-term prediction. The conventional wisdom is that a tectonic earthquake begins with an initial rupture at a point on the fault plane through a process known as nucleation. The scale of the nucleation zone is considered to be smaller than 100 m. Nucleation involves some sort of preparation. Foreshocks preceding more than 40 percent of major earthquakes are evidence of the preparation process. However, large earthquakes in the Chile-Peru region, such as the ($M_W = 8.8$) 2010 Maule earthquake, have no foreshocks. Table 3.15 and Figure 3.50 suggest that actually these earthquakes still have a preparation process, but carried out by a process of swarm and then a sharp drop of small earthquakes. This preparation process takes a longer time, more than five years, in this case. The good news is that we can monitor the evolution of the tectonic of the Chile-Peru region toward a major earthquake. The tell-tale signs of a coming major earthquake are mistakenly clear in the case of the 2010 Maule earthquake, as we can see in Figure 3.50b, showing the sharp drop in the number of small earthquakes. Let us describe Figure 3.50 a little bit more. We can monitor the increase and decrease in the number of earthquakes by computing this quantity,

$$N(t_k) = \text{Number of earthquakes in } [t_k - \Delta t, t_k] \ ,$$

where t_k is the date for which the count is carried out and $t_k - \Delta t$ defines how far back the count will start. We basically count all earthquakes with the same weight irrespective of their magnitudes and their locations, in this region, for a defined time period before t_k. Figure 3.50a shows the results of this count, with $\Delta t = 90$ days and $\Delta t = 365$ days for all earthquakes with magnitude-3 and greater. We can also limit this count to earthquakes of certain magnitudes, as in Figure 3.50b, in which the count is limited to earthquakes with magnitudes of between 3 and 4. As we can see in Figure 3.50b, the sharp drop in the number of earthquakes with magnitudes of between 3 and 4 clearly indicates more than a year in advance the coming of a very large rupture in this region. We can even discover a baseline under which the drop in small earthquakes implies a large seismic event in just a matter of days.

Note that we have not included the 1960 Valdivia earthquake

Table 3.15. The number of earthquakes on a yearly basis in the Chile-Peru region before and after the 2010 Maule earthquake. Notice the sharp decrease in the number of earthquakes, especially small earthquakes, before the earthquake. The variable M_W stands for earthquake magnitude.

		After		
Period	$M_W \geq 3$	$3 \leq M_W \leq 4$	$4 < M_W \leq 5$	$5 < M_W \leq 6$
2018-2019	1052	99	893	53
2017-2018	1260	181	987	88
2016-2017	1318	206	1029	75
2015-2016	2097	253	1686	133
2014-2015	1705	272	1281	136
2013-2014	939	99	776	60
2012-2013	795	75	656	54
2011-2012	795	65	636	82
2011-2010	2007	175	1585	228
		Before		
Period	$M_W \geq 3$	$3 \leq M_W \leq 4$	$4 < M_W \leq 5$	$5 < M_W \leq 6$
2010-2009	592	71	435	83
2009-2008	1360	719	577	58
2008-2007	2081	1154	784	129
2007-2006	2374	1610	677	81
2006-2005	1975	1218	692	59
2005-2004	1678	1117	500	57
2004-2003	2281	1782	438	61
2003-2002	2241	1631	534	72
2002-2001	2762	1708	935	112
2001-2000	1293	806	442	41
2000-1999	1638	1143	445	47
1999-1998	2103	1412	620	67
1998-1997	1121	622	517	66

in this study because less than 150 earthquakes are available in the USGS database between 1900 and 1960 for this region. Clearly, the distribution and quality of seismometers in the region were still under development or records are not well documented to reproduce figures and tables similar to those of the 2010 Maule earthquake for the 1960 Valdivia earthquake. Anyway, as we can see in Figure 3.50, the other

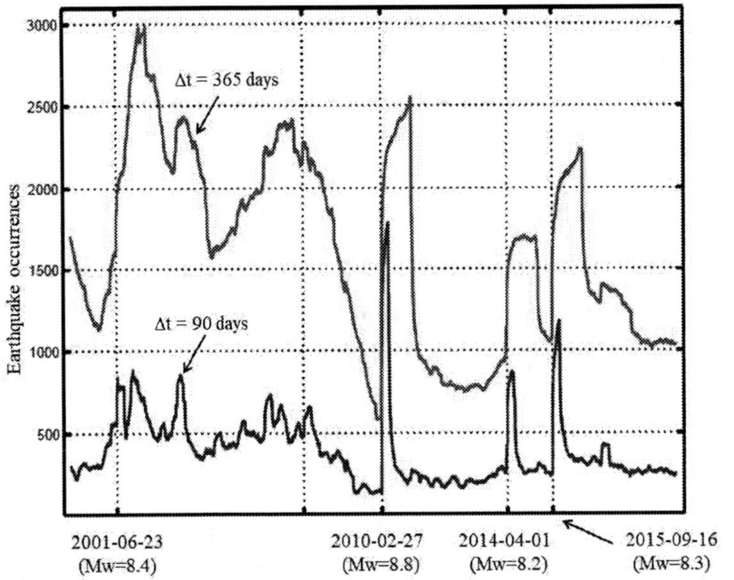

(a)

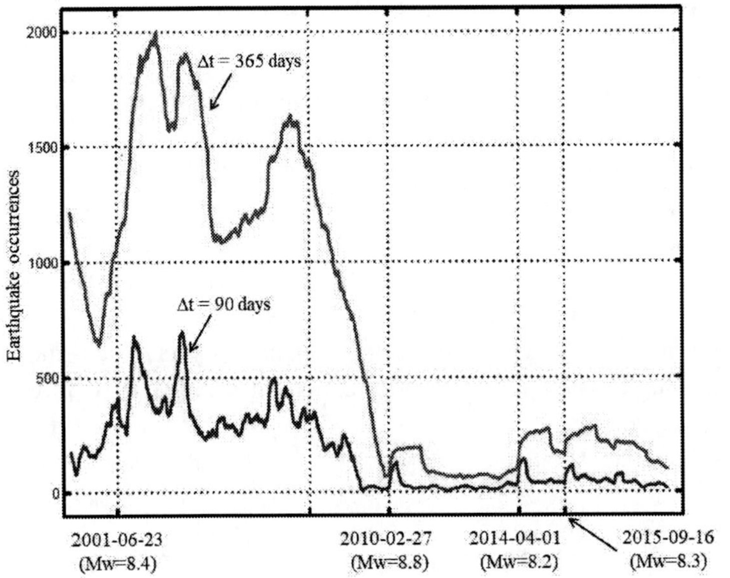

(b)

Figure 3.50. Variation of earthquake occurrences with time in the Chile-Peru region. At a given time, we compute all the earthquakes that occurred in the last Δt days with the same weight irrespective of their magnitudes and locations. (a) All earthquakes with magnitude 3 and greater. (b) All the earthquakes with a magnitude range between 3 and 4.

major earthquakes between 2000 and 2019 follow the same patterns as the 2010 Maule earthquake, with modest drops in small earthquakes. Moreover, some studies (e.g., Sieh, 1996; and Aki, 1979) suggest that large repeated earthquakes generally exhibit similar characteristics, including the magnitude, distribution, and style of slip on the fault. Therefore, it is very likely that the methodology of monitoring large earthquakes based on the increase and decrease of small earthquakes may hold for many other coming major earthquakes in the Chile-Peru area.

Arca major earthquake? As mentioned earlier, a number of studies have suggested that a major earthquake in the Arca area is imminent. As we were finishing drafting this paper, we could see that the number of small earthquakes was about 99, which is low (see Table 3.15). However, no obvious clusters of small earthquakes around Arca are yet visible, as in Figure 3.49b. Are we heading for another major earthquake in this region, and where, or have we started a slow evolution toward the next major earthquake between 2050 and 2070 in this region?

Box 3.5. Seismology and Reservoir Engineering

Seismology and reservoir engineering are two of the most important disciplines involved in the studies of water and hydrocarbon reservoirs. As far as reservoir description and characterization are concerned, these two disciplines are merging these days because they use common measurements and their objectives are the same, although they are generally presented as distinct disciplines, primarily due to the parallel development of these two disciplines over the last century or so. These measurements are taken on various scales and yet their shared objectives are to estimate the porosity, permeability, wettability, water saturation, organic-matter content, and capillary pressure of rock formations.

Note that the Chile-Peru trench includes a significant bend of more than 30 degrees in the Arca area. Fracture mechanics and multiple studies (e.g., King and Nablek, 1985; Wesnousky, 2006; and Ma and Elbanna, 2019) suggest that rupture often terminates at such a bend. Basically, when the rupture encounters a point that requires more fracture energy per unit area than it has, it will be stopped. The fracture energy required per unit area of the fault plane is less for points along a straight fault line without a bend or just with a slight bend (less than 10 degrees). In contrast, points along large bends sometimes require more energy than the fracture energy carried by the rupture. On the other hand, a bend like the Arca bend can act as a stress concentrator. In this case, the rupture may nucleate at the bend and terminate somewhere in south Chile or up in Peru when it runs out of energy.

3.5.2. Mexico Seismicity: Interplate Earthquakes

Plate tectonics. The area of Mexico under consideration is depicted in Figure 3.51. It includes parts of three large tectonic plates: the North America, Pacific, and Cocos Plates, and two small ones, the Caribbean and Rivera Plates. It also includes the boundaries between these plates. The Cocos Plate is subducting under North America, and more precisely, under Mexico. The Rivera Plate, which detached from the Cocos Plate about 10 million years ago (Demets and Traylen, 2000), is also subducting under North America, beneath Jalisco, Mexico. The subduction rate of the Rivera and Cocos Plates under North America varies between 69 mm/year and 75 mm/year. The motions of these plates have caused and continue to cause frequent and large earthquakes and occasionally volcanic eruptions. Most of them occur in the bending part of the subducting Cocos and Rivera Plates with an average depth of 20 km. Actually, the highest focal depth in Mexico is half that of the Chile-Peru and Japan regions, as we can see in Figure 3.52b. It is interesting to see, in Figure 3.52a, how the earthquake occurrences coincide with the Cocos-North American, Pacific-Cocos, Rivera-Cocos, Rivera-North American, and Caribbean-North American Plate boundaries.

The USGS database has more than 32,000 earthquakes of a magnitude greater than 3.0 between 1900 and 2019 in the area, as depicted in Figure 3.51, showing more than 26,000 earthquakes of a magnitude greater than 4.0 in the same period, and 270 earthquakes of a magnitude greater than 6.5, including the September 19, 1985,

Mw = 8.0 Michoacán earthquake, which caused about 10,000 fatalities. On an average, 260 earthquakes of a 4 magnitude occur in this area every year. Compared to the California region, which we will describe in the next subsection, with an average of 39 earthquakes of a 4 magnitude a year, the Mexico area is definitely a more seismic active region. Moreover, this region has registered five earthquakes of an 8 magnitude and greater in the last 120 years and none in the California region. Note that the number of magnitude-3 earthquakes in California is almost identical to that of Mexico, as we will see in the next subsection. This result is due to the slow implementation of advanced sensors capable of recording small earthquakes in Mexico. In the last 10 years, the Mexico region had 9,425 magnitude-3 earthquakes and California had 3166 magnitude-3 earthquakes. So the actual number of magnitude greater than 3.0 earthquakes in Mexico between 1900 and 2019 is probably around 96,000.

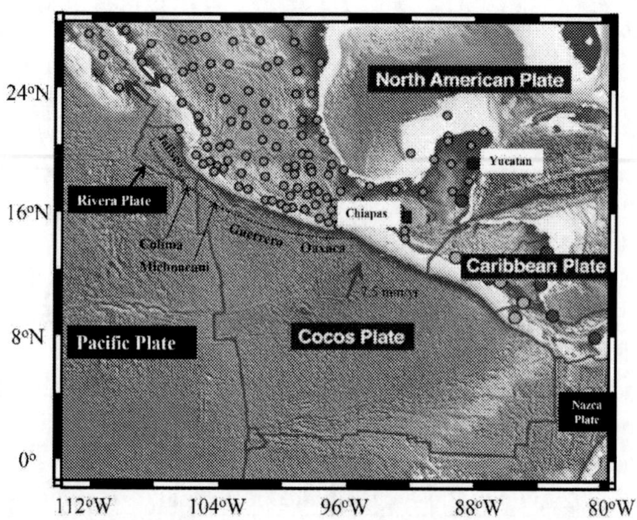

Figure 3.51. A Mexico area covering the collision zone, where the Rivera and Cocos Plates are subducting under North America at rates from 69 mm/year to 75 mm/year. The region of the Mexican subduction zone includes Jalisco, Colima, Michoacán, Guerrero, Oaxaca, and Chiapas. Red lines indicate plate boundaries.

Data analysis. There are a number of strange patterns of earth-

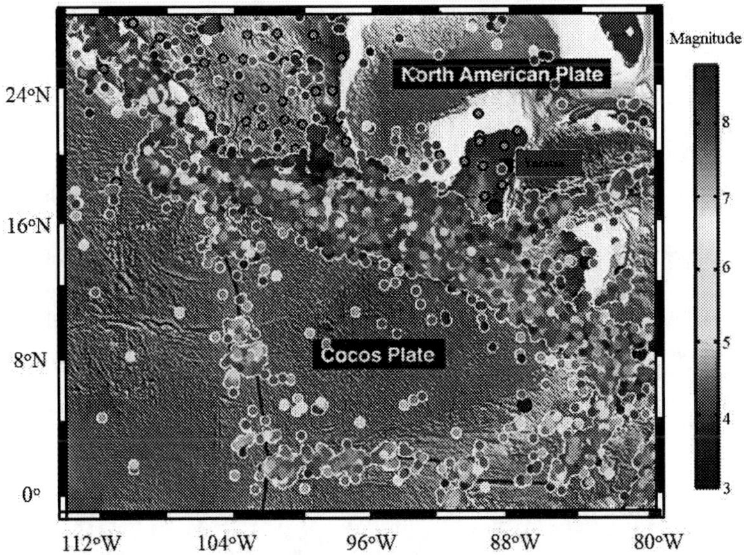

(a)

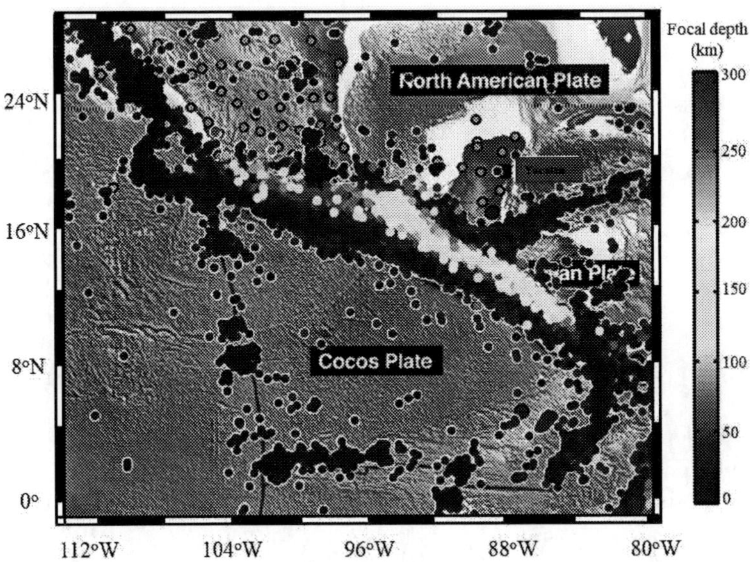

(b)

Figure 3.52. (a) Scatterplot of epicenters versus earthquake magnitudes of 3 or greater between 1900 and 2019 in Mexico. (c) Scatterplot of epicenters versus hypocenter depths of 3 or greater between 1900 and 2019 in Mexico.

quake occurrences in this region. For example, as pointed out by Segou and Parsons (2018), it is a common occurrence in Mexico of two magnitude-7 or greater earthquakes striking within a two-week or less interval more than 300 km apart, a distance that excludes the

Table 3.16. The largest earthquakes in Mexico on record in 2018 and with near-simultaneous earthquakes to them. Mexico has never had a magnitude-8.5 or even greater than a magnitude-8.5 earthquake in the last 120 years. Magnitudes of 7.5 or greater are considered major earthquakes in this region. Brackets mean [north latitude in degrees, south longitude in degrees].

Date	M_W	Depth (km)	Plate boundary	Location
2017-09-08	8.2	47.4	Cocos/North-American	Chiapas [15.0,-93.9]
2017-09-19	7.1	48	Cocos/Caribbean	Guatemala [11.4, -86.7]
2017-09-23	6.1	10	Cocos/North-American	Oaxaca [16.6, -95.1]
1932-05-21	7.1	15	Cocos/Caribbean	Guatemala [13.2, -88.1]
1932-06-03	8.1	15	Rivera/North-American	Jalisco/Colima [19.8, -103.8]
1932-06-18	7.8	15	Rivera/North-American	Jalisco/Colima [19.4, -103.9]
1932-06-22	7.7	25	Rivera/North-American	Jalisco/Colima [19.4 ,-104.2]
1985-09-15	6.0	64.8	Cocos/North-American	Oaxaca [18.0, -97.2]
1985-09-19	8.0	27.9	Cocos/North-American	Michoacán [18.2, -102.5]
1985-09-21	7.6	30.8	Cocos/North-American	Guerrero [17.8, -101.6]
1995-09-14	7.4	23	Cocos/North-American	Oaxaca [16.8 -98.6]
1995-10-09	8.0	33	Rivera/North-American	Jalisco/Colima [19.1 -104.2]
1995-10-21	7.2	159.3	Cocos/North-American	Chiapas [16.8, -93.5]

interpretation of one of them as foreshock or aftershock. Table 3.16 illustrates this phenomenon. On September 19, 2017, a magnitude-7.1 earthquake struck near Mexico City. This event was preceded by the magnitude-8.1 offshore Chiapas earthquake. The distance between these earthquakes exceeds 600 km. Also, there was a major earthquake on September 19, 1985; i.e., on the same day as the 2017 magnitude-7.1 earthquake, 32 years later. What is a coincidence!

Other features of earthquakes in Mexico include the so-called ground amplification and the long duration of some of the ground motion (e.g., Singh et al. 1980, 1985, 1988, and 1993; Bard et al., 1988; and Chávez-García, et al., 1995). Mexico City, for example, can be severely affected by earthquakes originated in the subduction zone of the Mexican Pacific coast more than 300 km away—that is, ground amplification, because we expect very small ground motion at 300 km away from the epicenter, even for a magnitude-8 earthquake. The damages caused by the earthquake of Michoacán in 1985 were due to the amplification of the ground motion and the duration of

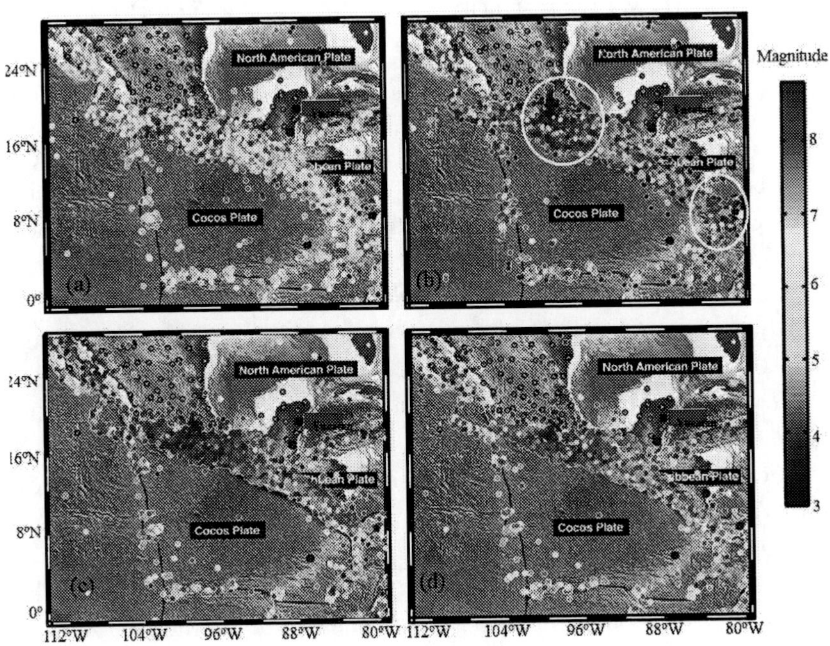

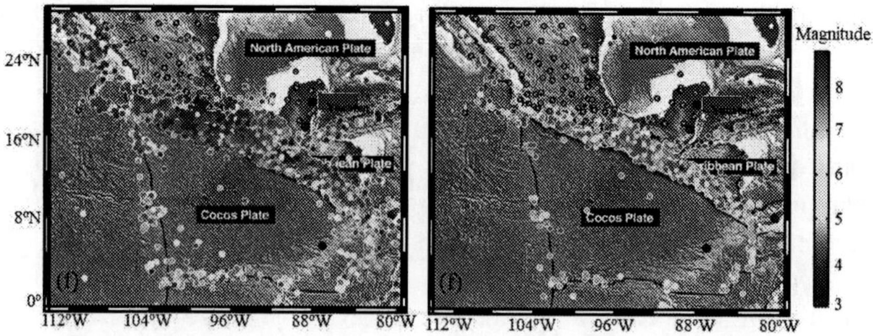

Figure 3.53. (a) Mexican seismicity from April 19, 1902 to December 31, 1994. There is an abundance of light blue and yellow dots because recordings of small quakes in the early part of the twentieth century are rare. (b) Mexico seismicity from January 1, 1995 to December 31, 2007, for quakes with a magnitude 3 and greater. (c) Mexico seismicity from January 1, 2003 to December 31, 2011, for quakes with a magnitude 3 and greater. (d) Mexico seismicity from January 1, 2007 to December 31, 2014, for quakes with a magnitude 3 and greater. (e) Mexico seismicity from August 9, 2003 to August 10, 2017, for quakes with a magnitude 3 and greater. (f) Mexico seismicity from October 10, 2017 to December 12, 2019, for quakes with magnitude 3 and greater.

ground motion and the regional amplification are two sides of the same coin. Both these phenomena are related to near-simultaneous earthquake occurrences, illustrated in Table 3.16. The constant tectonic adjustment due to near-simultaneous earthquake occurrences can produce constructive interferences of seismic waves that can lead to an increase in ground motion and the duration of these motions and provide paths of seismic wave propagation with little attenuation.

Notice that there are a multiplicity of gaps in this region with significant overlaps. For example, the last large earthquake (magnitude 7.9) in the Michoacán before 1985 was in 1911. This event was widely felt in southern Mexico, causing damage in the state of Michoacán and in Mexico City. In southern Oaxaca, there have been likewise no known large (a magnitude greater than 7) earthquakes since 1931. The occurrence of the great earthquake in Michoacán in 1985 suggests that there are seismic gap areas similar to the previous Michoacán gap, such as southernmost Oaxaca.

Figure 3.53 shows the variations in earthquake occurrences over time and space in this region. We can see, in Figure 3.53b, the two clusters of small earthquakes, thousands of kilometers, occurring in about the same period of time. These two clusters provide an indication of large earthquakes that occur nearly the same time. We can actually follow the evolution of an earthquake by just following the cluster of small earthquakes, which are characterized by a dark blue color. The cluster moves back and forth between Colima/Michoacán and Guerrero/Oaxaca/Chiapas with time. Then we hit geometric complexities at the junction of the Caribbean, North America, and Cocos Plates, the branch describing the Caribbean-North American boundary, and the bend describing part of the boundary between the Cocos and Nazca Plates. We can see, in Figures 3.53b, 3.53c, 3.53d, and 3.53e, significant blue dots characterizing these boundaries. In other words, fault bends, branches, and step overs, in this particular region, do not seem to arrest ruptures. Figure 3.53b even suggests a motion generation at the junction of the Cocos-Caribbean-Nazca Plates. The basic mechanism here is that the rupture motions cause the surrounding system of faults to move and the deformation is spread to branches. The rupture velocity can be determined by monitoring the evolution of maximum and minimum earthquake occurrences in this region. We can then determine the rupture velocity that allows the rupture to continue in some branches and/or on the main fault. So the rupture starts in the Colima/Michoacán area and propagates north toward the Gulf of California and south toward Guerrero, Oaxaca, and Chiapas.

Because of the large tectonic variations over a short time, the near-simultaneous occurrences of earthquakes, and the nonexistence of very large earthquakes, like the 2010 Maule earthquake, in this region, it is unrealistic to construct an earthquake cycle for the entire region. However, we can construct one for the magnitude- 8.2 earthquake on in 2017 in Chiapas. We define the Chiapas region as [13.84° N, 19.31° N; 99.7° W, 91.41° W].

Table 3.17 provides variations of earthquake variations on a yearly basis and even more details on the above observations. A year before the earthquake, there were only 26 earthquakes of a magnitude of between 3 and 4 for the entire year in the Chiapas region. It is clear that the solid Earth was evolving toward a large earthquake in Chiapas. The rise and then sharp drop in earthquake count, in Figure 3.54, confirms this evolution. We can discover the coming of this earthquake years in advance.

Table 3.17. The number of earthquakes on a yearly basis in the Chiapas region before and after the 2017 earthquake. Notice the sharp decrease in the number of earthquakes, especially small earthquakes, before the earthquake.

		After		
Period	$M_W \geq 3$	$3 \leq M_W \leq 4$	$4 < M_W \leq 5$	$5 < M_W \leq 6$
2018-2019	201	21	171	10
2017-2018	327	39	267	22
		Before		
Period	$M_W \geq 3$	$3 \leq M_W \leq 4$	$4 < M_W \leq 5$	$5 < M_W \leq 6$
2016-2017	197	26	157	15
2015-2016	233	38	181	15
2014-2015	230	36	182	14
2013-2014	431	160	253	16
2012-2013	591	195	376	20
2011-2012	597	212	361	24
2010-2011	320	108	202	11
2009-2010	212	56	145	12

The effects of the 2010 Maule earthquake in the Mexico region: monitoring the Chile mega-earthquake from different hemisphere. Figure 3.54, of the earthquakes in the Chiapas region, shows two major periods of a rise and then a sharp drop in earthquake count. One of the periods is related to the 2017 Chiapas earthquake, with a magnitude of 8.2. The other period, which is February 2010, cannot be linked to any significant event in Chiapas. In fact, the USGS database shows no earthquakes of magnitude 6 or greater between October 16, 2008, and June 30, 2010, in the area. Moreover, the patterns of earthquake occurrences after the drop in the occurrences of small earthquakes were not maintained in February 2010. So is this event coming from another part of the Mexico region? The answer is no. The count of earthquake occurrences for the entire region also shows the same rise and then sharp drop in February, 2010. Yet, there is no event between 2008 and 2011 in the

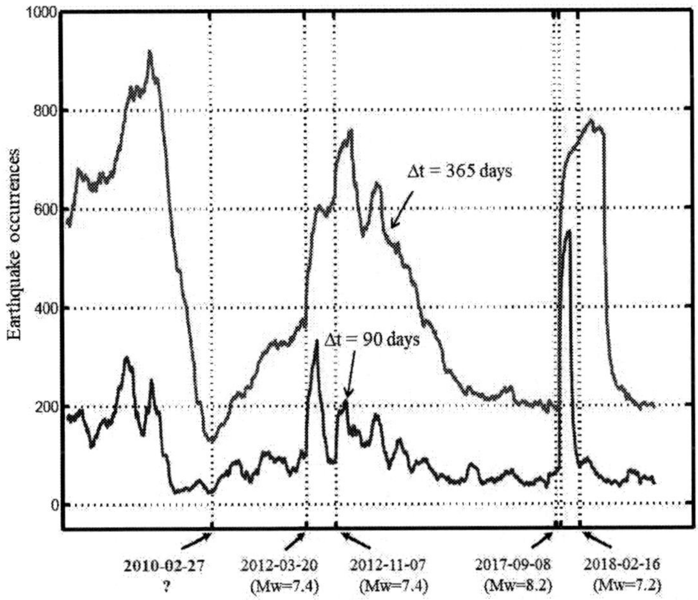

(a)

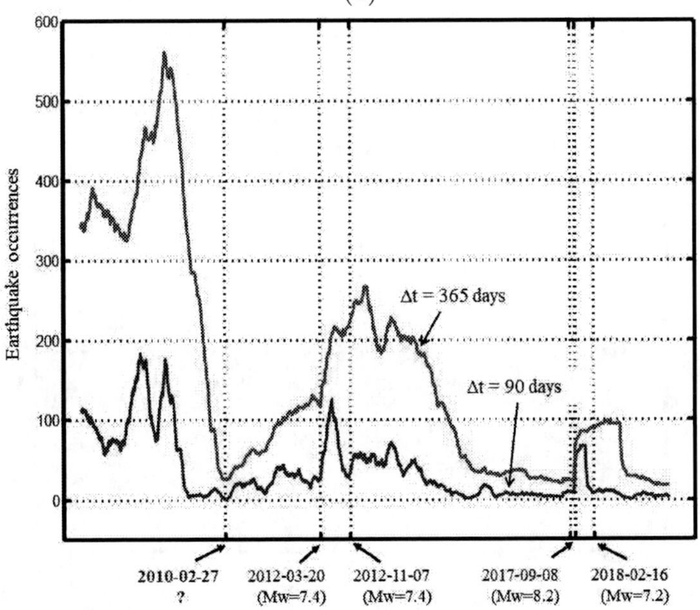

(b)

Figure 3.54. Variations in quake occurrences with time in the Chiapas region (i.e., [13.84, 19.31; -99.7 -91.41]). At a given time, we compute all the quakes that occurred in the last Δt days with the same weight irrespective of their magnitudes. (a) All quakes with $M_W \geq 3$. (b) All quakes for $3 \leq M_W \leq 4$.

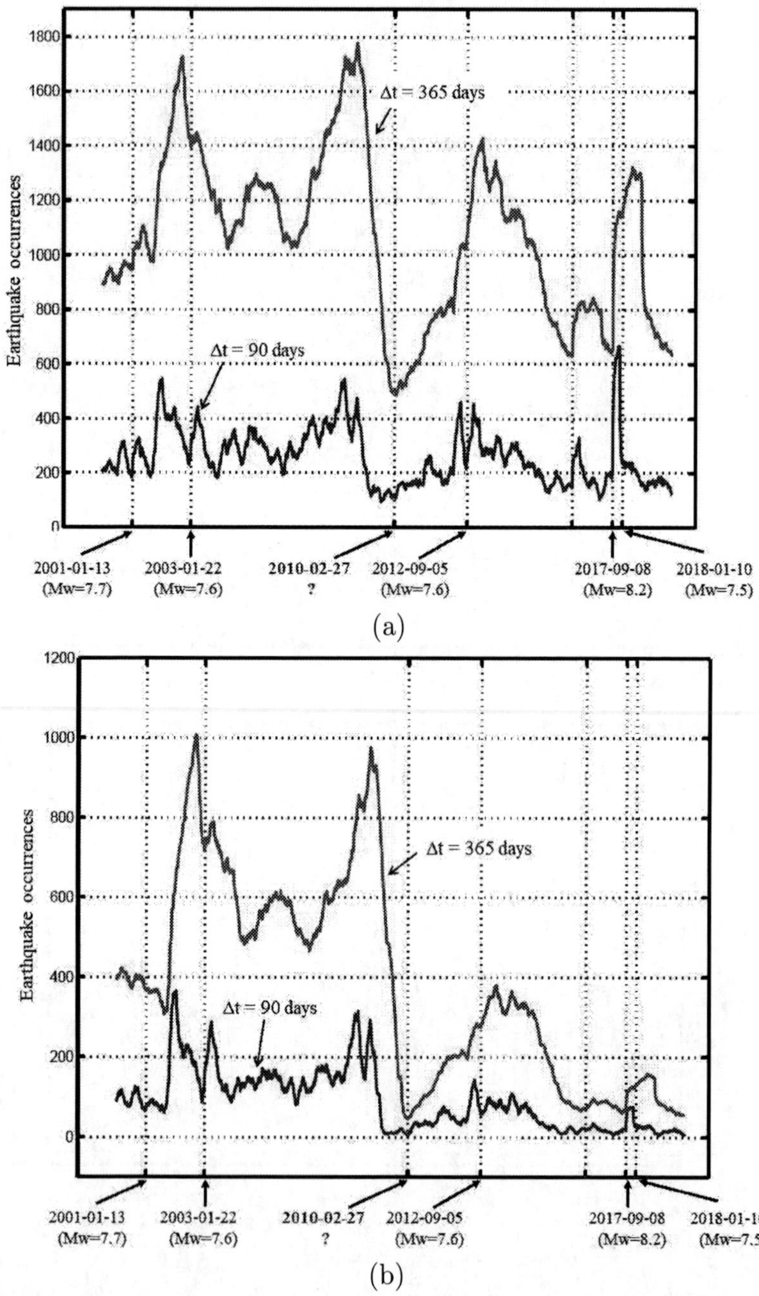

Figure 3.55. Same as Figure 3.54 for the entire Mexico region in Figure 3.52a. (a) All quakes with magnitude 3 and greater. (b) All quakes with a magnitude range between 3 and 4.

entire region of Mexico, depicted in Figure 3.51, with a magnitude more than 7.0 to justify the dominant peak first in Figure 3.55. Moreover, the count in the entire Mexico region shows this event is by far the most dominant tectonic event of the last two decades in the region.

By enlarging the region of study to include South America (i.e., [28.6° N, 51° S; 116.3° W, 63.3° W]), we can see that this strange event in the Mexico region is actually the consequence of the 2010 Maule earthquake in Chile with a magnitude of 8.8 on February 27, 2010. In other words, large earthquakes have effects far larger than 100 kilometers around the epicenter; very large earthquakes, like the 2010 Maule earthquake, are global tectonic events and can be monitored thousands of kilometers away from the eventual epicenter. However, by examining Figure 3.55, there is no indication of another eminent Maule-style earthquake with effects beyond Chile. In other words, the earthquake predicted to occur in Arca at any moment is not as large as the 2010 Maule earthquake, or it is not imminent.

3.5.3. Japan Seismicity

Plate tectonics. As shown in Figure 3.56a and discussed in Shimizu and Itaya (1993), Taira (2001), Grossi (2012), and Murotani et al. (2014), the Japanese islands (Hokkaido, Honshu, Shikoku, and Kyushu) sit at the intersection of four tectonic plates, the Pacific Plate, the Philippine Sea Plate, the North American Plate, and the Eurasian Plate, with the northern part of Japan belonging to the North American Plate. Their tectonics is driven by three subduction-related boundaries marked by deep oceanic trenches or troughs: the Sagami trough, at the interface of the Philippine Sea and the North American Plates, the Nankai trough, between the Philippine Sea Plate and the Eurasian Plate, and the Japanese trench, between the North American and Pacific Plates. The three boundary zones along the eastern coast of Japan have high earthquake activity rates and have historically produced very large damaging earthquakes (e.g., the 1944 Tonankai and 1946 Nankai events, along the Nankai trough, and the 1923 Great Kanto earthquake, along the Sagami trough), as described in Grossi (2012) and Murotani et al. (2014).

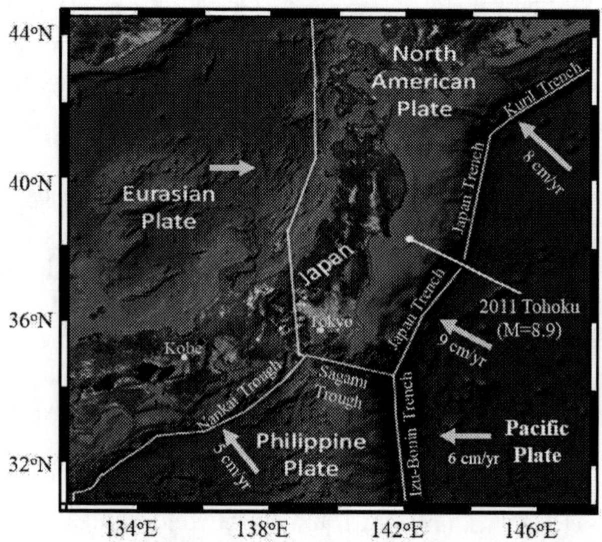
44°N
40°N
36°N
32°N
134°E
138°E
142°E
146°E
North American Plate
Kuril Trench
8 cm/yr
Eurasian Plate
Japan
Japan Trench
2011 Tohoku (M=8.9)
Tokyo
Kobe
9 cm/yr
Sagami Trough
Nankai Trough
Izu-Bonin Trench
Philippine Plate
3 cm/yr
Pacific Plate
6 cm/yr

(a)

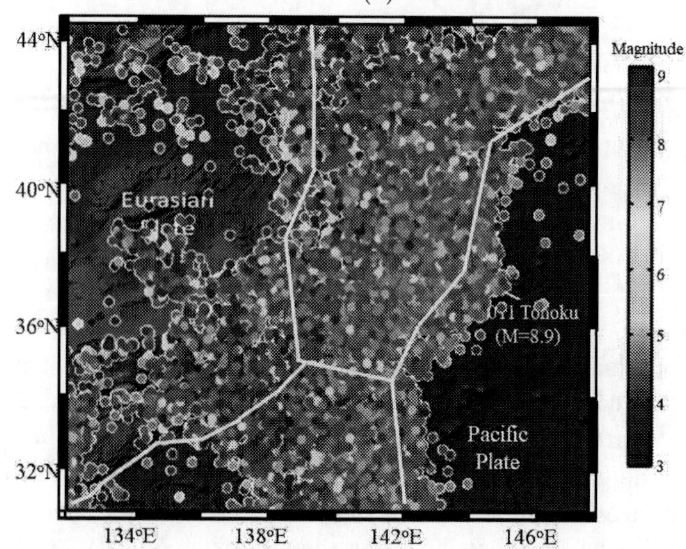
Magnitude
9
8
7
6
5
4
3
44°N
40°N
36°N
32°N
134°E
138°E
142°E
146°E
(M=8.9)
Pacific Plate

(b)

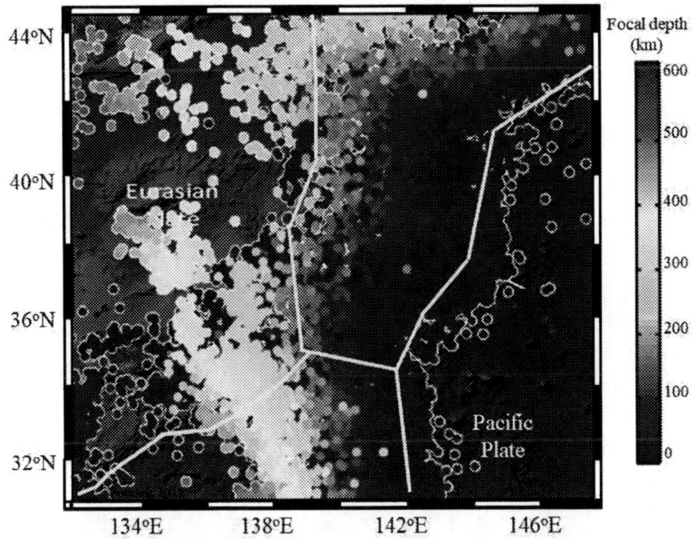

(c)

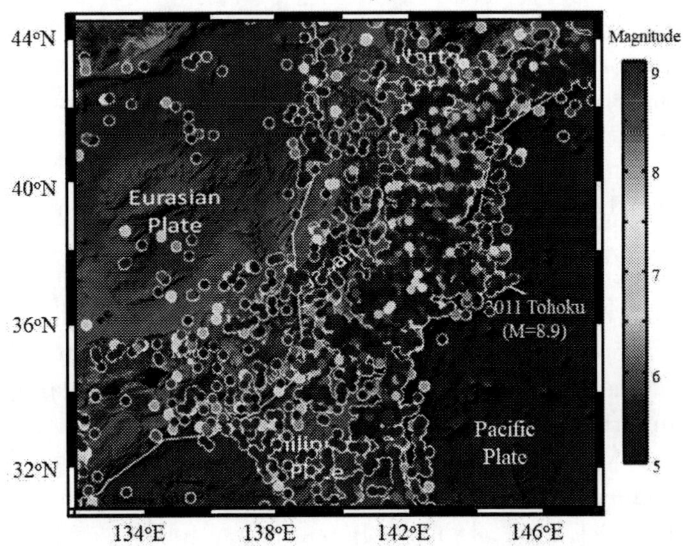

(d)

Figure 3.56. (a) Japan's tectonic setting, illustrating the Nankai trough, the Sagami trough, and the Japan trench. The yellow arrows indicate subduction. (b) Scatterplot of epicenters versus magnitudes of 3 and greater between 1900 and 2019. (c) Scatterplot of epicenters versus hypocenter depths of magnitudes of 5 for the same period. (d) Scatterplot of epicenters versus magnitudes of 5 or greater for the same period.

Table 3.18. The strongest earthquakes on record in the Japanese islands region in 2018

Date	M_W	Depth (km)	[lat (° N), lon (° E)]
2011-03-11	9.1	29	[38.3, 142.4]
1963-10-13	8.5	35	[44.9, 149.5]
1933-03-02	8.4	15	[39.2, 144.6]
1994-10-04	8.3	14	[43.8, 147.3]
1946-12-20	8.3	15	[33.1, 135.9]
1958-11-06	8.3	35	[44.5, 148.5]
1969-08-11	8.2	30	[43.4, 147.9]
1968-05-16	8.2	30	[40.9, 143.4]
2003-09-25	8.1	27	[41.8, 143.9]
1944-12-07	8.1	15	[33.7, 136.2]
1952-03-04	8.1	45	[42.1, 143.9]
1923-09-01	8.1	15	[35.4, 139.3]
1960-03-20	8	15	[39.9, 143.2]

In the Japanese islands region, depicted in Figure 3.56a, we recovered 34,000 earthquakes of magnitude 3 and greater between 1900 and 2019 and 41 earthquakes of magnitude greater than 7.5 from the USGS database. As we can see in Figure 3.56b, there is no place to hide, as there are earthquakes everywhere. Every resident has felt at least a magnitude-3 earthquake in their lifetime. We added a picture of earthquakes of a magnitude greater than 5 between 1900 and 2019 (Figure 3.56d) to improve the reading of these seismicity pictures. The east coast of the Japanese islands, along the Japan, Kuril, and Izu-Bonin trenches, has the highest amount of earthquakes of magnitude 5 and greater. We can see that the Japan Trench Plate boundary is the most active seismic zone in the Japanese islands. It also contains the largest number of earthquakes with a magnitude greater than 7.5 in the last 120 years or so. Table 3.18 shows the largest earthquakes in the last 120 years of Japan's history.

The March 11, 2011, Tohoku, Japan, magnitude-9.1 earthquake and the ensuing tsunami near the east coast of Honshu caused nearly 20,000 deaths and more than 300 billion dollars in damage, ranking as one of the worst natural disasters ever recorded (e.g., Geller, 2011; Hayes et al., 2011; Simons et al., 2011; Stein and Okal, 2011; Fujii et al., 2011; Hayes, 2011; Ide et al., 2011; Lee et al., 2011;

Pollitz et al., 2011; Shao et al., 2011; Kagan and Jackson, 2013; and Xu et al., 2014). The earthquake ruptured an area approximately 450 km long and 150 km wide. The tsunami was responsible for most of the damage and most of the lives lost. It was 15 m high when it hit the Fukushima Nuclear Power Plant. The 6-m seawall was insufficient. It was preceded by a magnitude-7.2 event on March 9, approximately 40 km from the main shock epicenter and was followed by a significant increase in seismicity in the immediate region of the magnitude-9.0 rupture (Grossi, 2012; and Murotani et al., 2015). The two largest earthquakes following this event, measuring 7.9 and 7.7 occurred on March 11, 2011 (Grossi, 2012 and Murotani et al., 2015).

Data analysis. Let us use the 2011 magnitude-9 Tohoku-Oki earthquake to analyze the earthquake cycle in the Japanese islands region. Figure 3.57 shows magnitude-3 and greater earthquake distributions

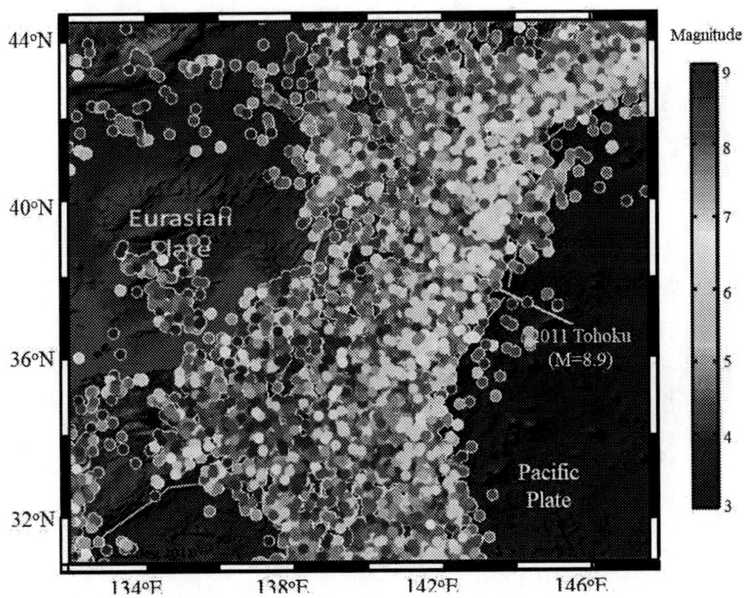

(a)

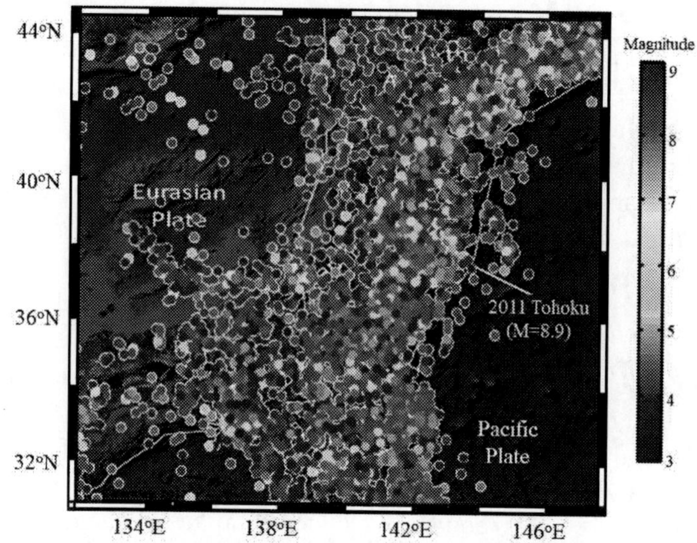

(b)

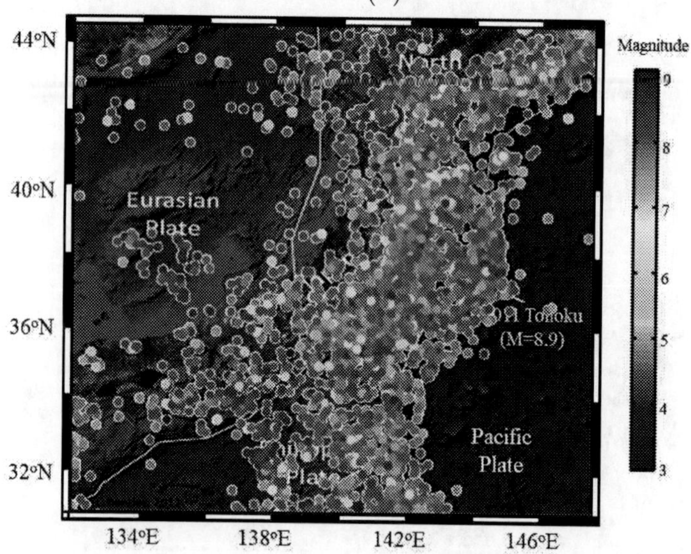

(c)

Figure 3.57. (a) Japan's seismicity from January 02, 2001, to February 10, 2011. (b) Japan's seismicity from April 10, 2011, to November 10, 2019. (c) Japan's seismicity from July 18, 2007, to February 10, 2011.

from January 2, 2001, to February 10, 2011 (i.e., a duration of 3,691 days, more than 10 years before the 2011 magnitude-9 Tohoku-Oki earthquake, until 30 days before this earthquake), and the same distribution from April 10, 2011, to November 10, 2019 (i.e., a duration of 3,136 days, about nine years after the earthquake, until 30 days after the earthquake). Notice how the small earthquake occurrences dominate the seismicity of the entire Japanese islands region before the 2011 magnitude-9 Tohoku-Oki earthquake. A clear cluster of small earthquakes appeared around Tohoku-Oki about five years before the 2011 Tohoku-Oki earthquake, as we can see in Figure 3.57c. Actually, the cluster of small earthquakes containing the Tohoku-Oki area captured the size and length of the ruptured area projection at the Earth's surface. In other words, the cluster of small earthquakes, which includes the eventual epicenter, was apparent about five years before the 2011 Tohoku-Oki earthquake. The preparation for the 2011 Tohoku-Oki earthquake was underway.

We can also monitor the preparation of the 2011 Tohoku-Oki earthquake by monitoring the increase and decrease in the number of earthquakes—that is, by computing $N(t_k)$, as defined in (3.3). Notice the low seismic activity in Figure 3.58, after the 2011 Tohoku-Oki earthquake. This low activity continues more than nine years after the earthquake, as we can see in Table 3.19 and Figure 3.58. This low activity suggests that an earthquake of the magnitude of the 2011 Tohoku-Oki earthquake or greater is not imminent. It is also in line with the common observation that such an earthquake returns to this area within a period of 40 to 400 years. Furthermore, the last time such a major earthquake occurred off the Tohoku-Oki region, before the 2011 Tohoku event, was in the year 869 (Minoura et al., 2001). The earthquakes have been primarily offshore since the 2011 Tohoku-Oki earthquake, as we can see in Figure 3.56c. They are also generally deep. Between 1900 and 2019, the Japanese islands have registered more than 2,000 earthquakes with a depth greater than 300 km.

Notice also that the occurrences of the other large earthquakes are hardly visible, in Figure 3.58, because of the large number of events that occurred that day and a few days after the 2011 Tohoku-Oki earthquakes. More than 4,700 earthquakes of magnitude 4.0 or greater occurred in the region of the Tohoku-Oki rupture through December 2011 (Grossi, 2012). Again, as we can see in Table 3.19, this seismic activity has significantly decayed with time since the event.

Table 3.19. The number of earthquakes on a yearly basis in the Japanese islands region before and after the 2011 magnitude-9 Tohoku-Oki earthquake. Notice the sharp decrease in the number of earthquakes, especially small earthquakes, before the earthquake. The variable M_W stands for earthquake magnitude.

		After		
Period	$M_W \geq 3$	$3 \leq M_W \leq 4$	$4 < M_W \leq 5$	$5 < M_W \leq 6$
2018-2019	706	41	613	52
2017-2018	595	26	534	34
2016-2017	935	38	818	74
2015-2016	794	40	704	46
2014-2015	912	35	829	52
2013-2014	817	17	740	60
2012-2013	979	18	878	74
2011-2012	2311	49	2087	156
		Before		
Period	$M_W \geq 3$	$3 \leq M_W \leq 4$	$4 < M_W \leq 5$	$5 < M_W \leq 6$
2011-2010	400	24	337	37
2010-2009	323	21	267	31
2009-2008	1156	447	648	55
2008-2007	877	331	502	42
2007-2006	724	254	441	30
2006-2005	941	230	657	48
2005-2004	1003	157	755	79
2004-2003	764	102	560	94
2003-2002	575	147	385	39
2002-2001	534	110	370	52
2001-2000	1265	304	844	111
2000-1999	453	122	279	51
1999-1998	550	160	341	49

Finally, notice that after the sharp drop in small events, in Figure 3.58, it took more than a year before the 2011 Tohoku-Oki earthquake. This observation suggests that after the materials reached their elastic limit their failure did not follow immediately. These

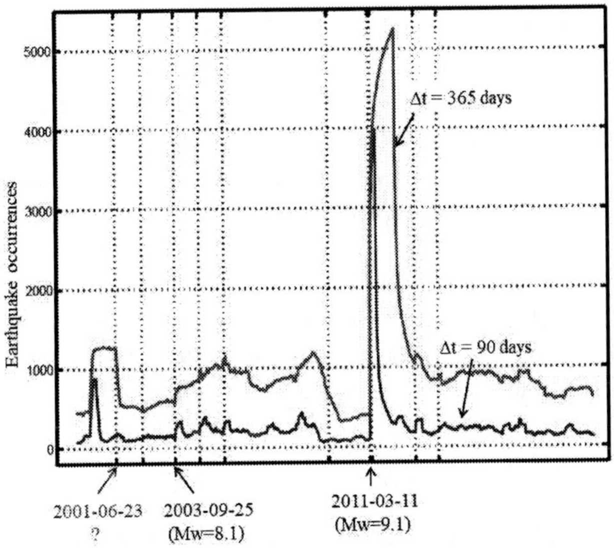

(a)

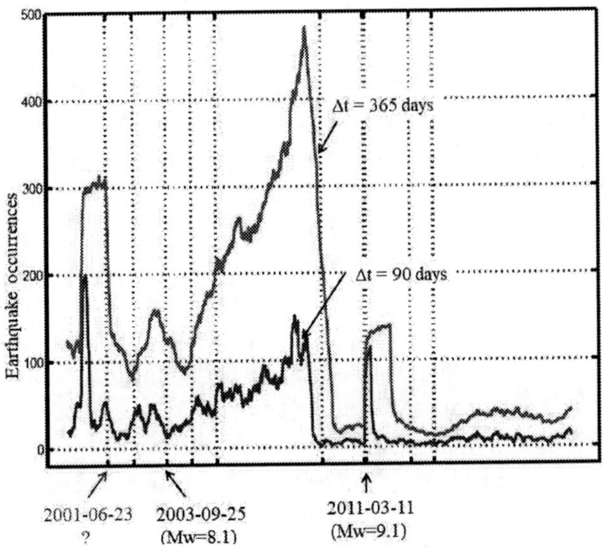

(b)

Figure 3.58. Variations in quake occurrences with time in the Japanese islands region. At a given time, we compute all the quakes that occurred in the last Δt days with the same weight irrespective of their magnitudes and locations. (a) All quakes with magnitude 3 and greater. (b) All quakes with a magnitude range between 3 and 4.

materials behave in a ductile manner rather than in a brittle manner. The assumption here is that rocks are weakened by small earthquakes until the elastic breaking limit of rocks. Once this point is passed, nucleation take about a year toward a major earthquake.

Note that we have not included the 1963 Kuril Islands earthquake ($M_W = 8.5$) and 1933 Sanriku earthquake ($M_W = 8.4$), indicated in Table 3.19, in this study because fewer than 200 earthquakes with a magnitude smaller than 4 are available in the USGS database between 1900 and 1963 for this region. Again, the distribution and quality of seismometers in the region were still under development or records are not well documented to reproduce figures and tables similar to those of the 2011 Tohoku-oki earthquake for the 1963 Kuril Islands earthquake. Anyway, as we can see in Figure 3.58, the other major earthquakes between 2000 and 2019 follow the same patterns as the 2011 Tohoku-oki earthquake, with modest drops in small earthquakes.

In summary, the monitoring process, based on the occurrences of small earthquakes, for large earthquakes is also possible for the Japanese islands. Not only can we know about one year in advance by using Figure 3.58 that a mega-earthquake is coming, we can even roughly identify its location as discussed in Ikelle (2022d).

The effects of the June 2001 Peru earthquake in the Japanese islands region. Figure 3.58a actually shows two major periods of a rise and then a sharp drop in earthquake count. One of the periods is related to the 2011 magnitude-9 Tohoku-Oki earthquake. The other period, which is June 23, 2001, cannot be linked to any significant event in Japan. In fact, the USGS database shows no earthquake with a magnitude of 7 or greater between April 8, 1999, and June 27, 2002, in the area. Moreover, the patterns of earthquake occurrences after the drop in occurrences of small earthquakes are not the same as on June 23, 2001. So where is this event coming from? It turns out that this strange event in the Japanese islands region is actually the consequence of the June 23, 2001, southern Peru earthquake, with a magnitude of 8.4. In other words, large earthquakes have effects far larger than 100 kilometers around the epicenter. Very large earthquakes, like the June 23, 2001, Peru earthquake, are global tectonic events.

Remember the June 23, 2001, Peru magnitude 8.4 earthquake was a major quake (e.g., Giovanni et al., 2018), with the epicenter located at 16.26° S latitude and 73° W longitude, deep in the Southern Hemisphere, while the Japanese islands are located in the Northern Hemisphere. This earthquake damaged southern Peru extensively, particularly the Arequipa-Moquegua-Tacna area. Its ground motions were so strong that even in Peru's capital, Lima (600 km away), homes collapsed and injured several people and were felt as far as Bolivia and northern Chile.

So lessons learned from the 2011 Tohoku-Oki 9.1 earthquake clearly suggest that the differences between the physics of small earthquakes and that of large earthquakes can be used to monitor the next subduction mega-earthquakes, especially in Japan, Chile, and Mexico. We had a period of a rise and then a sharp drop in the small earthquake count before the 2011 magnitude-9 Tohoku-Oki earthquake and the 2010 Maule earthquake of magnitude 8.8 in South Chile. Moreover, the natural preparation process for these very large earthquakes can be monitored in faraway places. They are global tectonic events.

3.5.4. California seismicity: intraplate earthquakes

In the California area, depicted in Figure 3.59a, more than 31,000 earthquakes of magnitude 3.0 and greater between 1900 and 2019 are available in the USGS database, more than 3,900 earthquakes of a magnitude greater than 4.0 in the same period, and 26 earthquakes of a magnitude of more than 6.5, including the magnitude-7.9 San Francisco earthquake (latitude 37.75° N and longitude 122.25° W) on April 18, 1906. Remember, only earthquakes greater than about 3 are commonly felt. Figure 3.59b shows the entire seismicity of this area from 1900 to 2019. This area includes part of the boundary between two of the Earth's largest tectonic plates: the Pacific and North America Plates. The Pacific and North American Plates are sliding past each other (horizontal displacements) along the San Andreas transform fault system at an average rate of about 5 cm/year. More than half of the relative plate motion between the Pacific and

North American Plates occurs along the San Andreas Fault. Most of the fault zone is now on land. The tectonic motion along this boundary (i.e., the San Andreas Fault) makes the region, depicted in Figure 3.59a, very susceptible to earthquakes and is a significant factor in creating the geology and geomorphology of the region. Note that the San Andreas Fault System has a historical record of small to moderate earthquakes, and rare massive earthquakes.

Actually, the San Andreas Fault is not a discrete fault strand, as depicted in Figure 3.59a, but rather a broad zone of faults as much as 120 km wide. Along this zone, earthquakes occur to depths of about 7 km. Most of the recent large earthquakes occur on land; in the interior of the North American Plate (i.e., they are intraplate earthquakes). The June 28, 1992, magnitude-7.3 Landers earthquake occurred in a remote area of the Mojave Desert in Southern California. The July 6, 2019, magnitude-7.1 Ridgecrest Sequence earthquake also occurred deep in the interior of the North American Plate, in Northeast California. Atwater and Stock (1988) have suggested

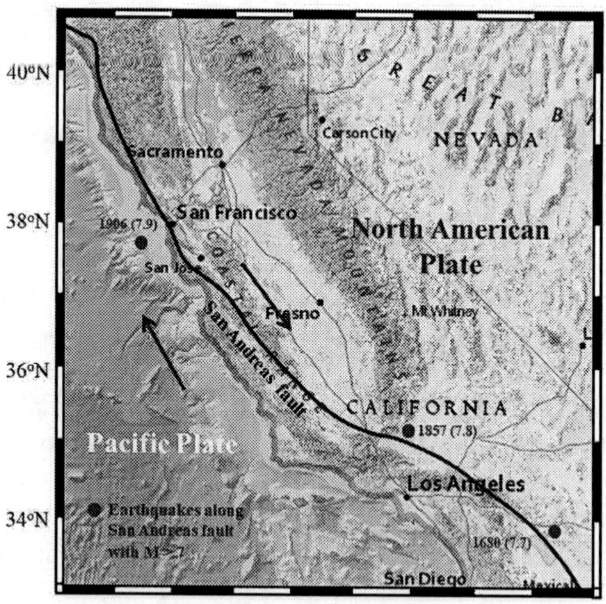

(a)

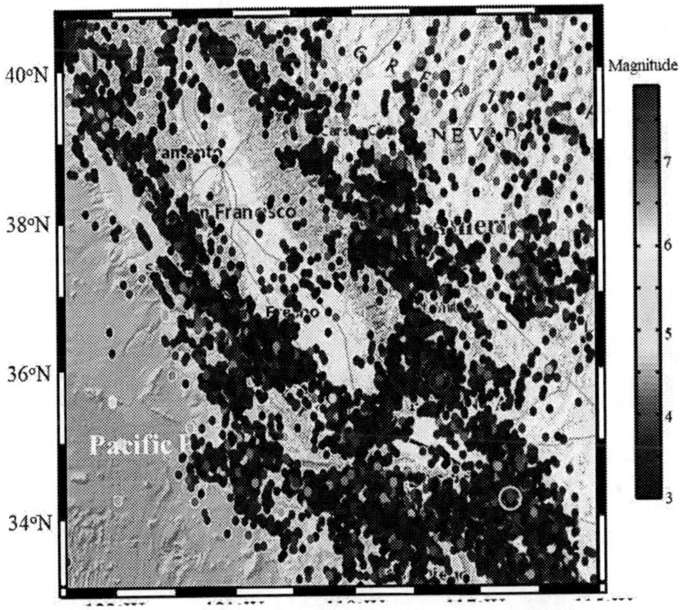

(b)

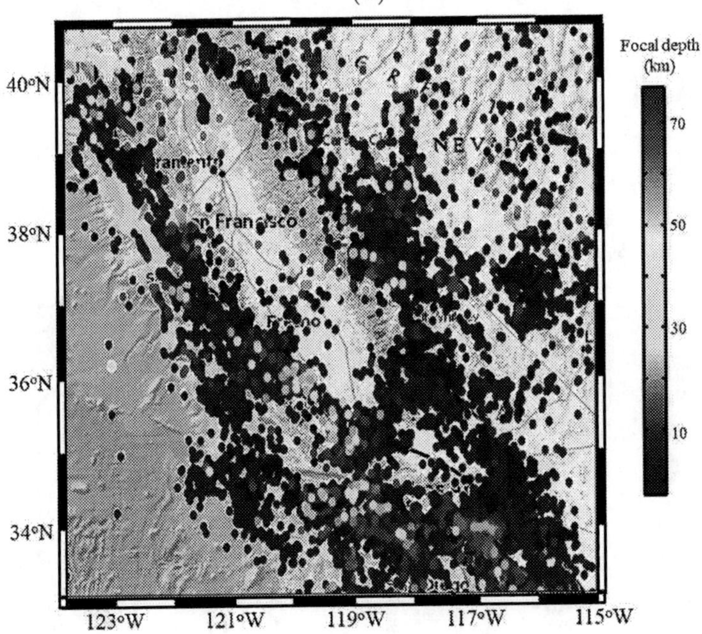

(c)

Figure 3.59. (a) The Pacific and North American Plates slide past each other along the San Andreas fault. (b) Scatterplot of epicenters versus magnitudes of 3 or greater between 1900 and 2019. (c) Scatterplot of epicenters versus hypocenter depths for the same quakes.

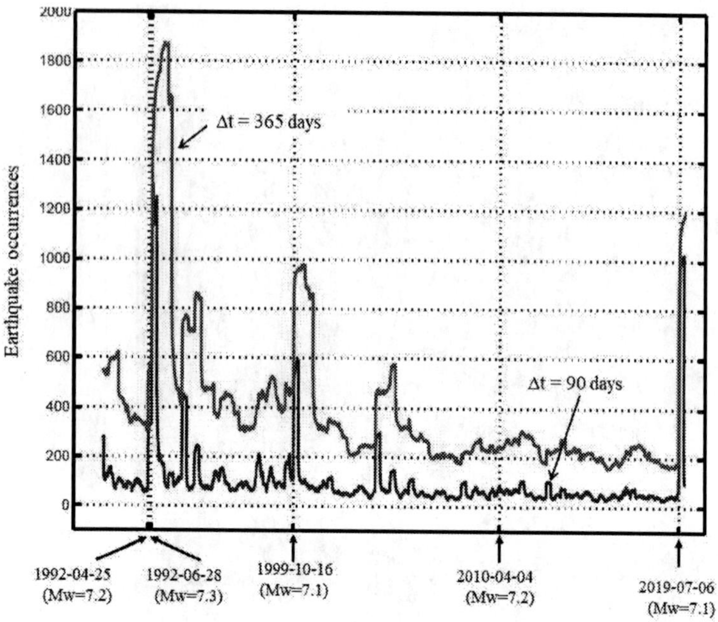

(a)

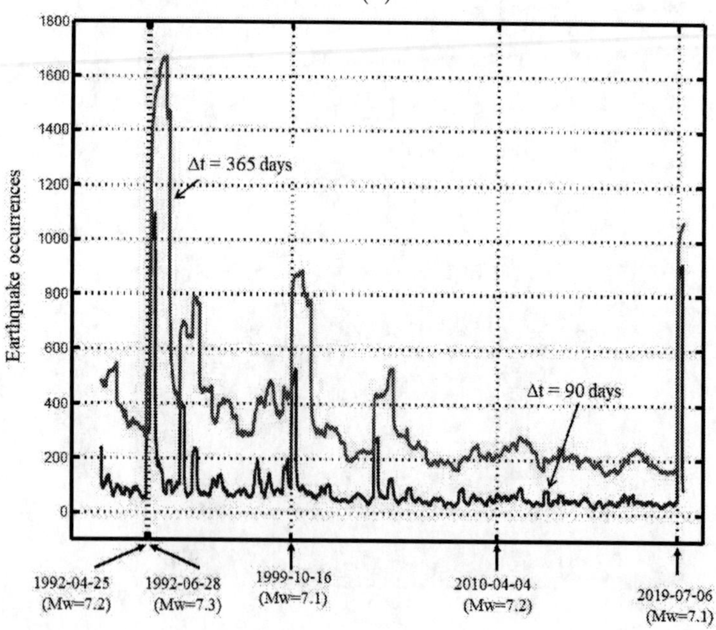

(b)

Figure 3.60. Variations in quake occurrences with time in the region, in Figure 3.59a. At a given time, we compute all the quakes that occurred in the last Δt days with the same weight irrespective of their magnitudes. (a) All quakes with $M_W \geq 3$. (b) All quakes with $3 \leq M_W \leq 4$.

that the San Andreas Fault has been stepping inland, transferring tectonic motion eastward over time.

Again, note that offshore strike-slip faults, such as the San Andreas Fault, do not produce large tsunamis because motions along these faults are primarily horizontal and would not cause the large vertical displacements of the seafloor needed to produce a tsunami. Therefore, it was not surprising that there was no significant tsunami observed during the 1906 San Francisco earthquake.

Data analysis. Large intraplate earthquakes are generally not prominent global tectonic events like the large interplate earthquakes described in the previous two subsections. Their importance stems from the fact that their epicenters are sometimes located in heavily populated areas like California.

As shown in Figure 3.59b, small earthquakes provide us with insight into the location of intraplate earthquakes. We can see clusters of earthquakes in some areas. Unfortunately, in contrast to subduction earthquakes in the previous sections, there is no clear evolution toward earthquakes, as in the previous sub-sections that we can capture by observing the evolution of small earthquakes. Changes in small earthquakes are almost identical to that of large earthquakes, as we can see in Figure 3.60. In other words, the physics of small intraplate earthquakes is identical to that of large intraplate earthquakes, even days before the large earthquake occurrences. This observation is valid even for earthquakes with magnitudes smaller than 3.

Quiz: Explain why the variation in the amount of small earthquakes cannot be interpreted as a precursor to large earthquakes in California.

Appendix A

Periodic Table

The periodic table of the elements is probably the greatest table in the sciences. Most texts dealing with some chemistry or material sciences contain some form of this table. We are here including another form of this table because we are emphasizing the electronegativity of the elements in addition to their atomic number (i.e., the number of protons in an atom, which also equals the number of electrons). Again, electronegativity refers to the ability of an atom to attract shared electrons in a covalent bond. The higher the value of the electronegativity, the more strongly that element attracts the shared electrons.

In Table A.1, we use the Pauling scale, in which fluorine (F) has the highest electronegativity, while francium (Fr) is one the least electronegative. Helium (He), neon (Ne), argon (Ar), krypton (Kr), and radon (Rn) are not listed in the Pauling electronegativity scale, although helium may have the highest electronegativity of all the chemical elements. Again, the difference in electronegativity between two bonded elements determines what types of bond they will form, as described in Chapter 2.

Table A.1. A periodic table with electronegativity in Pauling unit. Data taken from Emsley (1998)

Z — Atomic number
A — Symbol (element)
EN — Electronegativity

1 **H** 2.1																	2 **He**
3 **Li** 0.98	4 **Be** 1.57											5 **B** 2.04	6 **C** 2.5	7 **N** 3.0	8 **O** 3.44	9 **F** 3.98	10 **Ne**
11 **Na** 0.93	12 **Mg** 1.31											13 **Al** 1.61	14 **Si** 1.9	15 **P** 2.19	16 **S** 2.58	17 **Cl** 3.16	18 **Ar**
19 **K** 0.82	20 **Ca** 1.0	21 **Sc** 1.36	22 **Ti** 1.54	23 **V** 1.63	24 **Cr** 1.66	25 **Mn** 1.5	26 **Fe** 1.83	27 **Co** 1.88	28 **Ni** 1.91	29 **Cu** 1.9	30 **Zn** 1.65	31 **Ga** 1.81	32 **Ge** 1.8	33 **As** 2.18	34 **Se** 2.55	35 **Br** 2.96	36 **Kr**
37 **Rb** 0.82	38 **Sr** 0.95	39 **Y** 1.22	40 **Zr** 1.33	41 **Nb** 1.6	42 **Mo** 2.16	43 **Tc** 1.9	44 **Ru** 2.2	45 **Rh** 2.28	46 **Pd** 2.2	47 **Ag** 1.93	48 **Cd** 1.69	49 **In** 1.78	50 **Sn** 1.96	51 **Sb** 2.05	52 **Te** 2.1	53 **I** 2.66	54 **Xe** 2.6
55 **Cs** 0.79	56 **Ba** 0.89	57 **La** 1.1	72 **Hf** 1.3	73 **Ta** 1.5	74 **W** 2.36	75 **Re** 1.9	76 **Os** 2.2	77 **Ir** 2.2	78 **Pt** 2.28	79 **Au** 2.54	80 **Hg** 2.0	81 **Tl** 2.04	82 **Pb** 2.33	83 **Bi** 2.02	84 **Po** 2.0	85 **At** 2.2	86 **Rn**
87 **Fr** 0.7	88 **Ra** 0.89	89 **Ac** 1.1	104 **Rf**	105 **Db**	106 **Sg**	107 **Bh**	108 **Hs**	109 **Mt**	110 **Ds**	111 **Rg**	112 **Cn**						

← **Give electrons**

Acquire electrons →

58 **Ce** 1.12	59 **Pr** 1.31	60 **Nd** 1.14	61 **Pm** 1.13	62 **Sm** 1.17	63 **Eu** 1.2	64 **Gd** 1.2	65 **Tb** 1.1	66 **Dy** 1.22	67 **Ho** 1.23	68 **Er** 1.24	69 **Tm** 1.25	70 **Yb** 1.1	71 **Lu** 1.27
90 **Th** 1.3	91 **Pa** 1.50	92 **U** 1.38	93 **Np** 1.36	94 **Pu** 1.28	95 **Am** 1.3	96 **Cm** 1.3	97 **Bk** 1.3	98 **Cf** 1.3	99 **Es** 1.9	100 **Fm** 1.3	101 **Md** 1.3	102 **No** 1.3	103 **Lr** 1.3

Liquid	Solid	Gas

Appendix B

Answers to Some of the Quizzes and Exercises

Chapter 1:

Where does an atmosphere end?
Atmosphere gradually fades away as density and pressure decrease. By 60 km above Earth's surface, the sky is black (little air to scatter). Atmosphere extends up to 350 km (thermosphere). Between 350 km and 3000 km, we have exosphere. Above the exosphere, we are in interplanetary space which made of helium and hydrogen.

How did we prove that there is life deep in the ground?
Scientific drilling opens the window into environments hundreds of meters deep, then it became clear that life exists deep under the ground.

What is the difference between evaporation and precipitation?
Evaporation takes water from oceans and land into the atmosphere, and precipitation takes water vapor from the atmosphere to land and oceans.

Lands and oceans constitute the earth's surface. What percentage of the earth's surface is occupied by water? The earth's surface can be divided into the Northern Hemisphere and the Southern Hemisphere. Which hemisphere has the larger mass of ocean?
Water covers 71% of Earth's surface; water covers 61% (ocean) of Northern Hemisphere; water covers 81% (ocean) of Southern Hemisphere.

Define the mass density.
Density = mass/volume. It measures how much mass is in a given volume. Its unit is kg/m^3. It is also commonly expressed in g/cm^3 ($g/cm^3 = 10^{-3} \times kg/m^3$). Typical rocks at the surface of the earth have a density between 2.0 and 2.5 g/cm^3.

What are differences between continental crust and oceanic crust?
Continental crust underlies the continents with an average rock density of about about 2700 kg/m^3 (or 2.7 g/cm^3), an average thickness between 35 and 40 km, and granite as the common rock type. Oceanic crust underlies the the ocean basins with an average rock density of about about 3000 kg/m^3 (or 3 g/cm^3), an average thickness between 7 and 10 km, and basalt and Gabbro as the common rock types.

What are differences between crust and lithosphere?
Figure B.1

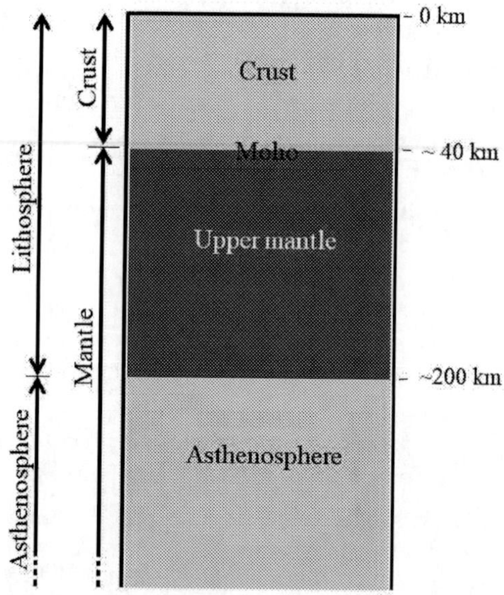

Figure B.1. An illustration of the crust, mantle, lithosphere, and asthenosphere.

Is the earth homogeneous?
No, it is very physically and chemically heterogeneous.

What is the temperature at the base of a 40-km-thick granitic crust with a geothermal gradient of 16 degrees Celsius per kilometer?
T = 40 km x 16 = 640 degrees Celsius. This temperature is not high enough to melt crustal rocks.
What is the pressure at the base of 35-km thick granitic crust with a density of 3000 kg/m^3*?*
Figure 1.8 shows that the pressure in the solid earth varies linearly with depth in the first 1000 km. The gradient of this linear relationship is $g \times \rho$, where g is the gravity on Earth and ρ is the density. The pressure is

$$\begin{aligned} p &= g \times \rho \times z \\ &= 9.8\,(\mathrm{m/s^2}) \times 3000\,(\mathrm{kg/m^3}) \times 35000\,(\mathrm{m}) \\ &= 1 \times 10^9\,\mathrm{Pa}\ [\mathrm{Pa} = \mathrm{kg/(ms^2)}] = 1\,\mathrm{GPa} \end{aligned} \tag{B.1}$$

What is the radius of the solid earth?
The radius of the solid earth is approximately 6371 km.

What the differences between minerals, glasses, and rocks?
Rocks [igneous rocks are formed from cooled from a liquid (melt); sedimentary rocks are formed from debris cemented from pre-existing rock; Metamorphic rocks are formed from rock alteration by pressure and temperature] are aggregates of minerals, minerals are inorganic crystalline solids, and glasses are non-crystalline minerals.

List two tools (methods) used look beneath our feet, inside the solid Earth.
The tools used to look beneath our feets, inside the solid Earth, are seismicity (Chapters 3), gravity and isostasy (see Ikelle, 2022c), heat flow (Ikelle, 2022c), and magnetism (Ikelle, 2022c and 2022d).

(1) Determining the state of water at the following temperatures and pressures: ($-10°C$ and 50 kPa), ($25°C$ and 90 kPa), ($80°C$ and 5 kPa), and ($50°C$ and 0.3 kPa).
(2) Determine the state of CO_2 at the following temperatures and pressures: ($-30°C$ and 2000 kPa), ($-60°C$ and 1000 kPa), ($-60°C$ and 100 kPa), and ($20°C$ and 0.3 kPa).
(1) Using the phase diagram for water, (−10°C and 50 kPa) is a solid; (25°C and 90 kPa) is a liquid; (80°C and 5 kPa) is a gas; (50°C and 0.3 kPa) is a gas.
(2) Using the phase diagram for CO_2, (−30°C and 2000 kPa) is a liquid,

(−60°C and 1000 kPa) is a solid, (−60°C and 100 kPa) is a gas, and (20°C and 100 kPa) is a gas.

(1) What are the terrestrial planets? What do they have in common?
(2) What does the Earth have lots of oxygen compared to, say, Venus?
(3) Why is Mercury's surface similar to that of the Moon's? What feature do they share that causes this?
(3) Which of the Jovian planets have solid surfaces? (i) Jupiter and Saturn, (ii) Uranus and Neptune, (iii) All four planets, and (iv) None of them. Select one answer.
(1) Terrestrial planets. Mercury, Venus, Earth, and Mars. Theses planets have higher densities, heavier elements, rocky surfaces, and iron cores. (2) Plants and other organic bodies are able to free the oxygen locked in CO2. (3) Both the Moon and Mercury have no atmosphere. This means they both have little protection from meteorites and are thus heavily crated by such objects. Other bodies, such as the Earth, also experience impacts, but are somewhat more protected by their atmosphere. If these bodies are active as well, these craters can be resurfaced. Unlike the Moon, however, Mercury has a somewhat larger iron core and is much denser. (4) None of them.

(1) The Mars Science Laboratory (MSL) rover, Curiosity, landed on Mars and began operations on August 6, 2012. When did rover mark its one Mars year?
(2) Many of the planets are similar. List two planets of our solar system which have both volcanoes, craters, and water features at their surfaces.
(1) On June 25, 2014, the rover marked one Mars year (687 Earth days) of operations on the surface.
(2) We know Earth and Mars both have volcanoes, craters, and water features.

Anywhere inside a planet (or star), the pressure is the weight per unit area of the material above that point. On Earth, a column of air 2.54 cm on a side, extending from sea-level to the top of the atmosphere, weighs 6.6678 kg. Show that the corresponding pressure is 1 bar.

$$\text{Pressure} = 6.6678/(2.54 \times 2.54) = 1.03\,\text{kg/cm}^2 = 1.03\,\text{bar}\ .$$

So less mass, less internal pressure.

Why are the inner and outer planets of our Solar System so different?
The terrestrial planets—Mercury, Venus, Earth, and Mars—all formed in the inner, hotter part of our Solar System. It was so hot that much of the

gas and ice in the solar system could not condense at normal temperatures and pressures could not condense. They evaporated. However, metals and silicates could withstand the high temperatures and it is these heavier materials that the rocky inner planets are made. In the outer, cooler portion of the Solar System materials such as water ice and gases were able to accumulate onto the giant Jovian planets. These planets have small icy rocky cores surrounded by thick accumulations of gases.

Suppose that the mass of the Sun is made up of hydrogen (73%), helium (25%), and heavy elements (which includes everything else, 2%). Compare the heavy elements in the Solar System that are found in the Sun to those in the planets.
We can observe that

$$0.02 \times m_{\text{sun}} > \sum_i m_{\text{planet}}^{(i)} , \tag{B.2}$$

with

$$m_{\text{sun}} = 1.989 \times 10^{30} \text{ kg} , \tag{B.3}$$

where $m_{\text{planet}}^{(i)}$ is the mass of the i-th planet of the Solar System. Most of the heavy elements in the Solar System are found in the Sun. Their mass is greater than the mass of all planets combined (see Table 10.2).

(1) What are differences and similarities between stars and giant (gaseous) planets?
(2) Venus spins backwards compared to most of the other planets of our solar system. Why?
(1) Stars are spheres of gas in hydrostatic equilibrium (i.e., gravitational force is balanced by pressure gradients). Densities and temperatures in a given star cover many orders of magnitude, as discussed in Ikelle (2022b). Giant (gaseous) planets are similar to stars. They are large spheres of gas, albeit with a rocky core. Contrary to stars, though, the gas is typically very dense and cold.
(2) Venus spins or rotates in the opposite direction that Earth rotates. This means that on Venus the Sun rises in the west and sets in the east. On Earth the Sun rises in the east and sets in the west. Scientists have not figured out why.
The common hypothesis is that Venus originally rotated counter-clockwise—like Earth and the other planets still do—but were struck at some point by massive objects (perhaps other planets) or a number of smaller collisions that sent them spinning in different directions.

Note that Oumuamua is the first known interstellar object detected passing through the Solar System. It was discovered October 19, 2017 by the University of Hawaii's. Oumuamua is still in our Solar System and is spinning in the same direction that Venus rotates. Maybe we have to look at interstellar medium for the reason why Venus rotates in the opposite direction that Earth rotates.
Most of the objects in our solar system, including the Sun, planets, and asteroids, all rotate counter-clockwise. This is due to the initial conditions in the cloud of gas and dust from which our solar system formed. These initial conditions must almost be included in the explanation of Venus rotation.

Chapter 2:

What is the number of protons in a 70-kg body under the assumption that protons and neutrons are nearly equally numerous in our bodies?

$$N = \frac{35\,\mathrm{kg}}{1.673 \times 10^{-27}\,\mathrm{kg}} \approx 10^{28}\,\mathrm{protons}\,.$$

Same amount of neutrons and electrons.

Write the nuclide notation, including charge, if applicable, for atoms with the following characteristics: (a) 20 protons, 20 neutrons, and 18 electrons, (b) 19 protons, 20 neutrons, and 18 electrons, (c) 26 protons, 28 neutrons, and 23 electrons; and (d) 90 protons, 142 neutrons, and 90 electrons.
All atoms except hydrogen contain more than one electron. (a) 20 protons, 20 neutrons, 18 electrons: ${}^{40}_{20}\mathrm{Ca}^{+2}$.
(b) 19 protons, 20 neutrons, 18 electrons: ${}^{39}_{19}\mathrm{K}^{+1}$.
(c) 26 protons, 28 neutrons, 23 electrons: ${}^{54}_{26}\mathrm{Fe}^{+3}$.
(d) 90 protons, 142 neutrons, 90 electrons: ${}^{232}_{90}\mathrm{Th}$.

(1) Determine Z/A of Hydrogen, ${}^{4}_{2}\mathrm{He}$, ${}^{60}_{27}\mathrm{Co}$, *and* ${}^{235}_{27}\mathrm{U}$. *Explain the variation of this ratio.*
(2) Calculate the average binding energy per nucleon of ${}^{93}_{41}\mathrm{Nb}$.
Predict the nuclear stability of ${}^{36}_{17}\mathrm{Cl}$. *(3) What is the binding energy of the oxygen isotope oxygen-18?*

$$B == (Z \times m_p + N \times m_n - M) \times c_0^2 - c2 = 139.8\ \mathrm{MeV}$$

$Z = 8$, $N = 10$, $m_p = 1.007825 \times 1.66 \times 10^{-24}$ g, $m_n = 1.008665 \times 1.66 \times 10^{-24}$ g, and $M\left({}^{1}8\mathrm{O}\right) = 17.99916 \times 1.66 \times 10^{-24}$

g. Remember that 1 amu $= 1.66 \times 10^{-24}$ g.

Assume we have a box of 5×10^8 ^{24}Na *atoms sitting on the table. How many will be left in the box after 3 hours and 30 minutes?*
From Table 3.4, we see that the half-life of ^{24}Na is 15 minutes. So we have 14 half-lives in 3 hours and 30 minutes; i.e.,

$$n = \frac{(3 \times 60) + 30}{15} = 14 \,.$$

The number of Ca atoms left after 3 hours and 30 minutes is

$$N_{\text{left}} = \frac{N_0}{2^n} = \frac{5 \times 10^8}{2^{14}} = 30517 \,.$$

We can alternatively calculate the number of radioactive nuclei left as

$$N = N_0 \exp(-\lambda t_0) = N_0 \exp\left[-\frac{0.6931}{\tau} \times (14\tau)\right] = (5 \times 10^8) \exp(-0.6931 \times 14) = 30517 \,,$$

where τ is the half-life of ^{24}Na and $t_0 = 14\tau$ corresponds to 3 hours and 30 minutes.

(1) Why are the properties of CO_2 *and* SiO_2 *different? (2) Again, an acid is a substance that gives up a proton. A base is a substance that accepts a proton. Using these definitions, rank the following molecules based on their acid strengths:* H_2SO_4, H_2O, HCl, H_3O^+, *and* $NH_4{}^+$.
Similarly, rank the following molecules based on their base strengths: $HSO_4{}^+$, OH^-, Cl^-, H_2O, *and* NH_3 .
(3) Using the periodic table rank the bonds in each set in order of decreasing bond length and decreasing bond strength: (a) Bond strength: S-F, S-Br, S-Cl and (b) C=O, C-O.
(1) Their properties are very different because CO_2 is a three-atom molecule gas and SiO2 is a solid compound.
(2) An acid is a substance that gives up a proton. A base is a substance that accepts a proton. The strength of an acid depends on its tendency to give up a proton, the strength of a base on its tendency to accept a proton.
Strength of Acid H_2SO_4, HCl (strongest) > H_3O^+ > $NH_4{}^+$ > H2O (weakest)
Strength of Base: $HSO_4{}^+$, Cl^- (weakest) < NH_3 < OH^- (strongest)
(3) Use bond Length and bond Energy table in Chapter 2 and/or Table A.1. Bond strength: S-F > S-Cl > S-Br and Bond length: S-Br > S-Cl > S-F. (b) Bond length: C-O > C=O and Bond strength: C=O > C-O.

Is CCl_4 polar or nonploar? Justify your answer.
Which bond is more polarized, the H−F or H−Br bond? A symmetric molecule (all bonds identical and no lone pairs) is nonpolar even if individual bonds within the molecular are polar. The individual dipoles on the bonds *cancel out* due to the symmetrical shape of the molecule. Carbon tetrachloride (CCl_4) is symmetrical molecule. The four atoms bonded to the carbon atom are identical and project symmetrically out from the carbon atom. Thus,the symmetry of the molecule causes the bond dipole moments to cancel; i.e., carbon tetrachloride is nonpolar. Dipole moment of H−F is 1.7 D and that of H−Br is 0.8 D.

Identify the intermolecular forces in Figure 2.17.
See Figure B.2.

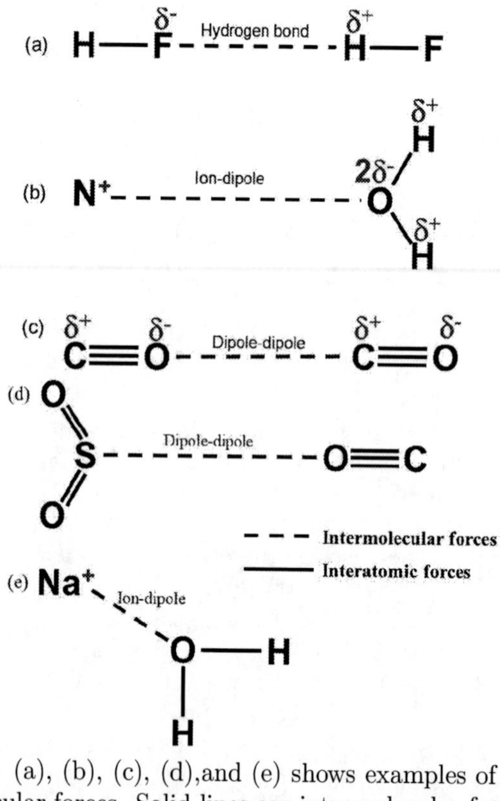

Figure B.2. (a), (b), (c), (d),and (e) shows examples of intramolecular and intermolecular forces. Solid lines are intramolecular forces and dashed lines are intermolecular forces. The slightly positive charges are denoted by δ^+ and the slightly negative charges are denoted by δ^-.

What is the number of atoms in the following molecules: argon, hydrogen gas, nitrogen gas, oxygen gas, carbon monoxide, carbon dioxide, and steam?
Argon (one-atom molecule), hydrogen gas (two-atom molecule), nitrogen gas (two-atom molecule), oxygen gas (two-atom molecule), carbon monoxide (tow-atom molecule), carbon dioxide (three-atom molecule), and steam (three-atom molecule).

Figure 2.28 shows a sea anemones. Is this creature an animal or a plant? Justify your answer.
It is not always easy to recognize an animal. For a very long time, people thought that's sea anemones were plants, because they tend to stay in one place and their tentacles look rather kike petals. Now we know that they are animals. One of the best way to tell if an organism is an animal is to look at its cells under the microscope. Animal cells never have cell walls.

(1) The large molecules of all living things fall into just four main classes. Name them.
(2) What are isomers?
(1) Carbohydrates, Lipids, Proteins, and Nucleic Acids. Carbohydrates, Proteins, Nucleic Acids are called macromolecules. Macromolecules are extremely large on the molecular scale, sometimes consisting of thousands of atoms.
(2) Isomers are compounds that have the same molecular formulas but different structural formulas.

Why viruses are not transmitted through cooked food?
Because cooked foods do not contain living cells for they replicate.

Does silicon-based life exist in our universe as an alternative to carbon-based life?
Carbon and silicon are chemically very similar. Silicon has a valency of 4 (the same as carbon) and so each atom could in principle form bonds with up to 4 other atoms and can therefore be a building block of life like carbon. Silicon is also widely available in the biosphere in our universe in large quantities. Beyond this observation, the answer is for now no.

Chapter 3:

Determine whether the compositions of rock samples associated with the columns in Table 3.3 correspond to felsic, intermediate, mafic, or ultramafic.

Oxide	Felsic (acidic) (%)	Intermediate (%)	Mafic (basic) (%)	Ultramafic (%)
SiO_2	71.3	57.94	49.2	42.26
TiO_2	0.31	0.87	1.84	0.63
Al_2O_3	14.32	17.02	15.74	4.23
Fe_2O_3	1.21	3.27	3.79	3.61
FeO	1.64	4.04	7.13	6.58
MnO	0.05	0.14	0.2	0.41
MgO	0.71	3.33	6.73	31.24
CaO	1.81	6.79	9.47	5.05
Na_2O	3.68	3.48	2.91	0.49
K_2O	4.07	1.62	1.1	0.34
H_2O	0.77	1.17	0.95	3.91
CO_2	0.05	0.05	0.11	0.30
P_2O_5	0.12	0.21	0.35	0.10

Determine the rock types in Figure 3.9. Select your answers among these three types: igneous, metamorphic, and sedimentary.
Figure 3.9 shows (a) igneous, (b) sedimentary, and (c) metamorphic rocks. The textures of these rocks provide us clues of the processes of their formations as we can see in here. For example, metamorphism is a change of form that includes the growth of new minerals, typically in response to a rise in temperature and pressure, and normally without the involvement of a magma.

(1) What are the four types of sedimentary rocks? Give two examples of each.
(2) How much of the Earth's crust is made up of sedimentary rock?
(3) In sedimentary rock types, what is the difference between the terms clastic and detrital?
(4) Sand produces sandstone and silt produces siltstone. What does clay produce?
(1) a. Clastic sedimentary rocks: Conglomerate, breccia, Sandstone, and shale; b. Chemical sedimentary rocks: rock gypsum, rock salt, travertine, and dolostone; c. Biochemical sedimentary rocks: limestone, chert, coquina, Oolite, and chalk; d. Organic sedimentary rocks: peat, lignite, bituminous coal, anthracite, and amber.
(2) Sediments and sedimentary rocks are the most commonly encountered Earth materials. They cover 75% of the continents and nearly all of the ocean floor except at ocean ridges.

(3) clastic = detrital
(4) clay (¡1/256 mm) produced mostly clay claystone, including shale which is actually a fissile claystone or mudstone (splits along many closely spaced planes).

We can reliably and precisely define the time cycle of moonquakes and not the time cycle of earthquakes. Why? The source (i.e., the cause) of moonquakes is periodic while that of earthquake is not.

Use the artificial-intelligence description in the previous chapter to propose a classification of the three key rocks in oil and gas exploration: sandstone, shale, and coal. Hint: Define input vector and features of classification.
We can classify these rocks using measurements that include colors, mineral type, non-minerals, grain size, and hardness of the rock (see Table below). Q. sandstone stands for quartz sandstone, A. sandstone for arkose sandstone, G. sandstone for graywacke sandstone.

Rocks	Hardness	Grain size	Composition	Color
Q. sandstone	hard	coarse	100% quartz	white to brown
A. sandstone	hard	coarse	feldspar and quartz	white to brown
G. sandstone	hard	coarse	sand, silt, and clay	white to brown
Shale	soft	fine	Silt and clay	black, gray, brown, green
Coal	soft	fine	carbon	black (Plants and skeletons)

Explain why the variation of the amount of small earthquakes cannot be directly used as a precursor of large earthquakes in California.
see Ikelle (2022d)

References

Abe, Y., 1993, Physical state of the very early Earth, *Lithos*, **30**, 223–235.

Adams, J. A. S. and Weaver, C. E., 1958, Thorium to uranium ratios as indications of sedimentary processes: Example of concept of geochemical facies, *American Association of Petroleum Geologists Bulletin*, **42**, 387–430.

Aki, K., 1966, Generation and propagation of G waves from the Niigata earthquake of June 14, 1964. Part 2. Estimation of earthquake moment, released energy and stress-strain drop from G wave spectrum, *Bulletin of the Earthquake Research Institute*, **44**, 73–88.

Aki, K., 1979, Characterization of barriers on an earthquake fault, *Journal of Geophysical Research*, **84**, 6140–6148.

Allegre, C. J., and Schneider, S. H., 1994, The Evolution of the Earth, *Scientific American*, **271**, 66–75.

Anderson, D. L., Miller, W. F., Latham, G. V., Nakamura, Y., Toksoz, M. N., Dainty, A. M., Duennebier, F. K., Lazarewicz, A. R., Kowach, R. L., Knight, T. C., 1977a, Seismology on Mars, *J. Geophys. Res.*, **82**, 4524–4546.

Anderson, D. L., Duennebier, F. K., Latham, G. V., Toksoz, M. N., Kovach, R. L., Knight, T. C. D., Lazarewicz, A. R., Miller, W. F., Nakamura, Y., Sutton, G. H., 1977b, Viking seismology experiment, *Bull. Am. Astron. Soc.* **9**, 447.

Anderson, D., Salm, S., and Allen, D., 2019, *Nester's microbiology: a human perspective;*, 9th ed., McGraw Hill, New York.

Atwater, T., and Stock, J., 1998, Pacific-North America Plate Tectonics of the Neogene Southwestern United States: An Update, *International Geology Review*, **40**, 375–402.

Audi, G., and Wapstra, A. H., 1995, The 1995 update to the atomic mass evaluation, *Nuclear Physics A*, **595**, 409–480.

Avdeiko, G. P., Savelyev, D. P., Palueva, A. A., and Popruzhenko, S. V., 2007, Evolution of the Kurile-Kamchatkan volcanic arcs and dynamics of the Kamchatka-Aleutian junction, *Geophysical Monograph Series*, **172**, 37–54.

Babcock, H.W., 1939, The rotation of the Andromeda Nebula, *Lick Observatory Bulletin*, **19**, 41–51.

Bannister, M. T., et al., 2017, Col-OSSOS: Colors of the Interstellar planetesimal 1I/Oumuamua, *Astrophys. J.*, **851**, L38.

Bard, P. -Y., Campillo, M., Chávez-García, F. J., and Sánchez-Sesma, F. J., 1988, A theoretical investigation of a large- and small scale amplification effects in the Mexico city valley: Earthquake spectra, **4**, 609–633.

Barazangi, M., Isacks, B. L., 1976, Spatial distribution of earthquakes and subduction of the Nazca plate beneath South America, *Geology*, **4**, 686–692.

Barnes, R., and Raymond, S. N., 2004, Predicting planets in known extrasolar planetary systems. I. Test particle simulations, *The Astrophysical Journal*, **617**, 569–574.

Bar-On, Y. M., Phillips, R., and Milo, R., 2018, The biomass distribution on Earth, *PNAS*, **115**, 6506–6511.

Bates, J. R, Lauderdale, W. W., and Kernaghan, H., 1979, *ALSEP Termination Report: NASA Reference Publication Series*, NASA-RP-1036, S-480, 914-40-73-01-72.

Beck, S., Barrientos, S., Kausel, E., and Reyes, M., 1998, Source Characteristics of historic earthquakes along the central Chile subduction Zone, *J. South Amer. Earth Sci.*, **11**, 115–129.

Belcher, C. M., Collinson, M. E., and Scott, A. C., 2013, A 450-million-year history of fire *Fire Phenomena and the Earth System: an Interdisciplinary Guide to Fire Science*, Oxford Wiley, 229–49.

Bennett, M. D., 1977, The time and duration of meiosis, *Phil. Trans. R. Soc. Lond. B*, **277(955)**, 201–226.

Berner, R., 1999, Atmospheric oxygen over Phanerozoic time, *Proc. Natl. Acad. Sci. USA*, **96**, 10955–10957.

Bianconi, E., Piovesan, A, Facchin, F., Beraudi, A., Casadei, R., Frabetti, F., Vitale, L., Pelleri, M. C., and Tassani, S., 2013, An estimation of the number of cells in the human body, *Annals of Human Biology*, **40(6)**: 463–471.

Bird, R. B., Stewart, W. E., and Lightfoot, E.N., 1960, *Transport phenomena*, Wiley, Inc. New York.

Bish, D. L. and Post, J. E., editors, 1989, *Modern Powder Diffraction: Reviews in Mienralogy*, Mineralogical Society of America, Washington D.C.

Bolin, B.T., et al., 2018, APO Time-resolved color photometry of highly elongated interstellar object 1I/Oumuamua, *Astrophys. J. Lett.*, **852**, L2.

Brewer, B. J., Chlebowicz-Sledziewska, E., and Fangman, W. L., 1984, Cell cycle phases in the unequal mother/daughter cell cycles of Saccharomyces cerevisiae, *Molecular and cellular biology*, **4(11)**, 2529–2531.

Bridges, B. A., Foster, P. L., and Timms, A. R., 2001, Effect of endogenous carotenoids on “adaptive” mutation in Escherichia coli FC40, *Mutat. Res.*, **473(1)**: 109–119.

Brooks, B. A., Bevis, M., Smalley Jr., R., Kendrick, E., Manceda, R., Lauria, E., Maturana, R., and Araujo, M. (2003), Crustal motion in the Southern Andes (26°–36° S): Do the Andes behave like a microplate?, *Geochem. Geophys. Geosyst.*, 4(10), 1085.

Brönsted, J. N., 1923, Einige Bemerkungen über den Begriff der Säuren und Basen: (Some observations about the concept of acids and bases), *Recueil des Travaux Chimiques des Pays-Bas*, **42(8)**, 718–728.

Buchanan, R. E., 1918, Life phases in a bacterial culture, *The Journal of Infectious Diseases*, **23**, 109–125.

Budyko, M. I., Ronov, A. B., and Yanshin, A. L., 1987, *History of the Earth's Atmosphere*, Springer Verlag, Heidelberg, Germany.

Cahill, T., Isacks, B. L., 1992, Seismicity and shape of the subducted Nazca plate, *J. Geophys. Res.*, **97**, 17503–17529.

Campos, J., and Kausel, E., 1990, The large 1939 intraplate earthquake of Southern Chile, *Seis. Res. Lett.*, **61**, 1.

Canup, R., and Asphaug, E., 2001, Origin of the Moon in a giant impact near the end of the Earth's formation, *Nature*, **412**, 708–712.

Canup, R. M., and Ward, W. R., 2009, *Origin of Europa and the Galilean Satellites*, University of Arizona Press.

Canup, R., and Salmon, J., 2018, Origin of Phobos and Deimos by the impact of a Vesta-to-Ceres sized body with Mars, *Science Advances*, **4**, 1–6.

Carlson, D. H., Plummer, C. C., and Hammersley, L., 2008, *Physical geology, earth revealed*, McGraw-Hill. New York, NY.

Chen, C. J., 1993, *Introduction to scanning tunneling microscopy*, Oxford University Press.

Chown, M., 2009, Cannibalistic Jupiter ate its early moons, *New Scientist* (March issue).

Clayton, D. D., 1983, *Principles of stellar evolution and nucleosynthesis*, The University of Chicago Press.

Coker, E., and Davis, P., 1999, Experiments with zeolites at the secondary school level: experience from the Netherlands, *Journal of Chemical Education*, **76**, 1417–1419.

Comiso, J. C., Gersten, R., Stock, L., Turner, J., Perez, G., and Cho, K., 2017, Positive trends in the Antarctic sea ice cover and associatedchanges in surface temperature, *J. Clim.*, **30**, 2251–2267.

Degens, E. T., Michaelis, W., Paluska, A., 1981, *Principles of petroleum source bed formation, in Energy - Present and Future Operations*, John Wiley and Sons, New York.

de Pater, I., and Lissauer, J. J., 2010, *Planetary Sciences*, Cambridge University Press.

DeMets, C., Gordon, R. G., Argus, D. F., and Stein, S., 1990, Current plate motions, *Geophys. J. Int.*, **101**, 425–478.

DeMets, C., Gordon, R. G., Argus, D. F., and Stein, S., 1994, Effect of recent revisions to the geomagnetic reversal timescale on estimates of current plate motions, *Geophys. Res. Lett.*, **21**, 2191–2194.

DeMets, C., and Traylen, S., 2000, Motion of the Rivera plate since 10 Ma relatives to the Pacific and North American plates and the mantle, *Tectonophysics*, **318**, 119–159.

De Silva, S. L. and Zielinski, G. A., 1998, Global influence of the AD1600 eruption of Huaynaputina, Peru, *Nature*, **393**, 455–458.

Einasto, J., Kaasik, A., and Saar, E., 1974, Dynamic evidence on massive coronas of galaxies, *Nature*, **250**, 309–310.

Emsley, J., 1998, *The Elements (Oxford Chemistry Guides)*: Oxford University Press, UK.

Enz, C. P., 1981, 50 years ago Pauli invented the neutrino, *Helvetica Physics Acta*, **54**, 411–418.

Erni, R., Rossell, M. D., Kisielowski, C., and Dahmen, U., 2009, Atomic-resolution imaging with a sub-50-pm electron Probe, *Physical Review Letters*, **102 (9)**: 096101.

Eskola, P., 1939, Die metamorphen Gesteine, in T. F. W. Barth, C. W. Correns, and P. Eskola, eds., *Die Entstehung der Gesteine*: Springer-Verlag, Berlin.

Faure, G., and Mensing, T. M., 2007, *Introduction to planetary science: the geological perspective*, Springer, The Nerthlands.

Fox, M. A., and Whitesell, J.K., 1997, *Organic Chemistry*, Jones & Bartlett Pub; 2nd edition.

French, B. M., 1998, *Traces of a Catastrophe: A Handbook of Shock-Metamorphic Effects in Terrestrial Meteorite Impact Structures*, LPI Contribution No. 954. Houston: Lunar and Planetary Institute.

Fujii, Y., Satake, K., Sakai, S., Shinohara, M., and Kanazawa, T., 2011, Tsunami source of the 2011 off the Pacific coast of Tohoku Earthquake, *Earth Planets Space*, **63**, 815–820.

Garcia, R. F., Gagnepain-Beyneix, J., Chevrot S, Lognonné, P., 2011, Very preliminary reference Moon model, *Phys Earth Planet Inter*,

188, 96–113.

Geller, R. J., 2011, Shake-up time for Japanese seismology, *Nature*, **472**, 407–409.

Gillespie, R. J., and Nyholm, R. S., 1957, Inorganic stereochemistry, *Quart. Rev. Chem. Soc.*, **11**, 339–380.

Giovanni, M. K., Beck, S. L., and Wagner, L., 2018, The June 23, 2001 Peru earthquake and the southern Peru subduction zone, *Geophysical Research Letters*, **29**, GL015774.

Glasspool, I. J., Edwards, D., and Axe, L., 2004, Charcoal in the Silurian as evidence for the earliest wildfire, *Geology*, **32**, 381–383.

Glasspool, I. J., and Scott, A. C., 2010, Phanerozoic concentrations of atmospheric oxygen reconstructed from sedimentary charcoal, *Nat. Geosci.*, **3**, 627–630.

Goins, N. R, Dainty, A. M., Toksöz M. N. 1981, Lunar seismology: The internal structure of the Moon, *J Geophys Res*, **86**, 5061–5074

Goins, N. R., Toksöz, M. N., and Dainty, A.M., 1978, Seismic structure of the lunar mantle: An overview. In: Lunar and Planetary science conference, 9th. Honston. 3575–3588.

Grossi, P., 2012, The M9.0 Tohoku, Earthquake: Short-Term Changes in Seismic Risk, *RMS Special Report*, https://www.researchgate.net/publication/309173507.

Hanks, T. C., and Kanamori, H., 1979, A Moment Magnitude Scale, *Journal of Geophysical Research*, **84**, 2348–2350.

Harlov, D. E., Austrheim, H., 2013, *Metasomatism and the Chemical Transformation of Rock: Rock-Mineral-Fluid Interaction in Terrestrial and Extraterrestrial Environments* Berlin: Springer.

Harrington, J., de Pater, I., Brecht, S.H., Deming, D., Meadows, V.S., Zahnle, K., and Nicholson, P.D., 2004, in Jupiter: The Planet, Satellites, and Magnetosphere, ed. F. Bagenal, T. E. Dowling, and W. B. McKinnon: Cambridge: Cambridge Univ. Press.

Hartmann, W. K. Phillips, R.J., and Taylor, G.J., eds., 1986. *Origin of the Moon*, Lunar and Planetary Institute, Houston.

Hartmann, W. K., and Miller, R., 1991, *The History of Earth*, Workman Publishing Co., New York.

Hartmann, W. K., 1997, A Brief History of the Moon, *The Planetary Report*, **17**, 4–11.

Haworth, W., 1937, The structure of carbohydrates and of vitamin C, *Nobel Lecture*, December 11, 1937.

Hayes, G. P., Earle, P. S., Benz, H. M., Wald, D. J., and Briggs, R. W. and the USGS/NEIC Earthquake Response Team, 2011, 88 Hours: The U.S. Geological Survey National Earthquake Information Center Response to the 11 March 2011 Mw 9.0 Tohoku Earthquake, *Seismol. Res. Lett.*, **82**, 481–493.

Hayes, G., 2011, Finite Fault Model Updated Result of the Mar 11, *2011 Mw 9.0 Earthquake Offshore Honshu*, Japan: U.S. Geological Survey.

Hood, L. L., 1986, Geophysical constraints on the lunar interior, *Origin Moon*, **31**, 361–410.

Hood, L. L., Herbert, F., Sonett, C. P., 1982, The deep lunar electrical conductivity profile: Structural and thermal inferences, *J Geophys Res*, **87**, 5311–5326.

Hooft, G., 1997, *In search of the ultimate building blocks*, Cambridge University Press.

Hurlbut, C. S., 1963, *Dana's manual of mineralogy*, 17th edition, John Wiley and Sons, New York.

Ide, S., Baltay, A., and Beroza, G.C., 2011, Shallow Dynamic Overshoot and Energetic Deep Rupture in the 2011 Mw 9.0 Tohoku-Oki Earthquake, *Science*, **332**, 1426–1429.

Ikelle, L. T., and Amundsen, L., 2005, *Introduction to Petroleum Seismology*: first edition, SEG, Tulsa, Okhlahoma.

Ikelle, L. T., 2009, *Coding and Decoding: Seismic Data*, First edition, Elsevier Science.

Ikelle, L. T., 2017a, *Coding and Decoding: Seismic Data, The concept of Multishooting*, Second edition, Elsevier Science.

Ikelle, L. T., 2017b, *Introduction to Earth sciences: a physics ap-*

proach, first edition, World Scientific, Singapore.

Ikelle, L. T., and Amundsen, L., 2018, *Introduction to Petroleum Seismology*, second edition, SEG, Tulsa, Okhlahoma.

Ikelle, L.T., 2020, *Introduction to Earth sciences: a physics approach*, Second edition, World Scientific, Singapore.

Ikelle, L.T., 2022a, *Introduction to Multidisciplinary Science at Artificial-Intelligence Age: Chemical, Nuclear, and Thermonuclear Reactions, and Oxygenic and Anoxygenic Photosyntheses*, Nova Science Publishers, New York.

Ikelle, L.T., 2022b, *Introduction to Multidisciplinary Science at Artificial-Intelligence Age: Elasticity, Permeability, Porosity, Viscosity, and Wettability*, Nova Science Publishers, New York.

Ikelle, L.T., 2022c, *Introduction to Multidisciplinary Science at Artificial-Intelligence Age: Geodesy, Geotherms, Spectroscopy, and Spontaneous Electricity*, Nova Science Publishers, New York.

Ikelle, L.T., 2022d, *Introduction to Multidisciplinary Science at Artificial-Intelligence Age: A Binary, Cyclical, Spectroscopic, and Multiple-Time Universe*, Nova Science Publishers, New York.

Isacks, B., and Molnar, P., 1971, Distribution of stresses in the descending lithosphere from a global survey of focal-mechanism solutions of mantle earthquakes, *Rev. Geophys.*, **9**, 103–174.

Isacks, B., Oliver, J., and Sykes, L. R., 1968, Seismology and the new global tectonics, *J. Geophys. Res.*, **73**, 5855–5899.

Jacqué, D., 2006, *APS X-rays reveal secrets of Mars' core*, Argonne National Laboratory.

Jett, J. H., 2015, How long does it take a cell to divide?, *Cytometry Part A*, **87(5)**, 383–384.

Jewitt, D., et al., 2017, Interstellar Interloper 1I/2017 U1: Observations from the NOT and WIYN Telescopes, *Astrophys. J.*, **850**, L36.

Joliot-Curie, I., 1935, *Nobel Lecture: Artificial Production of Radioactive Elements.*

Joliot, F., 1935, *Chemical Evidence of the Transmutation of Elements.*

Jones, L. M., and Molnar, P., 1979, Some characteristics of foreshocks and their possible relationship to earthquake prediction and premonitory slip on faults, *Journal of Geophysical Research*, **84(B7)**, 3596–3608.

Kagan, Y. Y., and Jackson, D. D., 2013, Tohoku earthquake: a surprise?, *Bulletin of the Seismological Society of America*, **103(2B)**, 1184–1194.

Kahn, F. D., and Woltjer, L. 1959, Intergalactic matter and the Galaxy, *Astrophysical Journal*, 130, 705–717.

Kanamori, H., and Brodsky, E. E., 2004, The physics of earthquakes, *Reports on Progress in Physics*, **67**, 1429–1496.

Kant, I., 1755, *Allgemeine Naturgeschichte und Theorie des Himmels* (Universal natural history and theory of the Heavens): SUB Göttingen, Leipzig, Germany.

Karttunen, H., Kroger, P., Oja, H., Poutanen, M., and Donner, K. J., 2017, *Fundamental astronomy* (6th edition): Springer, New York, NY.

Kasting, J. F., 1993, Earth's Early Atmosphere, *Science*, **259**, 920–926.

Keeling, R. F., Manning, A. C., McEvoy, E. M., and Shert, S. R., 1998, Methods for measuring changes in atmospheric O_2 concentration and their application in southern hemisphere air, *Journal of Geophysical Research*, **103**, 3381–3397.

Kekulé, A, 1857, Über die s. g. gepaarten Verbindungen unddie Theorie der mehratomigen Radicale, *Annalen der Chemie und Pharmacie*, **104 (2)**, 129–150.

Kekulé, A., 1858, Ueber die Constitution und dieMetamorphosen der chemischen Verbindungen und über diechemische Natur des Kohlenstoffs, *Annalen der Chemie und Pharmacie*, **106(2)**, 129–159.

Kelleher, J. A., 1972, Rupture zones of large South American earthquakes and some predictions, *J. Geophys. Res.*, **77**, 2087–2103.

King, G., and Nabelek, J., 1985, Role of fault bends in the initiation and termination of earthquake rupture, *Science*, **228**, 984–987.

Kious, W. J., and Tilling, R. I., 2001, *This dynamic earth: the story of plate tectonics*, United States Goverment Printing.

Kivelson, M. G., Kivelson, M. G., and Russell, C. T., 1995, *Introduction to space physics*, Cambridge University Press, U.K.

Klotz, J., Khazaradze, G., Angermann, D., Reigber, C., Perdomo, R., Cifuentes, O., 2001. Earthquake cycle dominates contemporary crustal deformation in Central and Southern Andes, *Earth and Planetary Science Letters*, **193**, 437–446.

Knapmeyer-Endrun, B., and Kawamura, T., 2020, NASA's InSight mission on Mars—first glimpses of the planet's interior from seismology, *Nature Communications*, **11**, 1451.

Knight, M. M., et al., 2017, On the rotation period and shape of the hyperbolic asteroid 1I/Oumuamua (2017 U1) from its lightcurve, *Astrophys. J. Lett.*, **851**, L31.

Kumar, A., Gadre, S. R., Mohan, N., and Suresh, C. H., 2014, Lone pairs: An electrostatic viewpoint, *The Journal of Physical Chemistry A.*, **118 (2)**, 526–532.

Kump, L. R., Kasting, J. F., and Robinson, J., 1991, Atmospheric oxygen variation through geologic time—introduction, *Global and Planetary Change.* **5**, 1–3.

Landis, G.A., 2001, Origin of Martian Moons from Binary Asteroid Dissociation: American Association for the Advancement of Science Annual Meeting; Boston, MA.

Laplace, P.-S., 1799, *Exposition du systeme du Monde*, Fayard, France.

Latham, G., Ewing, M., Dorman, J., Lammlein, D., Press, F., Toksöz, N., Sutton, G., Duennebier, F., Nakamura, Y., 1972, Moonquakes and lunar tectonism, *Moon*, **4**, 373–382.

Lee, S.-J., Huang, B.-S., Ando, M., Chiu, H.-C. , and Wang. J.-H., 2011, Evidence of large scale repeating slip during the 2011 Tohoku-Oki earthquake, *Geophys. Res. Lett.*, **38**, L19306.

Le Maitre, R. W., 1989, *A classification of igneous rocks and glossary of terms*: Blackwell Scientific Publications.

Lenton, T. M., 2013, Fire feedbacks on atmospheric oxygen *Fire Phe-*

nomena and the Earth System: an Interdisciplinary Guide to Fire Science: Oxford Wiley, 289–308.

Le Pichon, X., 1968, Sea-floor spreading and continental drift: *J. geophys. Res.*, **73**, 3661.

Lognonné, P., et al., 2019, SEIS: Insight's seismic experiment for internal structure of Mars, *Space Sci Rev*, **215:12**, 1–170.

Lorenz, R. D., and Nakamura, Y., 2013, *Viking Seismometer Record: Data Restoration and Dust Devil Search: 44th Lunar and Planetary Science Conference.*

Lowry, T. M., 1923, The uniqueness of hydrogen, *Journal of the Society of Chemical Industry*, **42 (3)**, 43–47.

Ma, X., and Elbanna, A., 2019, Dynamic rupture propagation on fault planes with explicit representation of short branches, *Earth and Planetary Science Letters*, **523**, 115702.

Madariaga, R., 1977, High-frequency radiation from crack (stress drop) models of earthquake faulting, *Geophys. J. Roy. Astron. Soc.*, **51**, 625–651.

Masterton, W., Hurley, C., and Neth, E., 2011, *Chemistry: principles and reactions:* Cengage Learning.

McKenzie, D. P., and Parker, R. L., 1968, The North Pacific: an example of tectonics on a sphere, *Nature*, **216**, 1276–1280.

McMurry, J., 2000, *Organic chemistry*, 5th edition.

McSween H. Y., 1994, *Meteoritics*, **29**, 757–779.

Meech, K. J., et al., 2017, A brief visit from a red and extremely elongated interstellar asteroid, *Nature* **552**, 378–381.

Meyer, E., Hug, H. J., and Bennewitz, R., 2004, *Scanning Probe Microscopy*, Springer, Berlin.

Minoura, K., Imamura, F., Sugawara, D., Kono, Y., and Iwashita, T., 2001, The 869 Jogan tsunami deposit and recurrence interval of large-scale tsunami on the Pacific coast of northeast Japan, *J. Natural Disaster Sci.*, **23**, 83–88.

Moore, D. M. and Reynolds, R. C., Jr, 1997, *X-Ray diffraction and the identification and analysis of clay minerals*, 2nd Ed. Oxford University Press, New York.

Moore, G. F., Shipley, T. H., Stoffa, P. L., Karig, D. E., Taira, A., Kuramoto, S., Tokuyama, H., and Suyehiro, K., 1990, Structure of the Nankai Trough accretionary zone from multichannel seismic reflection data, *J. Geophys. Res.*, **95**, 8753–8765.

Moore, J. C., et al., 1998, Consolidation patterns during initiation and evolution of a plate-boundary decollement zone: Northern Barbados accretionary prism, *Geology*, **26**, 811–814.

Morel, F. M. M., and Reinfelder, J.R., 1995, Growth limits on phytoplankton, *Nature*, **373**, 28–28.

Moreno, M. S., Klotz, J., Melnick, D., Echtler, H., and Bataille, K., 2008, Active faulting and heterogeneous deformation across a megathrust segment boundary from GPS data, south central Chile (36–39 S), *Geochem. Geophys. Geosyst.*, **9**, Q12024, doi:10.1029/2008GC002198.

Morgan, J., 1968, Rises, trenches, great faults, and ceustal blocks, *J. geophys. Res.*, **73**, 1959–1982.

Murotani, S., Shimazaki, K., and Koketsu, K., 2012, Rupture process of the 1946 Nankai earthquake estimated using seismic waveforms and geodetic data, *J. Geophys. Res. Solid Earth*, 120, 5677–5692.

Müller, H. S. P., Schlöder, F., Stutzki, J., and Winnewisser, G., 2005, The Cologne Database for Molecular Spectroscopy, CDMS: a useful tool for astronomers and spectroscopists, *Journal of Molecular Structure*, **742**, 215–227.

Myers, R., 2003, *The Basics of Chemistry*, Greenwood Publishing Group.

Nakamura Y., Latham, G. V., and Dorman, H.J., 1982. Apollo Lunar seismic experiment-final summary, *J. Geophys. Res.*, **87**, A117–123.

Nelson, D. L., and Cox, M. M., 2005, *Principles of Biochemistry*, (4th ed.), W. H. Freeman, New York, New York.

Newhall, C. G., and Self, S., 1982, The Volcanic explosivity index (VEI): An estimate of explosive magnitude for historical volcanism, *Journal of Geophysical Research*, **87**, 1231–1238.

Nimmo, F. and McKenzie, D., 1998, Volcanism and tectonics on venus, *Annual Review of Earth and Planetary Sciences*, **26**, 23–

51.

Novick, A., 1955, Growth of Bacteria, *Annual Review of Microbiology*, **9**, 97–110.

Nyquist, L. E., Bogard, D. D., Shih, C.-Y., Greshake, A., Stöffler, D., and Eugster, O., 2001, Ages and geologic histories of martian meteorites, *Space Sci. Rev.*, **96**, 105–164.

Olsen, N., et al.,2015, The Swarm Initial Field Model for the 2014 geomagnetic field, *Geophys. Res. Lett.*, **42**, 1092–1098.

Ostriker, J. P. and Peebles, P. J. E., 1973, A Numerical study of the stability of flattened galaxies: or, can cold galaxies survive?, *Astrophysical Journal*, **186**, 467–480.

Ostriker, J. P., Peebles, P. J. E., and Yahil, A., 1974, The size and mass of galaxies, and the mass of the Universe, *Astrophysical Journal*, **193**, L1–L4.

Pauli, W., 1934, Discussion du Rapport de M. Heisenberg. In: Institut International de Physique Solvay, Structures et proprietes des noyaux atomiques, *Rapports et discussions du septieme conseil de physique tenu a Bruxelles du 22 au 29 Octobre 1933* (pp. 324–325; pp. 214–215). Paris: Gauthier-Villars.

Pauli, W., 1961a, *Aufsatze und Vortrage über Physik und Erkenntnistheorie.* Braunschweig: Vieweg.

Pauli, W., 1961b, On the earlier and more recent history of the neutrino. In: *Winter* (2000), 1–22. Originally published in: *Pauli* (1961a), 156–180.

Pauli, W., 1964, Collected Scientific Papers (Vol. 2), *Interscience*, New York.

Pauling, L., 1932, The Nature of the Chemical Bond. IV. The Energy of Single Bonds and the Relative Electronegativity of Atoms, *Journal of the American Chemical Society*, **54(9)**, 3570–3582.

Peplow, M., 2007, *How Mars got its rust*, BioEd Online. MacMillan Publishers Ltd.

Pesonen, L. J., and Henkel, H., 1992a, Terrestrial Impact Craters and Craterform Structures with a Special Focus on Fennoscandia, *Tectonophysics*, **216**, 31–40.

Pesonen, L. J., Marcos, N., and Pipping, F., 1992b, Palaeomagnetism of the Lappajärvi impact structure, western Finland, *Tectonophysics*, **216**, 123–142.

Pesonen, L. J., Hietala, S., Poutanen, M., Moilanen, J., Mehtinen, M., and Ruotsalainen, H. E., 2005, *The Keurusselkä Meteorite Impact Structure*, Central Finland: Geophysical Data. 22nd Geophysics Days. Helsinki: Geophysical Society of Finland. 165–169.

Pollitz, F. F., Burgmann, R., and Banerjee, P., 2011, Geodetic slip model of the 2011 M9.0 Tohoku earthquake, *Geophys. Res. Lett.*, **38**, L00G08.

Pratt, L. M., Shaw, C. A., and Burruss, R. C. 1988. Thermal histories of the Hartford and Newark basins inferred from maturation indices of organic matter, in Froelich, A. J. and Robinson, G. R., eds., Studies of the Early Mesozoic Basins of the Eastern United States, *U. S. Geological Survey Bulletin*, 1776, pp.58–62.

Ravat, D., Wang, B., Wildermuth, E., and Taylor, P. T., 2002, Gradients in the interpretation of satellite-altitude magnetic data: an example from central Africa, *J. Geodyn.*, **33**, 131–142.

Raymond, S. N., Barnes, R., and Kaib, N. A., 2006, Predicting planets in known extrasolar planetary systems. III. Forming Terrestrial Planets, *The Astrophysical Journal*, **644**, 1223–1231.

Prandtl, L., 1904, *Uber Flussigkeitsbewegungen bei sehr kleiner Reibung,* in Verhandlungen des III, *Internationalen Mathematiker Kongresses*, pp. 484–491, Heidelberg, Germany.

Reynolds, S. J., Johnson, J. K., Kelly, M. M., Morin, P. J., Carter, C. M., 2010, *Exploring geology*, 2nd edition, McGraw-Hill Company Inc, New York, NY.

Richards, M. A., Duncan, R. A., and Courtillot, V. E., 1989, Flood basalts and hot spot tracks, *Science*, **246**, 103–107.

Rivoldini, A., Hoolst, T. V., Verhoeven, O., Mocquet, A., and Dehant, V., 2011, Geodesy constraints on the interior structure and composition of Mars, *Icarus*, **213**, 451–472.

Roberts, M. S., and Whitehurst, R. N., 1975, The rotation curve and geometry of M31 at large galactocentric distances, *Astrophysical Journal*, **201**, 327–346.

Rosman, K. J. R., and Taylor, P. D. P., 1998, Isotopic compositions of the elements 1997, *J. Phys. Chem. Ref. Data*, **27**, 1275–1287.

Rubin, V. C., and Ford, W. K. J., 1970, Rotation of the Andromeda Nebula from a spectroscopic survey of emission regions, *Astrophysical Journal*, **159**, 379–403.

Rudenberg, H. G., and Rudenberg, P. G., 2010, Origin and background of the invention of the electron microscope, *Advances in Imaging and Electron Physics*, **160**, 207–286.

Ruegg, J. C., Campos, J., Madariaga, R., Kausel, E., DeChabalier, J. B., Armijo, R., Dimitrov, D., Georgiev, I., Barrientos, S., 2002. Interseismic strain accumulation in south central Chile from GPS measurements, 1996-1999, *Geophys. Res. Lett.*, **29**, 11.

Schlesinger, W. H., 1997, *Biogeochemistry: An Analysis of Global Change*, 2nd edition, Academic Press, San Diego.

Scott, A. C., 2000, The pre-quaternary history of fire, *Palaeogeogr. Palaeoclimatol. Palaeoecol.*, **164**, 281–329.

Segou, M., and Parsons, T., 2018, Testing Earthquake Links in Mexico From 1978 to the 2017 M = 8.1 Chiapas and M = 7.1 Puebla Shocks, *Geophysical Research Letters*, **45**, 708–714.

Selley, R. C., and Morill, D. C., 1991, GL 101: Basic Concepts of Petroleum Geology, *IHRDC Video Library for Exploration and Production Specialists*, Boston, Ma.

Sellers, P. C., 1992, Seismic evidence for a low-velocity lunar core, *J. Geophys. Res.*, **97**, 11663–11672.

Shear, W. A., and Kukalová-Peck, J., 1990, The ecology of Paleozoic terrestrial arthropods: the fossil evidence, *Canadian Journal of Zoology*, **68**, 1807–1834.

Shao, G., Li, X., Ji, C., and Maeda, T., 2011, Focal mechanism and slip history of the 2011 Mw 9.1 off the Pacific coast of Tohoku earthquake, constrained with teleseismic body and surface waves, *Earth Planets Space*, **63**, 559–564.

Shimizu, S., and Itaya, T., 1993, Plio-Pleistocene arc magmatism controlled by two overlapping subducted plates, central Japan, *Tectonophysics*, **225**, 139–154.

Sidgwick and Powell, 1940, Foundations of valence shell electron pair repulsion theory, *Proc. R. Soc. London A*, **176**, 153–180.

Siegel, E., and Bang, A., 2018, This is how we know there are two trillion galaxies in the Universe, *Forbes* (Oct 18, 2018).

Sieh, K., 1996, The repetition of large-earthquake ruptures, *Proc. Natl. Acad. Sci. USA*, **93**, 3764–3771.

Simons, M., Minson, S.E., Sladen, A., Ortega, F., Jiang, J. L., Owen, S. E., Meng, L. S., Ampuero, J. P., Wei, S. J., Chu, R. S., Helmberger, D. V., Kanamori, H., Hetland, E., Moore, A. W., and Webb, F. H., 2011, The 2011 magnitude 9.0 Tohoku-Oki earthquake: mosaicking the megathrust from seconds to centuries, *Science*, **332**, 1421–1425.

Singh, S.K., Yamamoto, J, Haskov, J., Guzman, M., Novelo, D., and Castro, R., 1980, Seismic gap of Michoacán, Mexico, *Geophys. Res. Letters*, **7**, 69–72.

Singh, S.K., Ponce, L., and Nishenko, S.P., 1985, The great Jalisco, Mexico earthquakes of 1932: subduction of the Rivera plate, *Bull. Seism. Soc. Am.*, **5**, 1301–1314.

Singh, K., Mena, E., and Castro, R., 1988, Some aspects of the source characteristics and ground motion amplification in and near Mexico city from acceleration data of the September, 1985, Michoacán, Mexico earthquakes, *Bull. Seism. Soc. Am.*, **78**, 451–477.

Singh, K., and Ordaz, M., 1993, On the origin of the long coda observed in the lake-bed strong motions records of Mexico city, *Bull. Seis. Soc. Am.*, **83**, 1298–1306.

Solem, J. C., 1995, Cometary breakup calculations based on a gravitationally-bound agglomeration model: The density and size of Comet Shoemaker-Levy 9, *Astronomy and Astrophysics*, **302(2)**, 596–608.

Taira, A., 2001, Tectonic evolution of the Japanese island arc system, *Annual Review of Earth and Planetary Sciences*, **29**, 109–34.

Taylor, P. T., and Frawley, J. J., 1987, Magsat anomaly data over the Kursk region, *USSR Physics of the Earth and Planetary Interiors*, **45**, 275–265.

Tilling, R. I., 2009, Volcanism and associated hazards: the Andean perspective, *Advances in Geosciences*, **22**, 125–137.

Vaughan, D., et al., 2013, Observations: Cryosphere. climate change 2013, *The Physical Science Basis*, 317–382.

von Weizsäcker, C.F., 1935, Zur Theorie der Kernmassen: Zeitschrift für Physik (in German): **96**, 431–458.

Vigny, C., Rudloff, A., Ruegg, J. C., Madariaga, R., Campos, J., and Alvarez, M., 2009, Upper plate deformation measured by GPS in the Coquimbo Gap, Chile, *Physics of the Earth and Planetary Interiors*, **175**, 86–95.

Walter, F., Brinks, E., de Blok, W. J. G., Bigiel, F., Kennicutt Jr., R. C., Thornley, M.D., and Leroy, A.K., 2008, The HI Nearby Galaxy Survey (THINGS), *The Astronomical Journal*, **136**, 2563–2647.

Watson, A. J., and Lovelock, J. E., 2013, The dependence of flame spread and probability of ignition on atmospheric oxygen *Fire Phenomena and the Earth System: An Interdisciplinary Guide to Fire Science*, Oxford: Wiley, pp 273–87.

Weber, R. C., Lin, P. Y., Garnero, E. J., Williams, Q., and Lognonné, P., 2011. Seismic detection of the Lunar core. *Science*, **331**, 309–312.

Webster, G., Brown, D., and Napier, G., 2014, *Construction to begin on 2016 NASA Mars Lander NASA*, Retrieved April 2015.

Weinberg, S., 2011, *Dreams of a final theory: The Scientist's Search for the ultimate laws of nature*, Knopf Doubleday Publishing Group.

Wesnousky, S. G., 2006, Predicting the endpoints of earthquake ruptures, *Nature*, **444**, 358–360.

Wetherill, G. W., 1996, The formation and habitability of extra-solar planets, *Icarus*, **119**, 219–238.

Wieczorek, M. A., and Phillips, R. J., 1997, The structure and compensation of the lunar highland crust, *J Geophys Res*, **102**, 10933–10943.

Wiesendanger, R., 1994, *Scanning probe microscopy and*

spectroscopy-methods and applications, Cambridge University Press, UK.

Woese, C. R., and Fox, G. E., 1977, Phylogenetic structure of the prokaryotic domain: the primary kingdoms, *Proceedings of the National Academy of Sciences of the United States of America*, **74**, 5088–5090.

Woese, C. R., Magrum, L. J., and Fox, G. E., 1978, Archaebacteria, *J.Mol. Evol.*, **11**, 245–252.

Woese, C. R., 1987, Bacterial evolution, *Microbiol. Rev.*, **51**, 221–271.

Xu, L., Meng, X., and Xu, X., 2014, Natural hazard chain research in China: A review, *Natural Hazards*, **70**, 1631–1659.

Zwicky, F., 1933, Die Rotverschieb ung von extragalaktischen Nebeln, *Helvetica Phys. Acta*, **6**, 110–127.

Index

F

G

H

I

J

L

M

N

O

P

About the Author

Luc Thomas Ikelle
Imode Education & Energy,
College Station, Texas, USA
Email: luc.ikelle@gmail.com